UCLA Symposia on Molecular and Cellular Biology, New Series

Series Editor, C. Fred Fox

UCLA Symposia Board

C. Fred Fox, Ph.D., Director
Professor of Microbiology, University of California, Los Angeles

Molecular Strategies
for Crop Protection

Molecular Strategies for Crop Protection

Proceedings of a DuPont–UCLA Symposium on Molecular Strategies
for Crop Protection
Held in Steamboat Springs, Colorado
March 30-April 6, 1986

Editors

Charles J. Arntzen
Department of Plant Science and Microbiology
E.I. DuPont
Wilmington, Delaware

Clarence Ryan
Department of Biological Chemistry
Washington State University
Pullman, Washington

Alan R. Liss, Inc. • New York

Address all Inquiries to the Publisher
Alan R. Liss, Inc., 41 East 11th Street, New York, NY 10003

Library of Congress Cataloging-in-Publication Data

UCLA Symposium on Molecular Strategies for Crop
    Protection (1986 : Steamboat Springs, Colo.)
    Molecular strategies for crop protection.

    (UCLA symposia on molecular and cellular biology ;
new ser., v. 48)
    Includes index.
    1. Plants, Protection of—Congresses.   2. Micro-
organisms, Phytopathogenic—Control—Congresses.
3. Plant molecular biology—Congresses.   4. Molecular
biology—Congresses.   I. Arntzen, Charles J.   II. Ryan,
Clarence A.   III. Title.   IV. Series.
SB950.A2U25   1987   632   86-34284
ISBN 0-8451-2647-4

Contents

Contributors

Patricia P. Abel, Department of Biology, Plant Biology Program, Washington University, St. Louis, MO 63130 **[205]**

M.J. Adang, Agrigenetics Advanced Science Company, Madison, WI 53716 **[345]**

Paul Ahlquist, Biophysics Laboratory and Department of Plant Pathology, University of Wisconsin, Madison, WI 53706 **[335]**

Charles J. Arntzen, Department of Plant Science and Microbiology, E.I. DuPont, Wilmington, DE **[xix]**

Arthur R. Ayers, Department of Cellular and Developmental Biology, Harvard University, Cambridge, MA 02138 **[71]**

Ralph Baker, Department of Plant Pathology and Weed Science, Colorado State University, Fort Collins, CO 80523 **[115]**

Roger N. Beachy, Department of Biology, Plant Biology Program, Washington University, St. Louis, MO 63130 **[205]**

J.N. Bell, Plant Biology Laboratory, Salk Institute, San Diego, CA 92138 **[49]**

Michael W. Bevan, Department of Molecular Biology, Plant Breeding Institute, Cambridge CB2 2LQ, England **[215,307]**

J.M. Bonneville, Friedrich Miescher-Institut, CH-4002 Basel, Switzerland **[267]**

D. Borthakur, John Innes Institute, Norwich NR4 7UH, England **[169]**

Andrea D. Branch, Laboratory of Genetics, The Rockefeller University, New York, NY 10021 **[319]**

Clare L. Brough, Department of Pure and Applied Biology, Imperial College of Science and Technology, London SW7 2BB, England **[307]**

Kenneth W. Buck, Department of Pure and Applied Biology, Imperial College of Science and Technology, London SW7 2BB, England **[307]**

Allen Budde, USDA, ARS, Department of Plant Pathology, University of Wisconsin, Madison, WI 53706 **[95]**

L. Burhop, Agrigenetics Advanced Science Company, Madison, WI 53716 **[221]**

The numbers in brackets are the opening page numbers of the contributors' articles.

Luis F. Carbonell, Department of Bacteriology and Biochemistry, University of Idaho, Moscow, ID 83843 **[235]**

R.S. Chaleff, Department of Central Research and Development, Experimental Station, E.I. DuPont and Company, Wilmington, DE 19898 **[415]**

Forrest G. Chumley, Department of Central Research and Development, Experimental Station, E.I. DuPont and Company, Wilmington, DE 19898 **[83]**

Adrienne Clarke, Plant Cell Biology Research Centre, School of Botany, University of Melbourne, Victoria 3052, Australia **[35]**

R. James Cook, USDA, Agricultural Research Service, Washington State University, Pullman, WA 99164-6430 **[125]**

D.R. Corbin, Plant Biology Laboratory, Salk Institute, San Diego, CA 92138 **[49]**

Robert H.A. Coutts, Department of Pure and Applied Biology, Imperial College of Science and Technology, London SW7 2BB, England **[307]**

G.L. Creason, Department of Central Research and Development, Experimental Station, E.I. DuPont and Company, Wilmington, DE 19898 **[415]**

Catherine H. Daniels, Molecular Biology of Disease Resistance Laboratory, Department of Plant Pathology, Washington State University, Pullman, WA 99164-6430 **[13]**

S.D. Daubert, Department of Plant Pathology, University of Kentucky, Lexington, KY 40546 **[253]**

D. DeBoer, Agrigenetics Advanced Science Company, Madison, WI 53716 **[345]**

J.A. Downie, John Innes Institute, Norwich, NR4 7UH, England; present address: Division of Plant Industry, C.S.I.R.O., Act 2601, Canberra, Australia **[169]**

Theo W. Dreher, Biology Department, Texas A&M University, College Station, TX 77843-3258 **[295]**

Marilyn Ehrenshaft, Department of Botany and Plant Pathology, Oregon State University, Corvallis, OR 97331; present address: Department of Plant Pathology, University of Minnesota, St. Paul, MN 55108 **[135]**

D.A. Evans, DNA Plant Technology Corporation, Cinnaminson, NJ 08077 **[367]**

I.J. Evans, John Innes Institute, Norwich NR4 7UH, England; present address: Apcel Limited, Slough SL1 4EQ, Berkshire, England **[169]**

S.C. Falco, Department of Central Research and Development, Experimental Station, E.I. DuPont and Company, Wilmington, DE 19898 **[415]**

J.L. Firmin, John Innes Institute, Norwich NR4 7UH, England **[169]**

E. Firoozabady, Agrigenetics Advanced Science Company, Madison, WI 53716 **[345]**

Robert T. Fraley, Monsanto Company, St. Louis, MO 63198 **[205]**

Brian Fristensky, Molecular Biology of Disease Resistance Laboratory, Department of Plant Pathology, Washington State University, Pullman, WA 99164-6430 **[13]**

J. Fuetterer, Friedrich Miescher-Institut, CH-4002 Basel, Switzerland **[267]**

Burle G. Gengenbach, Department of Agronomy and Plant Genetics, University of Minnesota, St. Paul, MN 55108 **[107]**

K. Gordon, Department of Virology, Institut de Biologie Moleculaire et Cellulaire du C.N.R.S., Strasbourg, France **[267]**

Rebecca Grumet, Botany Department, Duke University, Durham, NC 27706 **[3]**

Lee A. Hadwiger, Molecular Biology of Disease Resistance Laboratory, Department of Plant Pathology, Washington State University, Pullman, WA 99164-6430 **[13]**

E. Halk, Agrigenetics Advanced Science Company, Madison, WI 53716 **[221]**

Timothy C. Hall, Biology Department, Texas A&M University, College Station, TX 77843-3258 **[295]**

William D.O. Hamilton, Department of Pure and Applied Biology, Imperial College of Science and Technology, London SW7 2BB, England **[307]**

B.D. Harrison, Department of Urology, Scottish Crop Research Institute, Invergowrie, Dundee DD2 5DA, Scotland **[215]**

Robert J. Hayes, Department of Pure and Applied Biology, Imperial College of Science and Technology, London SW7 2BB, England **[307]**

Michèle C. Heath, Botany Department, University of Toronto, Toronto, Ontario, Canada M5S 1A1 **[25]**

Renate Hellmiss, Institute for Molecular Biology, Department of Biology, Indiana University, Bloomington, IN 47405 **[145]**

H. Höfte, Plant Genetic Systems N.V., Ghent, Belgium 9000 **[355]**

T. Hohn, Friedrich Miescher-Institut, CH-4002 Basel, Switzerland **[267]**

David Holden, USDA, ARS, Department of Plant Pathology, University of Wisconsin, Madison, WI 53706 **[95]**

Barbara Howlett, Plant Cell Biology Research Centre, School of Botany, University of Melbourne, Victoria 3052, Australia **[35]**

Bruce Imrie, C.S.I.R.O., Division of Tropical Crops and Pastures, St. Lucia, Queensland 4067, Australia **[35]**

John Irwin, Department of Botany, University of Queensland, St. Lucia, Queensland 4067, Australia **[35]**

N. Jarvis, Agrigenetics Advanced Science Company, Madison, WI 53716 **[221]**

A.W.B. Johnston, John Innes Institute, Norwich NR4 7UH, England **[169]**

Stephen A. Johnston, Botany Department, Duke University, Durham, NC 27706 **[3]**

Mirek Kapuscinski, Plant Cell Biology Research Centre, School of Botany, University of Melbourne, Victoria 3052, Australia **[35]**

Michael Keil, Max-Planck-Institut für Züchtungsforschung, 5000 Köln 30, Federal Republic of Germany **[393]**

J.D. Kemp, Agrigenetics Advanced Science Company, Madison, WI 53716 **[345]**

David F. Kendra, Molecular Biology of Disease Resistance Laboratory, Department of Plant Pathology, Washington State University, Pullman, WA 99164-6430 **[13]**

Thomas Kinscherf, USDA, ARS, Department of Plant Pathology, University of Wisconsin, Madison, WI 53706 **[95]**

J. Klein, Agrigenetics Advanced Science Company, Madison, WI 53716 **[345]**

K. Krahn, Agrigenetics Advanced Science Company, Madison, WI 53716 **[221]**

Karl Josef Kunert, Faculty of Biology, University of Konstanz, 7750 Konstanz, West Germany; present address: Department of Virus Research, John Innes Institute, Norwich NR4 7UH, England **[401]**

S.A. Kut, DNA Plant Technology Corporation, Cinnaminson, NJ 08077 **[367]**

C.J. Lamb, Plant Biology Laboratory, Salk Institute, San Diego, CA 92138 **[49]**

J.W. Lamb, John Innes Institute, Norwich NR4 7UH, UK; present address: Mikrobiologisches Institute, ETHZentrum LFW, Zurich, Switzerland **[169]**

M.A. Lawton, Plant Biology Laboratory, Salk Institute, San Diego, CA 92138 **[49]**

J. Leemans, Plant Genetic Systems N.V., Ghent, Belgium 9000 **[355]**

Sally A. Leong, USDA, ARS, Department of Plant Pathology, University of Wisconsin, Madison, WI 53706 **[95]**

L.S. Loesch-Fries, Agrigenetics Advanced Science Company, Madison, WI 53716 **[221]**

W.H.-T. Loh, DNA Plant Technology Corporation, Cinnaminson, NJ 08077 **[367]**

Heather Macdonald, Department of Pure and Applied Biology, Imperial College of Science and Technology, London SW7 2BB, England **[307]**

Samuel W. MacDowell, Department of Pure and Applied Biology, Imperial College of Science and Technology, London SW7 2BB, England **[307]**

Susan F. Mackintosh, Department of Plant Pathology, Cornell University, Ithaca, NY 14853 **[59]**

Loren E. Marsh, Biology Department, Texas A&M University, College Station, Texas 77843-3258 **[295]**

J. Martinez-Izquierdo, Friedrich Miescher-Institut, CH-4002 Basel, Switzerland **[267]**

David E. Matthews, Department of Plant Pathology, Cornell University, Ithaca, NY 14853 **[59]**

C.J. Mauvais, Department of Central Research and Development, Experimental Station, E.I. DuPont and Company, Wilmington, DE 19898 **[415]**

B.J. Mazur, Department of Central Research and Development, Experimental Station, E.I. DuPont and Company, Wilmington, DE 19898 **[415]**

M.C. Mehdy, Plant Biology Laboratory, Salk Institute, San Diego, CA 92138 **[49]**

D.J. Merlo, Agrigenetics Advanced Science Company, Madison, WI 53716 **[221,345]**

R.W. Michelmore, Department of Vegetable Crops, University of California, Davis, CA 95616 **[193]**

Lois K. Miller, Department of Bacteriology and Biochemistry, University of Idaho, Moscow, ID 83843 **[235]**

Dallice Mills, Department of Botany and Plant Pathology, Oregon State University, Corvallis, OR 97331 **[135,199]**

E. Murray, Agrigenetics Advanced Science Company, Madison, WI 53716 **[345]**

J.B. Neilands, Department of Biochemistry, University of California, Berkeley, CA 94720 **[157]**

Richard S. Nelson, Department of Biology, Plant Biology Program, Washington University, St. Louis, MO 63130 **[205]**

S. Nelson, Agrigenetics Advanced Science Company, Madison, WI 53716 **[221]**

Peter Palukaitis, Department of Plant Pathology, Cornell University, Ithaca, NY 14853 **[243]**

Ernest G. Peralta, Institute for Molecular Biology, Department of Biology, Indiana University, Bloomington, IN 47405; present address: Genentech Inc., South San Francisco, CA 94080 **[145]**

Ian T.D. Petty, Department of Pure and Applied Biology, Imperial College of Science and Technology, London SW7 2BB, England **[307]**

P. Pfeiffer, Department of Virology, Institut de Biologie Moleculaire et Cellulaire du C.N.R.S., Strasbourg, France **[267]**

M. Pietrzak, Friedrich Miescher-Institut, CH-4002 Basel, Switzerland **[267]**

Alan Poplawsky, Department of Botany and Plant Pathology, Oregon State University, Corvallis, OR 97331; present address: Department of Plant Pathology, University of Wisconsin, WI 53706 **[135]**

Daryl R. Pring, USDA-ARS and Plant Pathology Department, University of Florida, Gainesville, FL 32611 **[107]**

Julie Ralton, Plant Cell Biology Research Centre, School of Botany, University of Melbourne, Victoria 3052, Australia **[35]**

K. Rashka, Agrigenetics Advanced Science Company, Madison, WI 53716 **[345]**

T.B. Ray, Department of Central Research and Development, Experimental Station, E.I. DuPont and Company, Wilmington, DE 19898 **[415]**

Walt Ream, Institute for Molecular Biology, Department of Biology, Indiana University, Bloomington, IN 47405 **[145]**

A. Reynaerts, Plant Genetic Systems N.V., Ghent, Belgium 9000 **[355]**

Eulian J.F. Roberts, Department of Pure and Applied Biology, Imperial College of Science and Technology, London SW7 2BB, England **[307]**

Hugh D. Robertson, Laboratory of Genetics, The Rockefeller University, New York, NY 10021 **[319]**

T.A. Rocheleau, Agrigenetics Advanced Science Company, Madison, WI 53716 **[345]**

Steven G. Rogers, Monsanto Company, St. Louis, MO 63198 **[205]**

L. Rossen, John Innes Institute, Norwich NR4 7UH, England **[169]**

Clarence Ryan, Department of Biological Chemistry, Washington State University, Pullman, WA **[xix]**

T.B. Ryder, Plant Biology Laboratory, Salk Institute, San Diego, CA 92138 **[49]**

Jose Sanchez-Serrano, Max-Planck-Institut für Züchtungsforschung, 5000 Köln 30, Federal Republic of Germany **[393]**

John C. Sanford, Department of Horticultural Science, Cornell University, Geneva, NY 14456 **[3]**

N. Sauer, Plant Biology Laboratory, Salk Institute, San Diego, CA 92138 **[49]**

Jeff Schell, Max-Planck-Institut für Züchtungsforschung, 5000 Köln 30, Federal Republic of Germany **[393]**

Arno Schmidt, Faculty of Biology, University of Konstanz, 7750 Konstanz, West Germany **[401]**

J.E. Schoelz, Department of Plant Pathology, University of Kentucky, Lexington, KY 40546 **[253]**

S.A. Sebastian, Department of Central Research and Development, Experimental Station, E.I. DuPont and Company, Wilmington, DE 19898 **[415]**

V. Sekar, Agrigenetics Advanced Science Company, Madison, WI 53716 **[345]**

C.A. Shearman, John Innes Institute, Norwich NR4 7UH, England **[169]**

R.J. Shepherd, Department of Plant Pathology, University of Kentucky, Lexington, KY 40546 **[253]**

Allan M. Showalter, Department of Biology, Washington University, St. Louis, MO 63130 **[375]**

H.D. Sisler, Department of Botany, University of Maryland, College Park, MD 20742 **[187]**

Marek J. Slomka, Department of Pure and Applied Biology, Imperial College of Science and Technology, London SW7 2BB, England **[307]**

Christopher Small, Department of Botany and Plant Pathology, Oregon State University, Corvallis, OR 97331 **[135]**

Michael Smart, Plant Cell Biology Research Centre, School of Botany, University of Melbourne, Victoria 3052, Australia **[35]**

M.J. Smith, Department of Biochemistry, University of California, Berkeley, CA 94720; present address: Department of Chemistry, Columbia University, New York, NY 10027 **[157]**

Timothy Smith, USDA, ARS, Department of Plant Pathology, University of Wisconsin, Madison, WI 53706 **[95]**

G. Staffeld, Agrigenetics Advanced Science Company, Madison, WI 53716 **[345]**

C. Stock, Agrigenetics Advanced Science Company, Madison, WI 53716 **[345]**

D. Sutton, Agrigenetics Advanced Science Company, Madison, WI 53716 **[345]**

L.S. Thomashow, USDA, Agricultural Research Service, Washington State University, Pullman, WA 99164-6430 **[125]**

M. Vaeck, Plant Genetic Systems N.V., Ghent, Belgium 9000 **[355]**

Barbara Valent, Department of Central Research and Development, Experimental Station, E.I. DuPont and Company, Wilmington, DE 19898 **[83]**

Hans D. VanEtten, Department of Plant Pathology, Cornell University, Ithaca, NY 14853 **[59]**

M. Van Montagu, Plant Genetic Systems N.V., Ghent, Belgium 9000 **[355]**

Joseph E. Varner, Department of Biology, Washington University, St. Louis, MO 63130 **[375]**

M.H. Walter, Plant Biology Laboratory, Salk Institute, San Diego, CA 92138 **[49]**

Jun Wang, ASDA, ARS, Department of Plant Pathology, University of Wisconsin, Madison, WI 53706 **[95]**

D.M. Weller, USDA, Agricultural Research Service, Washington State University, Pullman, WA 99164-6430 **[125]**

Janet Williams, Department of Botany and Plant Pathology, Oregon State University, Corvallis, OR 97331 **[135]**

Lothar Willmitzer, Max-Planck-Institut für Züchtungsforschung, 5000 Köln 30, Federal Republic of Germany **[393]**

Roger P. Wise, Plant Pathology Department, University of Florida, Gainesville, FL 32611 **[107]**

Keith L. Wycoff, Department of Cellular and Developmental Biology, Harvard University, Cambridge, MA 02138 **[71]**

N.B. Yadav, Department of Central Research and Development, Experimental Station, E.I. DuPont and Company, Wilmington, DE 19898 **[415]**

Nevin D. Young, Department of Plant Pathology, Cornell University, Ithaca, NY 14853 **[243]**

M. Zabeau, Plant Genetic Systems N.V., Ghent, Belgium 9000 **[355]**

Milton Zaitlin, Department of Plant Pathology, Cornell University, Ithaca, NY 14853 **[243,339]**

Preface

Scientists from two rather distinct scientific disciplines have traditionally worked together to prevent crop losses to pathogens, fungal parasites, insects and weeds. Crop breeders have empirically manipulated genetic variability to select cultivars that are resistant to insects and diseases. In parallel, agronomists have devised cultural practices (crop rotations, use of crop protection chemicals) to reduce populations of weeds, insects and pathogens. At present, more than thirteen billion dollars are spent each year for crop protection chemicals alone; the total cost of protecting crops (including genetically superior seed, field management practices, etc.) is much higher.

The availability of new molecular biological tools has recently begun to revolutionize our thinking about crop production systems. A new field of research specialists has emerged, many of whom were represented at this conference on Molecular Strategies for Crop Protection. These scientists are charting new paths toward understanding the molecular basis of host–parasite interactions and plant defensive responses; the results of these basic investigations are already being utilized for directed genetic alterations of crop plants to create resistances to insects and viruses. Gene transfer technology also is being utilized to map and manipulate genes controlling crop plant tolerance to pesticides, thereby opening new avenues to the use of low-cost, environmentally safe chemicals. The research programs which have led to these exciting developments, and other related studies, were the topics of our conference.

This UCLA Symposium was characterized primarily by the broad mix of research backgrounds of the individuals in attendance, all of whom shared an understanding of the need to apply molecular biology techniques to crop protection problems. The conference was held concurrently with the UCLA Symposium on Molecular Entomology; the groups held two joint sessions to emphasize research in the plant-insect interface. All participants recognized that basic research advancements in plant biology, virology, microbial biology, and insect physiology are now sharing common principles, and that basic understanding of these principles is opening dramatic new strategies for crop protection. Examples of success in utilizing these new strategies heightened the general enthusiasm and excitement of the conference (described as being a meeting "notable for its ebullience in light of recent triumphs," NATURE, Vol 320, p. 571).

The travel and subsistence expenses of the invited speakers were defrayed in large part by a major contribution to the meeting through sponsorship funding provided by E.I. du Pont de Nemours & Company. Additional funding was provided by: Zoecon Corporation, Agrigentics Corporation, Rohm and Hass Company, Agricultural Biotechnology, Monsanto Company, Pioneer Hi-Bred International, Inc., Calgenne, Campbell Institute for Research and Technology, Campbell Soup Company, FMC Corporation, Agriculture Chemical Group, Northrup King Company, and Ecogen, Inc.

We wish to thank Dr. C. Fred Fox for his total commitment and support of the meeting. Our thanks also go to the invaluable assistance of Robin Yeaton, Betty Handy, and their colleagues on the UCLA Symposia staff.

Charles J. Arntzen
Clarence Ryan

I. Protection Against Microorganisms

Molecular Strategies for Crop Protection, pages 3–12
© **1987 Alan R. Liss, Inc.**

A DEMONSTRATION OF PATHOGEN-DERIVED RESISTANCE USING
ESCHERICHIA COLI AND THE BACTERIOPHAGE, Qβ[1]

Rebecca Grumet,[2] John C. Sanford,[3] and
Stephen A. Johnston[2]

[2]Botany Department, Duke University, Durham,
North Carolina 27706
[3]Department of Horticultural Science, Cornell
University, Hedrick Hall, Geneva, New York 14456

ABSTRACT Pathogen-derived resistance is a new
approach to creating disease resistant organisms.
The idea is to transfer genes from the pathogen to
the host in order to block or attenuate infection.
By causing the host to produce pathogen-specific
products at the wrong time in the pathogen's life
cycle, in the wrong amount, or in an altered form, it
should be possible to interfere with the normal
pathogenic process. This approach has several
potential advantages over the more traditional use of
host-encoded resistance genes. Most importantly,
each pathogen would provide its own source of
resistance genes, and isolation of genes from a
relatively small pathogen genome should be simpler
than from a large host genome. In this paper we
demonstrate the successful application of this method
to developing resistance in a model system using
Escherichia coli and the bacteriophage, Qβ. E. coli
JM103 that were transformed with a high copy number
plasmid containing the Qβ coat protein gene fused to
the lacZ promoter, show 1000-fold greater resistance
to Qβ infection than do control bacteria containing a

[1]Support for this research was provided by the North
Carolina Consortium for Research and Education in Plant
Molecular Biology, sponsored by the North Carolina
Biotechnology Center (fellowship to R.G.) and by N.S.F.
Grant DCB-8502626 to S.A.J.

control plasmid without the coat protein gene. The
E. coli expressing the coat protein also showed
varying levels of resistance (0 to 10-fold) to
several other related and unrelated bacteriophages.
The relevance to, and potential application of this
approach to developing disease- resistant plants is
discussed.

INTRODUCTION

Pathogen-derived resistance is a new and potentially
advantageous approach to genetically engineering disease
resistant organisms (1). This approach, which utilizes
the genome of the pathogen as the source of resistance
genes, is based on the principle that in any set of host-
pathogen interactions there are certain pathogen-encoded
functions that are essential to the pathogen, but not the
host. If one of these pathogen-specific functions is
disrupted, the pathogenic process should be stopped. An
outline of this method of developing resistance is to
isolate a gene from the pathogen which encodes a function
essential for pathogenicity and then use that gene, or an
altered form of it, to transform the host. Disease
resistance would result by disrupting the pathogen's life
cycle with a pathogen product that is expressed at the
wrong time, in the wrong amount, or in a counterfunctional
form. Since such disruption would be aimed specifically
at a pathogen function, it should have little, if any,
effect on the host. Two obvious potential advantages of
this approach over the more conventional host-derived
resistance are: first, the source of resistance genes
would never be in question as the genome of each pathogen
would provide resistance genes, and second, since
pathogens have smaller genomes and shorter life cycles
than their hosts, the isolation of such genes should be
simpler.
        In this paper we describe the successful application
of this technique to developing disease resistance in a
simple, well-defined, model system: virus infection of
Escherichia coli by the positive, single-stranded RNA,
F-specific, group III (2) bacteriophage, Qβ. The Qβ
genome is completely sequenced and consists of only four
cistrons (3,4). In theory, each of the cistrons could be
manipulated to interfere with the normal Qβ life cycle
(1). For the experiments described here we chose the coat

protein gene. In addition to serving a structural function of forming the viral capsid, the coat protein acts as a negative regulatory molecule (4,5). Late in the Qβ life cycle the coat protein binds to a specific site on the Qβ genome; this prevents transcription and allows for viral assembly. Thus premature expression of coat protein would be predicted to inhibit phage replication. In this work we demonstrate that premature expression of coat protein by transformed E. coli results in a very high level of protection against Qβ infection. Like Qβ, a majority of the economically important plant viruses are also positive, single-stranded RNA viruses. Thus the Qβ work should also provide an indication of directions to pursue, or not to pursue, in developing virus resistant plants.

MATERIALS AND METHODS

E. coli JM103 [F', lacI^q, (lac pro), thi, strA, supE, endA, sbcB, hsdR-, F'traD36, proAB, ZDM15] (6) were transformed as per Maniatis (7) with three high copy number plasmids containing Qβ-derived cDNA inserts fused to a lacZ promoter. The plasmid inserts are shown in Fig. 1. pRW1169 contains a 1362 base pair (bp) coding sequence

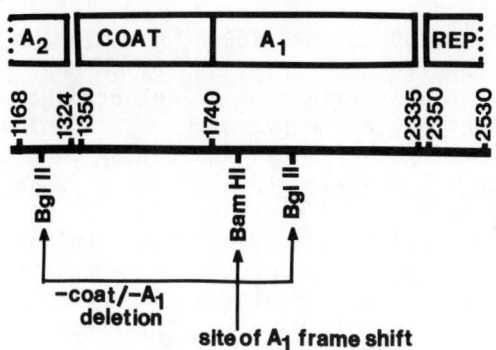

FIGURE 1. Structure of the Qβ cDNA insert of the RW1169 plasmid. Modifications to derive the pRWΔ and RW1169-A$_1$ plasmids were made at the indicated restriction sites. Base pair numbers refer to positions in the Qβ genome from the 5' terminus (3).

for the Qβ coat protein and its read-through protein A_1
(gift of Winter and Gold). pRWΔ was made as a control.
It has an approximately 700 bp Bgl II deletion including
the 390 bp coat protein coding sequence. A second type of
control plasmid, pRW1169-A_1, was used to test the effect
of the A_1 protein. A frame shift mutation was made using
a BamHl digest, Klenow (fill-in) reaction, and blunt end
ligation at the 5' end of the A_1 sequence while leaving
the coat protein gene intact. All plasmids contain a gene
for ampicillin resistance.

     The transformed bacterial strains were tested for
susceptibility to Qβ infection using plaque assays (7) on
YT media (7) with 25 ug/ml ampicillin, or in liquid
cultures as follows. Bacterial cultures were grown in
L-broth (7) with 25 ug/ml ampicillin to A_{550} = 0.4 and
then inoculated with Qβ at a multiplicity of infection
(m.o.i.) of 0.1, 0.2, 1, 2, 5, or 10. At various time
intervals after inoculation with Qβ, 0.5 ml samples were
removed from the bacterial cultures. The samples were
either centrifuged at 12,000 g for 10 min immediately
after removal or artificially lysed as per Model et al.
(8) prior to centrifugation. The supernatants (cell
lysates) were diluted and titered for plaque forming units
of Qβ/ml using JM103+pRWΔ bacteria.

                         RESULTS

     Both the pRW1169 and pRW1169-A_1 plasmids, but not the
pRWΔ plasmid, conditioned expression of Qβ coat protein as
visualized by SDS-polyacrylimide gel electrophoresis (data
not shown). Coat protein expression was inducible by IPTG
(isopropyl-B-D-thiogalactoside) indicating that the coat
gene expression was under control of the lacZ promoter.
There was no effect of the coat protein gene or IPTG on
growth of the bacterial strains in liquid culture.
Doubling times during log phase growth were: JM103+pRWΔ,
39 min; JM103+pRW1169-A_1, 39 min; JM103+pRW1169, 37 min.
     Depending on the multiplicity of infection, addition
of Qβ to the control JM103+pRWΔ cultures caused either a
reduction or cessation of growth, and visible cell lysis.
The titer of Qβ increased rapidly in these cultures (Fig.
2). In contrast, bacteria with the coat protein gene
(JM103+pRW1169 or JM103+pRW1169-A_1) continued to grow
normally and showed no evidence of lysis. Even at
m.o.i.'s as high as 10, replication of Qβ was completely

prevented (Fig. 2).

The difference in levels of Qβ found in the bacterial lysates of the JM103+pRW1169 and JM103+pRWΔ strains was not due to a difference in lysis between the two strains; artificial lysis did not alter the relationship between the two curves (Fig. 2). The A_1 protein was also not responsible for the difference in rate of Qβ replication; both the pRW1169 and pRW1169-A_1 plasmids were equally successful in conferring resistance to Qβ infection. The difference was also not due to a differential loss of F' among the strains; all strains were equally susceptible to infection by the pilus-specific phage, f2 (Table 1). We conclude that resistance to Qβ infection was due to expression of the coat protein gene.

We also tested whether expression of Qβ coat protein could protect against infection by phage other than Qβ

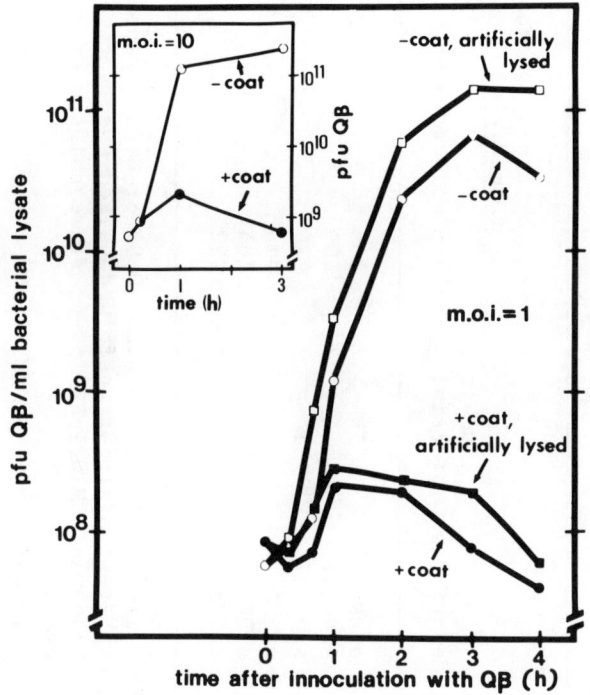

FIGURE 2.   Plaque forming units of Qβ per ml of bacterial lysate vs. time after inoculation. −Coat = JM103+pRWΔ E. coli; +coat = JM103+RW1169 or JM103+RW1169-A_1 E. coli. Results were equivalent whether the RW1169 or RW1169-A_1 plasmids were used.

TABLE 1

TEST FOR RESISTANCE TO INFECTION BY OTHER BACTERIOPHAGE

phage	description	number of trials	mean number of plaques formed/plate		ratio[b]	t-test
			JM103+pRWΔ	JM103+pRW1169[a]		
Qβ	ssRNA,F'	18	336	0.33	1019	18.6***
SP	ssRNA,F'	6	400	39.5	10.1	5.92**
f2	ssRNA,F'	14	220	165	1.3	1.63
f1	ssDNA,F'	9	314	362	0.9	0.97
St-1	ssDNA	11	184	117	1.6	1.84*
P1	dsDNA	14	382	210	1.8	3.85**

*,**,*** significant at P = 0.05, 0.01 or 0.001, respectively according to a one-tailed t-test of log-transformed paired observations.
[a]JM103+pRW1169 and JM103+pRW1169-A₁ frame shift bacteria responded equivalently.
[b]ratio of plaques formed on JM103+pRWΔ vs. JM103+pRW1169 bacteria.

(Table 1). The most striking result was that the
JM103+pRW1169 bacteria showed 10-fold resistance to
infection by the extensively homologous (9) phage, SP.
The JM103+pRW1169 bacteria were not resistant to infection
by the unrelated (4) RNA phage, f2, or the single-stranded
DNA phage f1. Surprisingly, though, there was a low level
of resistance (ca. 2-fold) to the DNA phages St-1 and P1.

DISCUSSION

The data presented above clearly indicate that
expression of Qβ coat protein gene confers a very high
level of resistance to Qβ infection. Since the Qβ coat
protein has been shown to be a negative regulatory
molecule that binds Qβ RNA and inhibits replication in
vitro (4) this is probably the mechanism by which
premature expression of coat protein prevented Qβ
replication in our experiments. Interestingly, even at
m.o.i.'s of 5 or 10 it was not necessary to add IPTG to
obtain resistance. This suggests that only a low level of
coat protein is necessary to confer a high level of
resistance, and is consistent with in vitro studies
showing that a very few molecules of Qβ coat protein are
necessary to repress Qβ replication (5). Leaky expression
of the coat protein gene in the absence of IPTG is likely
due to the high copy number of the plasmid. Similar
uninduced expression of the lac-controlled β-galactosidase
gene was observed by the appearance of light blue colonies
on X-gal (5-bromo-4-chloro-3-indolyl-β-D-galactoside)
indicator plates when the plasmid pUC9 (10) was used to
transform E. coli JM103.
The 10-fold level of cross protection against
infection by SP is especially interesting in light of work
by Coleman et al. (9). They tested for protection against
SP infection using antisense SP RNAs and obtained up to a
10- to 20-fold reduction in SP plaques. However, despite
strong sequence homology between SP and Qβ, the bacteria
expressing SP antisense RNA were not resistant to Qβ
infection. In contrast, the Qβ coat protein conferred
1000-fold resistance to Qβ infection and a 10-fold
inhibition of SP infection. Perhaps there is sufficient
homology between Qβ and SP to enable the Qβ coat protein
to repress SP replication in an attenuated, but analogous
manner to its repression of Qβ replication. On the other
hand, Qβ coat protein did not prevent infection by f2.

This is consistent with the inability of f2 and Qβ coat
proteins to reciprocally inhibit transcription in vitro
(4).   The low level protection against infection by the
DNA phages St-1 and P1 was unexpected. This may be due to
an interference effect of the capsid protein rather than a
regulatory effect.

These observations may provide insight into the
relative effectiveness of alternative approaches to
developing pathogen-derived resistance. For different
possible sources of resistance, (e.g., regulatory
proteins, non-regulatory proteins, or antisense RNAs),
there will be distinct advantages and disadvantages.
Possible differences include: the extent of knowledge of
the biology of the pathogen that is required, the required
level of expression of the pathogen gene, the level of
resistance afforded, and the pathogen specificity of the
resistance.

Pathogen-derived resistance should be especially
applicable to the development of disease resistant plants.
A primary limitation of the more traditional host-derived
resistance is identification of sources of resistance.
The desired resistance is not always available, it may be
associated with undesirable characteristics, it may exist
in a species that is not interfertile with the crop
species, or it may be multigenic and difficult to transfer
(11).   Furthermore, although recombinant DNA technologies
can relieve the limitation that sources of resistance be
confined to the host species or closely related
interfertile species, identifying and isolating resistance
genes from a large plant genome can be formidable.

This may be contrasted with using pathogen genes to
develop resistance. The source of resistance genes would
never be in question as each pathogen would provide its
own. Additionally, since pathogens have smaller genomes
and shorter life cycles than their hosts, the isolation of
such genes should be simpler. Another advantage, is that
in contrast to the natural pesticides of many resistant
plant taxa (12), the pathogen-derived resistance genes
should not produce substances harmful to man or grazing
animals. Pathogen-derived resistance genes should also
have a minimal effect on the host as they should not cause
a change in normal host functions. It has been argued
that host genes controlling susceptibility exist because
they are optimal for essential host functions (13). To
the extent that this is true, host-derived resistance
alleles will tend to disrupt optimal functioning of the

host; this is observed in the Qβ system where host-derived resistance is likely to be achieved by disrupting sex pili formation (14). With pathogen-derived resistance only pathogenic functions should be disrupted. At least in the case of Qβ, there was no negative effect on growth by introducing the coat protein gene.

To conclude, expression of Qβ coat protein by transformed strains of E. coli confers a very high level of resistance to Qβ infection. This clearly demonstrates that a pathogen's genes can be effectively turned against the pathogen and used as a source of resistance by the host. Although these results are not unexpected in light of Qβ biology and studies of translational repression of the Qβ replicase gene by Qβ coat protein (15), it is only recently that the possibility of using such pathogen-derived genes for the purpose of creating disease-resistant organisms has been considered or feasible (1,9,16). Since it has now been shown that it is possible to transform plants using pathogen-derived genes (17), and because at least in the case of virus infection there are direct analogies to the Qβ-E. coli system, we are confident that the approach of pathogen-derived resistance holds promise for the future development of disease resistant plants.

ACKNOWLEDGEMENTS

We especially thank R. Winter and L. Gold for generously providing the pRW1169 plasmid. We also thank M. A. Billeter for the Qβ sequence and R. Webster and M. Inouye for phage strains.

REFERENCES

1. Sanford JC, Johnston SA (1985). The concept of parasite-derived resistance — deriving resistance genes from the parasite's own genome. J Theor Biol 113:395.
2. Weber K, Konigsberg W (1975). Proteins of the RNA phages. In Zinder, ND (ed): "RNA Phages," New York: Cold Spring Harbor Laboratory, p. 51.
3. Meckler P (1981). PhD thesis. University of Zurich.
4. Robertson HD (1975). Functions of replicating DNA in cells affected by RNA bacteriophages. In Zinder ND

(ed) "RNA Phages," New York: Cold Spring Harbor
Laboratory, p. 112.

5. Robertson H, Webster RE, Zinder ND (1968).
   Bacteriophage coat protein as repressor. Nature
   218:533.

6. Messing J, Crea R, Seeburg PH (1981). A system for
   shotgun DNA sequencing. Nucl Acids Res 9:309.

7. Maniatis T, Fritsch EF, Sambrook J (1982). "Molecular
   Cloning a Laboratory Manual," New York: Cold Spring
   Harbor Laboratory.

8. Model P, Webster RE, Zinder ND (1968).
   Characterization of Op3, a lysis defective mutant of
   bacteriophage f2. Cell 18:235.
      9. Coleman J, Hirashima A, Inokuchi Y, Green PJ,
      Inouye M (1985). A novel immune system against
      bacteriophage infection using complementary RNA (mic
      RNA). Nature 351:601.

10. Viera J, Messing J (1982). The pUC plasmids, and
    M13mp7-derived system for insertion mutagenesis and
    sequencing with synthetic universal primers. Gene
    19:259.

11. Knott DR, Dvorak J (1976). Alien germplasm as a
    source of resistance to disease. Ann Rev Phytopathol
    14:211.

12. Ames BN (1983). Dietary carcinogens and
    anticarcinogens. Science 221:1256.

13. Vanderplank JE (1978). "Genetic and Molecular Basis
    of Plant Pathogenesis." New York: Springer-Verlag,
    p. 167.

14. Paranchych W (1975). Attachment, ejection, and
    penetration stages of the RNA phage infectious
    process. In Zinder ND (ed): "RNA Phages," New York:
    Cold Spring Harbor Laboratory, p. 85.

15. Campbell KM, Stormo GD, Gold L (1983).
    Protein-mediated translational repression. In
    Beckwith J, Davies J, Gallant JA (eds): "Gene
    Function in Prokaryotes." New York: Cold Spring
    Harbor Laboratory, p 185.

16. Hamilton RI (1985). Using plant viruses for disease
    control. Hort Sci 20:848.

17. Bevan MW, Mason SE, Goelet P (1985). Expression of
    tobacco mosaic virus coat protein by a cauliflower
    mosaic promoter in plants transformed by
    Agrobacterium. EMBO Journal 4:1921.

Molecular Strategies for Crop Protection, pages 13–24
© 1987 Alan R. Liss, Inc.

DISEASE RESISTANCE RESPONSE GENES IN PLANTS:
EXPRESSION AND PROPOSED MECHANISMS OF INDUCTION

David F. Kendra, Brian Fristensky, Catherine H. Daniels,
and Lee A. Hadwiger

Molecular Biology of Disease Resistance Laboratory
Department of Plant Pathology
Washington State University
Pullman, WA 99164-6430

ABSTRACT Several cloned pea genes are activated in
temporal correlation with the expression of disease
resistance observed cytologically in pea pod tissue
inoculated with Fusarium solani f. sp. phaseoli (non-
host resistance) or with races of Pseudomonas syringae
pv. pisi (race-specific resistance). Chitosan, a minor
component of the fungal cell wall is capable of inducing such genes. Some of the same genes were induced
in both the non-host and race-specific resistance,
suggesting that the single dominant Mendelian traits
reportedly involved in race-specific resistance may be
involved in the regulation of multiple response genes
which are closely associated with the expression of
resistance. Proposed mechanisms by which pathogens
induce these genes are discussed in reference to (i)
the in vitro effects of chitosan both on CD spectra of
DNA and on restriction enzyme digests of plasmid DNA;
and (ii) chitosan as a mutagen in the Ames test.

INTRODUCTION

Peas (Pisum sativum) express a "non-host" resistance
response against Fusarium solani f. sp. phaseoli but are
susceptible to Fusarium solani f. sp. pisi. The root rotting fungus f. sp. phaseoli infects beans but not peas and
f. sp. pisi infects peas but not beans. In another host/
parasite interaction five pea varieties express differential
"race-specific" resistance to races 1, 2 and 3 of Pseudomonas

syringae pv. pisi (1). Previously, we have utilized 2D-PAGE
of in vitro translation products to show that at least 20
pea mRNAs are induced in pea tissue during non-host resistance to f. sp. phaseoli (2). Treatments which interfere
with the expression of these mRNAs also cause the tissue to
be susceptible to f. sp. phaseoli (3).
      Recently we have cloned cDNAs corresponding to mRNAs
which were induced in pea tissue as it resisted F. solani
f. sp. phaseoli (4). We refer to the genes coding for
these mRNAs as "disease-resistance-response-genes." Selected clones from the cDNA library were also used to monitor
the accumulation of individual mRNA species in each of five
differential varieties of peas challenged with three races
of Pseudomonas syringae pv. pisi. Northern blot analyses
of pea RNA indicated that phenotypic resistance correlates
temporally with accumulation and susceptibility correlates
with depressed levels of, or absence of, some clone specific
RNAs (5). Thus, some of the same pea genes are expressed
both in non-host and in race-specific resistance.
      A current problem in understanding disease resistance
at the molecular level is determining how components released
from a wide variety of plant pathogenic organisms can activate the same multiple gene-controlled responses in the host.
Many of the multigene responses are associated with resistance to a number of different phytopathogenic organisms
which in some cases are believed to be controlled by a
single Mendelian trait. We have challenged pea tissue with
a broad array of compounds which have a known regulatory
function (6). In preliminary evaluations we assayed the
potential of these compounds to activate increases in the
synthesis, activity, or accumulation of both phenylalanine
ammonia lyase, a key enzyme in lignin and other secondary
metabolism, and pisatin, a pea phytoalexin. To date these
studies indicate that many candidates for regulatory molecules such as plant (or animal) hormones, cyclic AMP, cGMP
(or derivatives), Ca^{++} and calmodulin antagonists, membrane
specific or deteriorating components, wounding, heat shock,
etc., have no effect on enhancing these responses. The
classes of compounds which do have major positive effects
on these assays include polycationic molecules, compounds
which intercalate, alkylate, or crosslink DNA molecules,
enzymes such as RNase and pectic enzymes, base analogs,
heavy metals, X-ray, U.V. (260 nm), certain pharmaceuticals
and a few detergents (6,7,8,9). The most potent elicitors
within these classes were those that in some way affected
DNA. This initial screening assisted in the discovery that

the polycationic carbohydrate, chitosan, which is naturally present in Fusarium walls can induce complete resistance to F. solani f. sp. pisi (10). Thus, chitosan appeared to be the most suitable biotic inducer available for pursuing the mode by which disease resistance response genes are activated. In addition to inducing disease resistance, chitosan also induced the accumulation of mRNAs in peas which when translated in vitro or in vivo selectively synthesized the same 20+ proteins selectively enhanced by F. solani f. sp. phaseoli as it was resisted by pea tissue (11). Similarly, chitosan has been shown to induce the disease resistance response genes in a temporal pattern much like induction by F. solani f. sp. phaseoli (12). Two of the DNA specific abiotic elicitors, Actinomycin D and U.V. (366nm)-activated trimethylpsoralen, also mimicked induction of the 20+ proteins induced in pea tissue by the pea pathogen (10). Trimethylpsoralen functions as an inducer only when pods were pretreated with this compound prior to receiving the 366nm light required for the light induced linking of pyrimidine bases. [Other compounds such as quinacrine and $CdCl_2$ which also induce phytoalexin accumulation induce accumulation of mRNAs which translate into strikingly different protein patterns (11)].

The question arises, by what mechanism can such a diverse set of externally applied stimuli as a bean pathogen, chitosan, Actinomycin D or activated trimethylpsoralen influence the accumulation in peas of the same gene products? Compounds added to pea pod endocarp surfaces have no detectable cuticle structure to confront, but must traverse the plant cell wall and the plasma membrane to gain entrance to the cytoplasm. Externally applied ^{3}H-chitosan is detectable within the cytoplasm and the pea nucleus 20 minutes after treatment (13). Chitosan from Fusarium solani cell walls is detectable in surface pea cells by antichitosan-antisera 30 min following inoculation with F. solani. Alterations in pea chromatin structure are observed in electron micrographs of tissue cross-sections from pea pods treated with chitosan or the major abiotic elicitors. However, the period of chitosan traversion is not accompanied by loss of cell viability or ion leakage (14).

It is certainly a possibility that a route of secondary messenger elicitation exists in mediating the influence of externally applied chitosan. The alternating positive charges associated with the amino groups of the β 1,4 linked glucosamine sugars of the chitosan molecule enable it to aggregate various negatively charged proteinaceous or other

biological materials. Little is known about the membrane
alterations or specific affinities which might be involved
as this compound traverses the plasma membrane and localizes
in the cytoplasm and nucleus of the recipient cell. Although
polycations in high concentrations have been shown to influence membrane conformation in cultured cells (15) resulting in ion leakage and cell death, we have recently utilized
vital staining and conductivity measurements to demonstrate
that chitosan, when applied to endocarp tissue (14) did not
enhance leakage at concentrations which induce the resistance
response.

Chitosan as a polycationic molecule shares some similarities with spermine, protamine, poly-L-lysine and his-
tones and thus also has a high affinity for negatively
charged PO_4^-groups along the sugar-phosphate backbone of DNA
(16). We have proposed (14) that the elicitor action that
chitosan and other DNA specific factors have in common may
be explained via direct alterations in the DNA within
chromatin rather than as multiple ligands acting on a common
receptor molecule at the level of the plasma membrane or
cytosol. We are thus reporting preliminary observations on
chitosan as a mutagen and on the ·in vitro effects of
chitosan/DNA interactions.

Polyamines have been shown to bind preferentially to
double stranded nucleic acids in the minor groove of the
double helix where they interact with the phosphate groups
thereby stabilizing the double-helical structure (16). Also,
polyamines have been shown to enhance nucleic acid-associated
replication and enzymatic transcription reactions as well as
affect translation.

Pingoud, et al. (17) demonstrated that at low concentrations, polyamines stimulate the rate and accuracy of the
restriction endonuclease cleavage of DNA, while at higher
concentrations cleavage was inhibited. Shishido (18) showed
that spermine reduced the cleavage specificity of the single-
strand-specific nucleases, S_1 and Bal 31 while it stimulated
the exonuclease action of Bal 31.

EFFECTS OF CHITOSAN ON DIGESTION OF DNA
BY RESTRICTION ENDONUCLEASES

In order to assess the preference of chitosan for
specific types of sequences, we have performed partial restriction digests of pUC12 (19) supercoiled plasmid DNA,
using a range of chitosan concentrations and limiting concentrations of enzyme. All digests were carried out for 60

min in standard buffer and temperature conditions for each
enzyme (20). Crab-shell chitosan was partially hydrolyzed
with 6N HCl at 53°C for 2 hr, and neutralized by dialysis
with ddH$_2$0. Concentrations of chitosan in reaction mixtures
were 0, 75, 125 and 250 µg/ml. At 500 µg/ml chitosan usually
precipitated DNA from solutions. At 250 µg/ml chitosan will
also precipitate DNA, but this result is difficult to obtain
consistently.

Two effects were seen which were generally enzyme dependent. In the cases of Taq1, Hae3 and Mbo2 a reproducible
enhancement of digestion efficiency was seen at chitosan
concentrations of 75 µg/ml or above. Conversely, the
activity of Pvu1 in digesting pUC12 appears to be inhibited
by chitosan. As seen in Fig. 1, with increasing chitosan
concentration, the mass of DNA in the Taq1 digest is shifted
from lower mobility partial bands towards bands predicted
for a complete digest. With Pvu1, the two complete digest
bands seen in the absence of chitosan decrease in intensity
when chitosan is added, indicating an inhibition in digestion. Significantly, the Pvu1 digest shows that the inhibition affects both the supercoil to linear transition, as well
as further digestion of linear DNA. A final observation
worth noting is that not all bands in the digests are influenced equally by chitosan. This result could be explained
by hypothesizing that sequences adjacent to restriction recognition sites may play a role in the specificity of
chitosan's interaction with DNA. The isolation of genomic
clones for the disease resistance response genes should permit a more careful examination, by techniques such as DNA
footprinting, of sites on DNA for which chitosan may show a
strong preference.

In chitosan/DNA mixtures in which chitosan is added at
concentrations such that its positive charges approach
equivalency with the negative charges of the DNA in solution,
precipitation of the DNA occurs. At lower concentrations,
circular dichroism (CD) spectra reveal subtle effects of
this polycationic polymer on specific DNA oligomer duplexes.

EFFECT OF CHITOSAN AND OTHER POLYCATIONIC
ELICITORS ON CD SPECTRA OF DNA

The CD spectra in Figure 2A indicates that the negative
peak in the area of 270 nm wave length and the positive peak
near 260 nm diminish following the addition of chitosan (36
µg) to 50 µg of a Poly dG·Poly dC segment of DNA. There is
a reduction (Fig. 2D) in the positive 270 nm peak of native

FIGURE

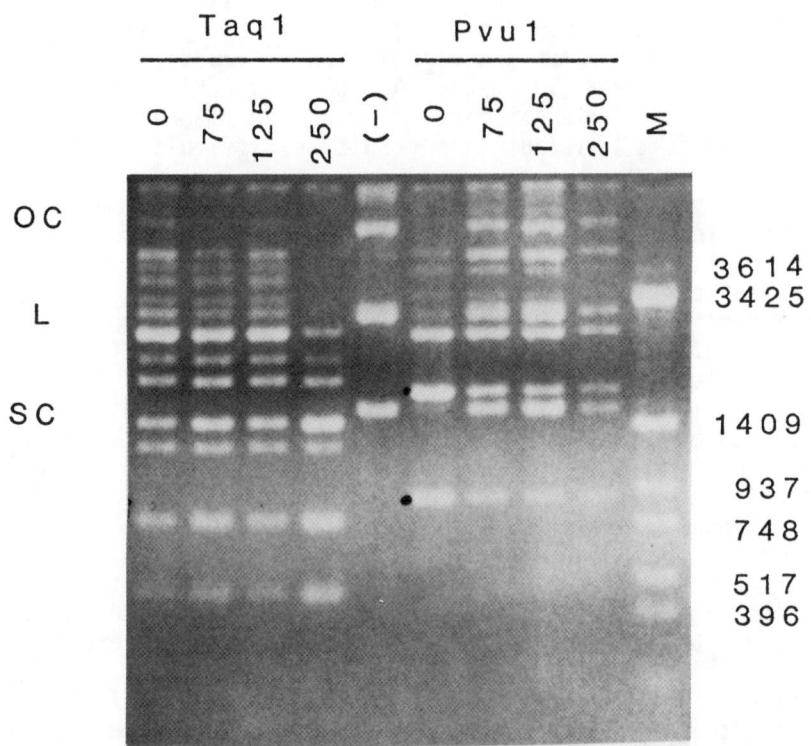

FIGURE 1. Effects of chitosan on partial restriction digests of pUC12. Reaction conditions are descibed in the text. Restriction enzymes used and chitosan concentrations (μg/ml) are shown above digest lanes. (-) undigested pUC12 DNA; SC - supercoiled plasmid; OC - open circular plasmid; L - linear plasmid. Bands expected in a complete digest are indicated by dots. Sizes of marker bands (M) are listed in bp.

pea DNA (50μg) following the addition of 12 or 36 μg of chitosan. Addition of chitosan extracted from walls of incompatible fungi or from crab shells gave similar results. Chitosan, which alone has a straight line spectra, also altered the CD spectra of Poly d[G-C] segments but had little or no effect on Poly d[A-T] or Poly dA·Poly dT segments. Other polycationic compounds such as spermine and protamine which elicit pisatin accumulation but not immunity to F. solani f. sp. pisi in pea pods also have unique effects on

FIGURE

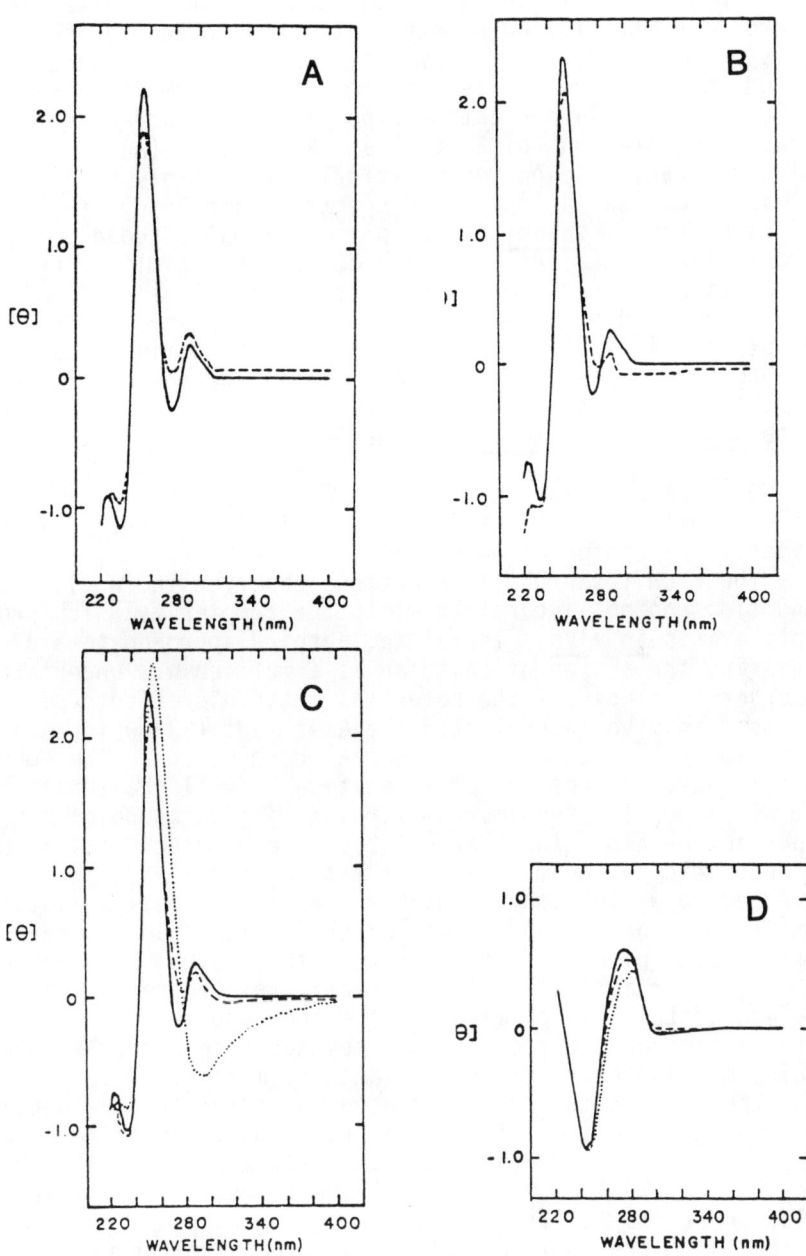

FIGURE 2. Conformational changes in synthetic and
native pea DNA indicated by CD spectra in 3 ml cuvettes in
Dichrograph Mark III CD spectrophotometer following the
addition of polycationic polymers. DNA [50 µg] was dissolved in 2.75 ml of buffer [15 mM NaCl, 1.5 mM Na citrate,
pH 7.0 (SSC x .1)] and cationic polymers were added in
stock concentrations of 2 mg/ml at 22°C (21). (A)
Polydeoxyguanydilic·polydeoxycytidylic acid (Poly dG·Poly
dC) spectrum (solid line), and spectrum with 36 µg of
chitosan from F. solani f. sp. pisi cell walls added
(broken line). (B) Poly dG·Poly dC spectrum (solid line)
and spectrum with 24 µg of spermine (broken line). (C)
Poly dG·Poly dC spectrum (solid line) and spectrum with 12
µg (broken line) and 24 µg (dotted line) protamine added.
(D) Native pea DNA spectrum (solid line) and spectrum with
12 µg (dashed line) and 36 µg (dotted line) of chitosan from
F. solani f. sp. phaseoli cell walls added.

the CD spectra of Poly dG-Poly dC segments (Fig. 2B,C).
These results indicate the potential of elicitors to change
slightly the conformation of DNA.
        These chitosan/DNA interactions observed in in vitro
conditions do not accurately mimic the complexity which probably exists in vivo. Therefore, a third approach toward
assessing the action of chitosan as a mutagen was undertaken.
In order to elucidate the potential mutagenic effects of
chitosan in vivo, we selected the Ames test (22) which has
been widely used to detect potential carcinogens. The Ames
test is based on the use of a genetically well characterized
set of Salmonella typhimurium strains with known point mutations in the histidine operon (22). These mutant strains can
produce colonies on minimal agar medium plates when they are
reverted to prototrophy. The his$^+$ colonies appear against a
thin film of background growth which develops due to the
small amount of histidine present in the medium. The number
of his$^+$ revertant colonies is directly related to the concentration of the test chemical if the compound is mutagenic.
Not all compounds tested in the Ames test are directly mutagenic, but instead must be activated by microsomal enzymes.
        The S. typhimurium stains used to screen for the poten-
tial mutagenic action of chitosan (generously supplied by B.
Ames) are: TA 97a (frameshift), TA98 (frameshift), TA100
(base pair), TA 102 (Ochre) and TA104 (nonsense). A chitosan
dilution series was assayed both with and without addition
of the S-9 microsomal activation mixture using the protocol
of Maron and Ames (22). Previously described positive con-

trols for each tester strain were also run.

The mutagenic action of chitosan was most noticeable in
S. typhimurium strains TA97a, TA98 and TA100 (Table 1). At
a concentration of 500 µg/ml, chitosan induces his$^+$ revertants in all strains (except TA104, data not shown) with a
frequency substantially greater than the level of spontaneous
revertants (0 µg/ml chitosan). From the results presented,
it appears that in strains 97a and 98 chitosan concentrations
of 300 µg/ml or greater are required for mutagenic action;
while for strain TA100, a chitosan concentration as low as
100 µg/ml can induce mutations. The concentrations of
chitosan required to induce frameshift mutations are much
higher than those which permanently eliminate fungal macroconidia viability (14). However, the minimum chitosan
concentration which stimulates base-pair susbtitutions (100

TABLE 1

Mutagenic activity of chitosan toward
Salmonella typhimurium stains TA97a, TA98,
TA100 and TA102

Chitosan Concentration (µg/ml)	His$^+$ revertants[a]							
	TA97a		TA98		TA100		TA102	
	+S9	-S9	+S9	-S9	+S9	-S9	+S9	-S9
500[b]	137	121	127	125	270	271	663	524
400[b]	101	100	96	81	274	273	-	-
300	96	79	58	65	220	222	-	-
200	54	48	33	22	219	208	-	-
100	60	50	33	28	198	199	-	-
50	55	60	31	23	166	154	-	-
8	57	57	33	19	142	168	543	395
2	59	57	30	23	151	152	528	390
0	60	59	21	28	155	163	462	423

a The number of revertant per chitosan concentrations are
the average between two separate experiments.
b Due to the agglutination properties of the 500 and 400
µg/ml chitosan solutions, some of the bacteria were agglutinated into small clumps; therefore the reported
number of revertants is probably an underestimate of the
true value.

μg/ml) closely mimics the minimum concentration which permanently eliminates fungal macroconidia viability (14). In all strains it appears that the mutagenic activity exhibited by chitosan does not require enzymatic activation.

CONCLUSION

The experiments described in this paper were not intended as an exhaustive study of the physical chemistry of DNA/chitosan interactions. Rather, they were pilot experiments intended to determine whether the hypothesis, that chitosan affects gene expression through interaction with DNA, merited further pursuit. We have presented evidence using several parameters suggesting an interaction of chitosan with DNA. As a whole the data indicate that under normal physiological conditions, given the opportunity, chitosan will interact with DNA, due to the physical properties inherent in the two molecules. Since chitosan does enter the plant cell during infection with Fusarium solani, and localizes in the nucleus, it is reasonable to expect that chitosan would interact with those portions of the chromatin that are most accessible, i.e., potentially transcriptionally active regions. The apparent sequence dependent effects of chitosan on restriction digests bolsters the possibility that genes whose activites are regulated by chitosan would have regulatory sequences for which chitosan had some preference. Future investigations which concentrate on the interactions between chitosan and cloned genomic copies of the disease resistance response genes, and the effects of chitosan on the chromatin conformations of these genes in vivo may provide more definitive conclusions.

ACKNOWLEDGEMENT

We acknowledge the support by Washington Sea Grant RX-13 and National Science Foundation Grant DMB 84-14870.

REFERENCES

1. Hadwiger LA, Webster DM (1984). Phytoalexin produc-duction in five cultivars of peas differentially resistant to three races of Pseudomonas syringae pv. pisi. Phytopathology 74:1312.
2. Hadwiger LA, Wagoner W (1983). Electrophoretic patterns of pea and Fusarium solani proteins synthesized in vitro or in vivo which characterize the com-

patible and incompatible interactions. Physiol Plant Path 23:153.

3. Hadwiger LA, Wagoner W (1983). Effect of heat shock on the mRNA-directed disease resistance response of peas. Plant Physiol 72:553.

4. Riggleman RC, Fristensky B, Hadwiger LA (1985). The disease resistance response in pea is associated with increased levels of specific mRNAs. Plant Mol Biol 4:81.

5. Hadwiger LA, Daniels C, Fristensky B, Kendra DF, Wagoner W (1986). Pea genes associated with the non-host resistance to Fusarium solani are also induced by chitosan and in race-specific resistance by Pseudomonas syringae. In Bailey JA (ed): "Biology and Molecular Biology of Plant Pathogen Interactions." In press.

6. Hadwiger LA, Jafri A, vonBroembsen S, Eddy R (1974). Mode of pisatin induction. Plant Physiol 53:52.

7. Schwochau ME, Hadwiger LA (1969). Regulation of gene expression by actinomycin D and other compounds which change the conformation of DNA. Arch Biochem Biophys 134:34.

8. Hadwiger LA (1972). Induction of phenylalanine ammonia-lyase and pisatin by photo sensitive psoralen compounds. Plant Physiol 49:779

9. Walker-Simmons M, Jin D, West CA, Hadwiger LA, Ryan CA (1984). Comparison of proteinase inhibitor- inducing activities and phytoalexin elicitor activities of a pure fungal endopolygalacturonase, pectic fragments and chitosan. Plant Physiol 76:833.

10. Hadwiger LA, Beckman JM (1980). Chitosan as a component of pea-Fusarium interactions. Plant Physiol 66:205.

11. Loschke DC, Hadwiger LA, Wagoner W (1983). Comparison of mRNA populations coding for phenylalanine ammonia-lyase and other peptides from pea tissue treated with biotic and abiotic phytoalexin inducers. Physiol Plant Path 23:163.

12. Fristensky BW, Riggleman RC, Wagoner W, Hadwiger LA, (1985). Gene expression in susceptible and disease resistant interactions of peas induced with Fusarium solani pathogens and chitosan. Physiol Plant Pathol 27:15.

13. Hadwiger LA, Beckman JM, Adams MJ (1981). Localization of fungal components in the pea-Fusarium interaction detected immunochemically with anti-

chitosan and antifungal cell wall antisera. Plant
Physiol 67:170.
14. Kendra DF, Hadwiger LA (1986). Cell death and membrane leakage are not associated with the induction of
disease resistance in peas by chitosan. Phytopathol.
(In Press).
15. Young DH, Kauss H (1983). Release of calcium from
suspension-cultured Glycine max cells by chitosan,
other polycations and polyamines in relation to effects
on membrane permeability. Plant Physiol 73:698.
16. Tabor CW, Tabor H (1984). Polyamines. Ann Rev Biochem
53:749.
17. Pingoud A, Urbanke C, Alves J, Ehbrecht H-J, Zabeau M,
Gualerzi C (1984). Effect of polyamines and basic
proteins on cleavage of DNA by restriction endonucleases. Biochemistry 23:5697.
18. Shishido K (1985). Effect of spermine on cleavage of
plasmid DNA by nucleases S_1 and Bal 31. Biochem
Biophys Acta 826:147.
19. Messing J (1983). New M13 vectors for cloning. Meth
Enzymol 101:20.
20. Fuchs R, Blakesley R (1983). Guide to use of Type II
restriction endonucleases. Meth Enzymol 100:3.
21. Zarlenga DS, Halsall HB, Day RA (1984). A polycationic amine that induces unique conformational changes
in poly(dA-dT) in low salt. Nucl Acids Res 12:6325.
22. Maron DM, Ames BN (1983). Revised methods for the
Salmonella mutagenicity test. Mutation Res 113:173.

Molecular Strategies for Crop Protection, pages 25–34

HOST VS. NONHOST RESISTANCE

Michèle C. Heath

Botany Department, University of Toronto, Toronto, Ontario, Canada M5S 1A1

ABSTRACT To be a plant pathogen, a fungus usually must possess attributes that enable it to 1) enter the tissue, 2) obtain from the tissue those factors necessary to support growth and reproduction, and 3) combat the plant's constitutive or induced defense mechanisms. The lack of any of these attributes with respect to a given plant will result in plant resistance. Conceptually, two types of resistance towards a given plant pathogen can be recognized: host resistance that is expressed only by certain genotypes of an otherwise susceptible plant species, and nonhost resistance that is exhibited by plant species that contain no susceptible genotypes. Studies with rust fungi suggest that compared with host resistance, nonhost resistance is expressed earlier, may involve different mechanisms, and is more difficult for the fungus to overcome. The possible reasons for the latter phenomenon, and their significance in designing novel and durable forms of disease resistance, are discussed.

INTRODUCTION

One of the more attractive ways of protecting crop plants against fungal parasites is to manipulate them genetically so that their previous state of susceptibility becomes one of resistance. Such an approach has been used successfully by plant breeders for many years, but the resulting "host" (or "cultivar") resistance often is not durable, and a new race of the pathogen may soon evolve that can overcome this resistance (20). In contrast, "nonhost" resistance (the resistance shown by all genotypes of a species not considered to be a host for the fungus in question) is extremely durable judging by the fact that the host species range of most plant pathogens has not changed noticeably over recorded history. Since cultivar resistance

is often conferred by the introduction of single genes into
a susceptible crop species from a resistant relative, it
seems highly amenable to precise manipulation using
recombinant DNA technology. Moreover, such technology allows
the tranfer of genes from a wider selection of donors than
can be achieved by conventional breeding, and it has often
been suggested that the use of donor species unrelated to
the crop plant will result in the introduction of more
durable, nonhost-type resistance (3,8). I believe this idea
to be overly simplistic. There is no doubt in my mind that
recombinant DNA technology does indeed provide the promise
for producing durable resistance in our crop plants, but not
from such an empirical approach. Rather, this technology
should be used to construct predictably durable resistance
based on a detailed understanding of what makes resistance
durable. To my mind, the clues to this understanding lie in
a consideration of the evolutionary history of both
parasitism and resistance with respect to vascular plants
and fungi.

THE EVOLUTION OF RESISTANCE AND SUSCEPTIBILITY

To stand any chance of survival to maturity, any
terrestrial vascular plant must be resistant to invasion by
most of the hundreds of extant species of terrestrial fungi
that it is likely to encounter during its lifetime. Given
the number of fungal species involved, it is highly unlikely
that each plant species possesses a defense mechanism
specific for each species of fungus (23,24). More likely, it
has acquired during its evolution a "battery" of nonspecific
passive and active defense mechanisms against fungal
invasion. In fact, deterrents against parasitism by
micro-organisms may pre-date the evolution of vascular
plants, since the ability to deter invasion and predation
must have been of strong selective advantage in the first
non-motile autotrophs that could not move to escape attack.
These defense mechanisms may not always have been as complex
as they are today since the resistance of a plant is only a
reflection of the lack of parasitic capabilities of the
fungus. This point will be returned to later, but to give a
simple example here, the plant cuticle will be an effective
defense mechanism for as long as the fungus lacks the
ability to penetrate it. One can see how, during the
co-evolution of plants and fungal parasites (as of plants
and herbivorous insects see ref. 4), defense mechanisms may
have become more complex as fungi acquired ways of

overcoming pre-existing deterrents against infection. It is also important to recognise that many defense mechanisms may have evolved as part of an integrated system of defense against micro-organisms, predation and mechanical damage, with many mechanisms serving more than one role. For example, high levels of tannins and other phenolics in the plant could have an adverse effect on predation and parasitism by mammals, insects and bacteria as well as fungi (10).

Resistance to parasitism by most fungi is, therefore, a characteristic of all extant plants and is exemplified by the relatively few fungal species that can successfully invade a given plant (26). Consequently, a parasitic fungus is one that has acquired, during its evolution, special attributes to combat this resistance (14), as well as the ability to attempt invasion and to obtain from the plant the necessary factors for growth and reproduction (18). Therefore, the host species specificity of a fungal pathogen reflects more the specific attributes of the fungus than the specific resistance mechanisms of the plant. Given the biochemical differences between plants, particularly in their constitutive and inducible antifungal compounds (10), many of these fungal attributes are likely to be specific for the plant in question, and each can be regarded as a pathogenicity factor. The number and type of such factors needed for parasitism is going to depend on the life style of the fungus; for example, a parasite that has adapted to penetrate through, natural openings is not going to require enzymes that degrade the plant cuticle. However, it is highly unlikely that a single attribute could confer complete pathogenicity on a previously saprophytic fungus. It seems reasonable to assume, therefore, that pathogenesis usually is based on a number of pathogenicity factors.

Although all plants have a battery of potential defense mechanisms against fungal invasion, not all of them will be called into play by a single fungus. To return to a previous example, the cuticle can only be a defense barrier if the fungus tries to penetrate directly. Similarly, a lignified mesophyll cell wall cannot act as a physical barrier if the fungus remains in the intercellular spaces (whether one regards the ability to avoid penetrating such walls as a pathogenicity factor is a moot point). Therefore, susceptibility has to be viewed in terms of the fungus being able to match (not necessarily on a 1:1 basis) a pathogenicity factor with any resistance mechanism that it would otherwise encounter. Thus, the possession of a

cutinase for a fungus that must enter via the cuticle is a
pathogenicity factor; the cuticle is the matching resistance
mechanism, since it prevents infection if the enzyme is
absent. It is important to note that a pathogenicity factor,
in the present context, could be the absence of a molecule
or process that might, for example, otherwise trigger active
defense responses. Similarly, the ability of the fungus to
respond to a plant product in some manner essential for
parasitism (e.g. formation of an infection structure or
directional growth toward the plant) also can be regarded as
a pathogenicity factor; the corresponding resistance
mechanism would be the lack of this product in the plant. It
also should be realized that not all pathogenicity factors
have to be matched by a resistance mechanism in the plant.
Certain features necessary for parasitism may be
constitutive in a given fungus (e.g. the ability to form
appressoria without exogenous stimuli) and would not
correspond to any feature in the plant that could act as a
resistance mechanism if the fungal feature was lost.

One can see, therefore, that the ease with which a
fungus can become a parasite of a specific plant species
depends on the number of pathogenicity factors that it has
to acquire. The degree of specificity associated with these
factors also is important; if the fungus combats, for
example, the effect of the plant's antifungal phytoalexins
by not eliciting their synthesis, this may merely require
the loss or modification of an eliciting molecule, and may
be achieved relatively easily by random mutation (14,20).
However, if the eliciting molecule is essential to fungal
survival, the fungus may have to solve the problem by
producing a specific enzyme that detoxifies the phytoalexin
(see ref. 22 and chapter by VanEtten in this volume). In
this case, random mutation is much less likely to produce
such a specific molecule (14,20) unless the fungus already
possesses a related enzyme of slightly different
specificity. Presumably, for a saprophytic fungus or one
that is a parasite of a different plant family, the number
of pathogenicity factors that must be acquired in order to
successfully attack a given plant species is going to be
quite large (17). Consequently, the chances of random
mutation producing the necessary variety of specific and
non-specific pathogenicity factors tailor-made to this
nonhost species becomes vanishingly small, and the plant's
resistance is correspondingly durable.

A COMPARISON OF HOST AND NONHOST RESISTANCE

When looked at in this light, it seems obvious why so many examples of man-bred host resistance are not durable. Such resistance is bred empirically into a species for which the given fungus already has the pathogenicity factors necessary for successful parasitism. The fact that such resistance may be based on defense mechanisms different from, and more easy to overcome than, those involved in nonhost resistance is exemplified by cytological studies of host and nonhost resistance towards rust fungi. Nonhost resistance towards these biotrophic fungi is usually expressed before the fungus manages to form the first haustorium inside a plant cell (11,12) and seems to involve several mechanisms (even in one plant) such as poor recognition of the leaf surface, antifungal compounds, and changes in the plant wall (16). In contrast, in its resistant host, a rust fungus usually begins to develop normally, presumably because it has the pathogenicity factors necessary to combat these prehaustorial defenses (11,12,16). Resistance is manifested by plant cell death after haustorium formation, a phenomenon usually seen infrequently in nonhost plants. Only rarely does host resistance involve prehaustorial defenses. Significantly, such resistance may be unusually durable in the field (2,20).

Judging by the gene-for-gene relationship that often exists between typical host resistance and fungal virulence and avirulence (5), overcoming this resistance relies on the fungus modifying a single allele. The common recessive nature of this allele strongly suggests that the pathogenicity factor in such cases is the lack or modification of a molecule that would otherwise trigger resistance in the plant, and it is not surprising that random mutation may produce such changes at relatively high frequency. However, if such a mutation was deleterious to the survival of the fungus, or if the pathogenicity factor had to have a high level of molecular specificity, then one would expect that resistance, even though controlled by a single gene, would be relatively durable. Therefore, the durability of resistance ultimately lies in the ease with which random mutation can produce a means to combat it, not whether it originated in host or nonhost plants or is controlled by one or many genes.

STRATEGIES FOR ENGINEERING DURABLE FORMS OF RESISTANCE

If these arguments are correct, then it seems unrealistic to expect to be able to take a completely susceptible host plant and turn it into a true nonhost of the pathogen in question; this requires changes in the plant's basic defense mechanisms so that they no longer match the pathogen's pathogenicity factors. Nevertheless, using the precision inherent in recombinant DNA technology, it should be possible to use the principles of durable resistance discussed above to create novel forms of resistance that are difficult for the fungus to overcome. One strategy is to introduce many new defense mechanisms into the plant that are unmatched by any pathogenicity factor in the fungus. As long a different pathogenicity factor is required to match each defense mechanism, the chance of a fungal individual gaining them all would be small even if the fungus could evolve some of the necessary pathogenicity factors fairly readily. Another, perhaps technically easier, strategy would be to introduce into a susceptible plant one or two defenses that are known to be particularly difficult for the given fungus to evolve a pathogenicity factor for.

There are two basic ways in which these "new" defense mechanisms can be engineered. The first is to change the nature or regulation of an existing "defeated" or unused defense mechanism so that it becomes effective against, or is triggered by, the pathogen. The second is to introduce a new defense mechanism (either from another plant or one that is totally novel) for which the fungus has no pathogenicity factor. Which approach is the easiest is going to depend on the plant - parasite system under consideration. An example of systems where the first approach may be more profitable are those involving rust fungi. Many rust fungi penetrate their host plants through stomata which they find and recognize by responding to topographical features of the leaf surface (25). On nonhost plants, particularly if their surfaces significantly differ in waxiness from that of the host plant, a given rust fungus will make more mistakes, forming fewer appressoria and/or more of these in places other than over stomata (11,12,25); consequently, fewer fungal individuals enter the leaf. In spite of various suggestions (reviewed in ref. 25), it is still not known for certain what surface feature the fungus responds to. When this feature has been identified, it may be possible to produce host plants with surfaces that fail to trigger fungal differentiation and are therefore protected against rust infection. Superficially, one might expect that the adaptation of the fungus to the new host surface would require specific attributes that are difficult to evolve. Whether this assumption is correct depends on the nature of the mechanism by which the fungus recognizes the plant surface and although progress is being made in this area (e.g. 21), much is still unknown. However, lab-induced mutants could be used to investigate the frequency with which adaptation to previously non-inducing surfaces might arise.

The second approach, to my mind, may be the more difficult in the long run despite it often appearing the most attractive. The first problem is to choose what defense mechanism to transfer to the susceptible plant, and to do this one has to be certain that the plant truely lacks the mechanism under consideration. For example, French bean plants respond to the cowpea rust fungus by depositing silica in and on adjacent mesophyll cell walls (13), and these silicified walls seem to be involved in preventing haustorium formation by the fungus (19). Susceptible cowpea plants do not respond in this way to the cowpea rust fungus and one might be tempted to regard silica deposition as a trait that could be transferred to the cowpea plant to confer rust resistance. However, when tested with other rust fungi nonpathogenic to cowpeas, cowpea plants readily deposit silica in their cell walls (Heath unpublished). Therefore, the susceptibility of cowpea to the cowpea rust fungus lies not in its lack of ability to form silica, but in the ability of the fungus to specifically suppress or not trigger this response. Indeed, further investigations revealed that silica deposition is a rather common response of legumes and other plants to damage (13, Heath unpublished). Like lignification (see 9) and callose formation (1), silica deposition may be a potential defense reaction that most, if not all, vascular plants can exhibit if given the right stimulus.

More promising defense reactions in this context are those based on species- or genus-specific molecules (e.g. toxins, phytoalexins). However, currently it is difficult to assess how easy it is going to be to introduce new biosynthetic pathways into a plant, or what effect this may have on its metabolic equilibrium. As discussed by Foard et al (7) (also see chapter by Ryan in this volume), defenses effected by primary gene products would seem to be the easiest to manipulate in this way, although depressingly few polypeptides have been demonstrated to have a direct role in plant resistance towards fungi (see 7).

Obviously, such a rational approach to crop protection against fungal parasitism requires a much better knowledge of the interactions between fungi and plants than is currently available. Apart from a few noticeable exceptions covered in other chapters in this volume, pathogenicity factors have not received wide attention among pathologists working with fungal pathogens. As indicated earlier, the idea of what is a pathogenicity factor may run into some conceptual difficulties. However, in practice, such factors could be defined, as suggested by Ellingboe and Gabriel (6), in terms of genes which when mutated affect pathogenicity while not affecting saprophytic growth. Nonhost resistance also has not been a popular field of study, although knowledge of the types of basic defense mechanisms possessed by a given species, and how the active forms are triggered by a nonpathogen, is essential if we wish to exploit such resistance. In my opinion, the more popular studies of host resistance are relatively unlikely to aid in the precise production of durable disease resistance unless they focus on comparisons of forms known from field studies to be particularly difficult, or particularly easy, for a given fungus to overcome.

REFERENCES

1.   Aist JR (1976). Papillae and related wound plugs of plant cells. Ann Rev Phytopathol 14:145.
2.   Clifford BC, Carver TLW, Roderick HW (1985). The implications of general resistance for physiological investigations. In Groth JV, Bushnell WR (eds): "Genetic Basis of Biochemical Mechanisms of Plant Disease", St. Paul: APS, American Phytopathological Society Symposium Book No.4, p 43.
3.   Day PR, Barrett JA, Wolfe MS (1983). The evolution of host-parasite interaction. In Kosuge T, Meredith CP, Hollaender A (eds): "Genetic Engineering of Plants. An Agricultural Perspective", New York: Plenum, p 419.
4.   Edwards PJ, Wratten SD (1980). "Ecology of Insect-Plant Interactions", Edward Arnold: London.
5.   Ellingboe AH (1976). Genetics of host-parasite interactions. In Heitefuss R, Williams PH (eds): "Encyclopedia of Plant Physiology Vol. 4. Physiological Plant Pathology", Berlin: Springer, p 761.

6.   Ellingboe AH, Gabriel DW (1977). Induced conditional
       mutants for studying host/pathogen interactions. In:
       "Induced Mutations against Plant Disease", Vienna:
       International Atomic Energy Agency, p. 35.
7.   Foard DE, Murdock LL, Dunn PE (1983). Engineering of
       crop plants with resistance to herbivores and
       pathogens: an approach using primary gene products. In
       "Plant Molecular Biology", New York: Alan R. Liss, p
       223.
8.   Gilchrist DG, Yoder OC (1984). Genetics of
       host-parasite systems: A prospectus for molecular
       biology. In Kosuge T, Nester EW (eds): "Plant-Microbe
       Interactions. Molecular and Genetic Perspectives. Vol
       1", New York: Macmillan Publishing Company, p 69.
9.   Hammerschmidt R, Bonnen AM, Bergstrom GC, Baker KK
       (1985). Association of epidermal lignification with
       nonhost resistance of cucurbits to fungi. Can J Bot
       63:2393.
10.   Harborne JB (1982). "Introduction to Ecological
       Biochemistry", 2nd. Edit. London: Academic Press.
11.   Heath MC (1974). Light and electron microscope studies
       of the interactions of host and non-host plants with
       cowpea rust- Uromyces phaseoli var. vignae . Physiol
       Plant Pathol 4:403.
12.   Heath MC (1977). A comparative study of non-host
       interactions with rust fungi. Physiol Plant Pathol
       10:73.
13.   Heath MC (1979). Partial characterization of the
       electron-opaque deposits formed in the non-host plant,
       French bean, after cowpea rust infection. Physiol
       Plant Pathol 15:141.
14.   Heath MC (1981). A generalized concept of
       host-parasite specificity. Phytopathology 71:1121.
15.   Heath MC (1981). The suppression of the development of
       silicon-containing deposits in French bean leaves by
       exudates of the bean rust fungus and extracts from
       bean rust-infected tissue. Physiol Plant Pathol
       18:149.
16.   Heath MC (1981). Resistance of plants to rust
       infection. Phytopathology 71:971.
17.   Heath MC (1985). Implications of nonhost resistance
       for understanding host-parasite interactions. In Groth
       JV, Bushnell WR (eds): "Genetic Basis of Biochemical
       Mechanisms of Plant Disease", St Paul: APS, American
       Phytopathological Society Symposium Book No.4, p 25.
18.   Heath MC (1986). Fundamental questions related to

plant-fungal interactions: can recombinant DNA
technology provide the answers? In Bailey JA (ed):
"Biology and Molecular Biology of Plant-Pathogen
Interactions", New York, Plenum (in press).

19. Heath MC, Stumpf MA (1986). Ultrastructural
observations of penetration sites of the cowpea rust
fungus in untreated and silicon-depleted French bean
cells. Physiol Plant Pathol (in press).

20. Parlevliet JE (1983). Models explaining the
specificity and durability of host resistance derived
from the observations on the barley- Puccinia hordei
system. In Lamberti F, Waller JM, Van der Graaff NA
(eds): "Durable Resistance in Crops", New York:
Plenum, p 57.

21. Staples RC, Hoch HC (1982). A possible role for
microtubules and microfilaments in the induction of
nuclear division in bean rust uredospore germlings.
Exp Mycol 6:293.

22. VanEtten HD (1982). Phytoalexin detoxification by
monooxygenases and its importance for pathogenicity.
In Asada Y, Bushnell WR, Ouchi S, Vance CP (eds):
"Plant Infection: the Physiological and Biochemical
Basis", Berlin: Springer, p 315.

23. Ward EWB, Stoessl A (1976). On the question of
"elicitors" or "inducers" in incompatible interactions
between plants and fungal pathogens. Phytopathology
66:940.

24. Wood RKS (1976). Specificity - an assessment. In Wood
RKS, Graniti A (eds): "Specificity in Plant Diseases",
New York: Plenum, p 327.

25. Wynn WK, Staples RC (1981). Tropisms of fungi in host
recognition. In Staples RC, Toenniessen GH (eds):
"Plant Disease Control: Resistance and
Susceptibility", New York: John Wiley and Sons, p 45.

26. Yarwood CE (1967). Response to parasites. Ann Rev
Plant Physiol 18:419.

Molecular Strategies for Crop Protection, pages 35–48
© **1987 Alan R. Liss, Inc.**

TISSUE-SPECIFIC RESPONSES TO INFECTION

Adrienne Clarke[1], Barbara Howlett[1], Bruce Imrie[2],
John Irwin[3], Mirek Kapuscinski[1], Julie Ralton[1],
and Michael Smart[1]

[1]Plant Cell Biology Research Centre, School of Botany
University of Melbourne, Parkville, Victoria 3052
AUSTRALIA

[2]CSIRO, Division of Tropical Crops and Pastures
St. Lucia, Queensland 4067, AUSTRALIA

[3]Department of Botany, University of Queensland
St. Lucia, Queensland 4067, AUSTRALIA

ABSTRACT   Aspects of the interaction between the stem-
rotting fungus Phytophthora vignae (Oomycetes) and near-
isogenic cultivars of cowpea, Vigna unguiculata, which
differ in a dominant resistance gene for race 2 of the
fungus, are described. The events of epidermal protoplast collapse and production of phenylalanine ammonia
lyase, hydroxyproline-rich glycoprotein and arabinogalactan protein have been mapped in relation to fungal
growth in the host. The potential of hybridization histochemistry for following gene expression within infected
tissues is demonstrated.

INTRODUCTION

Few generalizations regarding tissue-specific responses
to infection can be made because different pathogens establish
different nutritive and cellular relationships with their
hosts, and because of the varied growth patterns of pathogens
within host tissues. These patterns are dependent on factors
such as the stage of the life cycle of the fungus which provides the inoculum, and the maturity of the host tissue.
Resistance to particular pathogens may involve a number of
different strategies which may or may not be common to expression of host and non-host resistance (for reviews see

(1), (2), (3) and references therein). Implicit in our
thinking about host responses is the idea that the primary
event is an interaction of host resistance gene products with
fungal avirulence gene products. A scheme summarizing the
state of information regarding single gene resistance is presented in Figure 1. At present most of the available information concerns host reponses which can be observed cytochem-
ically (e.g. papilla formation, hypersensitive death) or
measured biochemically (e.g. PAL and elicitor activity). As
no product of a host resistance gene has been identified, we
cannot yet connect the information on the classical genetics
of host resistance with the cytochemical and biochemical observations.

FIGURE 1. Scheme summarizing state of information
regarding single gene resistance in host-fungal pathogen
interactions.

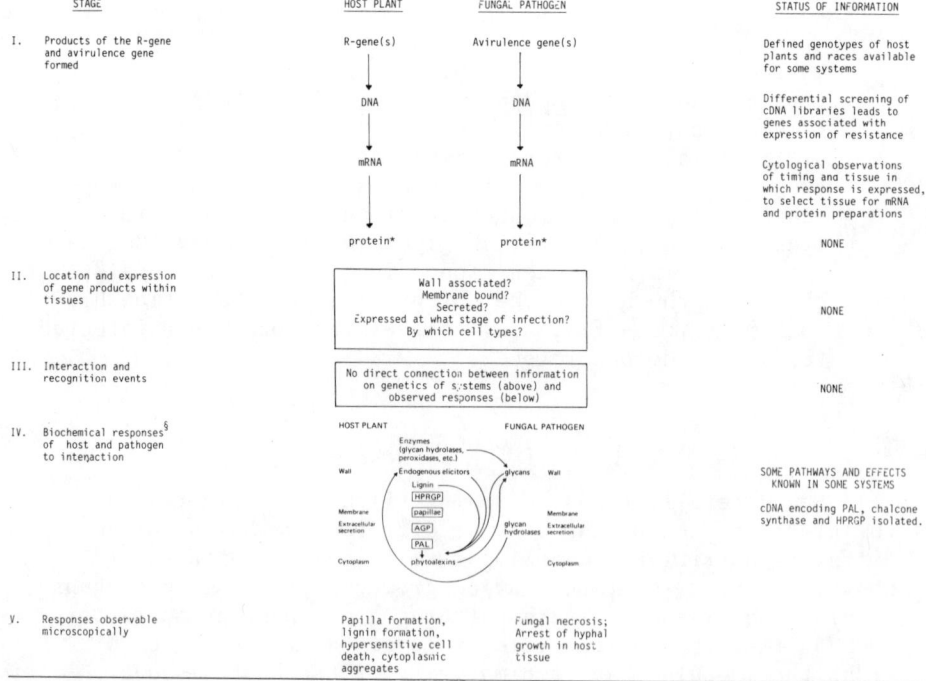

We have studied one interaction, that between a soil
pathogen Phytophthora vignae (Oomycetes) and its specific
host, cowpea (Vigna unguiculata) in detail. We have established the course of the interaction by observations at the
microscope level, and measured the timing of a number of
responses in relation to the position of the fungus within
the host tissue. These studies show that:

. Microscopy gives crucial information for planning
  molecular studies; this information allows selection
  of the most appropriate tissue and the time after inoculation to give optimum yield of particular host proteins or mRNA species;

. Specific molecular probes (cDNA or specific antibodies)
  derived from these studies can be used (hybridization
  histochemistry or immunoelectro microscopy) to unite
  the molecular and microscopic approaches.

RESULTS AND DISCUSSION

Experimental system: the cowpea - P. vignae interaction.

Phytophthora vignae is a soil-borne homothallic fungus
in the class Oomycetes. In contrast to some other pathogens
in the genus Phytophthora, which have a broad host range,
P. vignae is restricted to cowpea. The disease is apparently
limited to Australia and was first described in the North-
eastern state of Queensland by Purss (4). Purss (5) described
four races of the pathogen which differed in their ability
to cause disease of different cowpea cultivars. He also produced near-isogenic lines of cowpea differing in their resistance to P. vignae (race 2) by crossing a susceptible cultivar,
cv. Poona, with a resistant variety, Blackeye 5, and backcrossing the F1 generation to cv. Poona for three successive
generations, with selection for resistant progeny at each
generation. The new resistant cultivar is known as cv.
Caloona (6).
The disease reactions of the F1, F2 and back-cross seedlings derived from variety Blackeye 5 and cv. Caloona, indicate single major gene resistance to race 2 of the pathogen
with resistance dominant to susceptibility (Table 1). We
have started a detailed study of the interaction of the near-
isogenic cultivars cv. Poona and cv. Caloona with race 2 and
race 3 of the pathogen. The dominant resistance gene is apparently expressed in all tissues of the intact plant, as all
tissues express symptoms of resistance to experimental

inoculation. This resistance is independent of the age of the intact plant or the method of inoculation and is expressed in both intact plants and excised tissues (3).

TABLE 1
Interaction of cowpea varieties with races of
Phytophthora vignae

Cowpea Variety	P. vignae race		
	1	2	3
Blackeye 5[1]	R	R	S
Caloona	R	R	S
Poona	S	S	S
CP128215[2]	S	S	S

[R = resistant; S = susceptible]

[1]This line, referred to by Purss (5) as Blackeye 5, is a black seeded line but is not the cultivar California Blackeye 5.

[2]Commonwealth Plant Introduction number. Accession received from South Africa.

The combinations used in this study are indicated by the box.

Events associated with expression of resistance in roots of cowpea cultivars.

The response of roots of the resistant and susceptible cowpea cultivars to experimental inoculation with zoospore suspensions was recorded by careful direct observation (Table 2) (7). Seedlings were germinated under axenic conditions and the roots exposed to zoospore suspensions ($10^3 ml^{-1}$) for 0.5h. The roots were then removed by cutting at the root-hypocotyl interface and the entire root system placed on a slide for direct microscopic observation. The events in the host epidermis and first two to three cortical cell layers could be viewed distinctly. For each cultivar, events assoc-iated with sixty individual germinated cysts were followed in a number of separately inoculated roots.

The outstanding difference in response in the two cultivars is collapse of the protoplast of epidermal cells (Fig 2).A
high proportion (68%) of germinated cysts which penetrated the
epidermis of cv. Caloona were associated with collapse of the
protoplast of an epidermal cell. Curiously, usually only one
of the two cells in contact with the germ tube responded in
this way. The other usually remained apparently healthy although in some cases papillae or cytoplasmic aggregates were
formed (Figure 3).

FIGURE 2. Timing of observable events in roots of resistant and susceptible hosts to inoculation by P. vignae
(race 2). Observations were made from the time of inoculation (time 0); the arrow heads represent the spread
which includes 90% of all observations. The timing of
events is similar in susceptible and resistant hosts.
Epidermal protoplast collapse, a typical response of the
resistant host, occurs on average 1h after the fungus
has penetrated the epidermis.

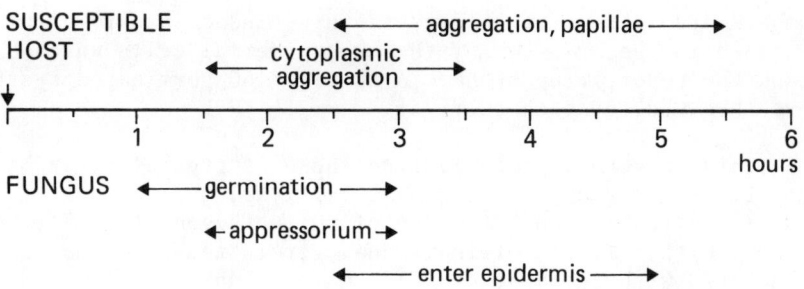

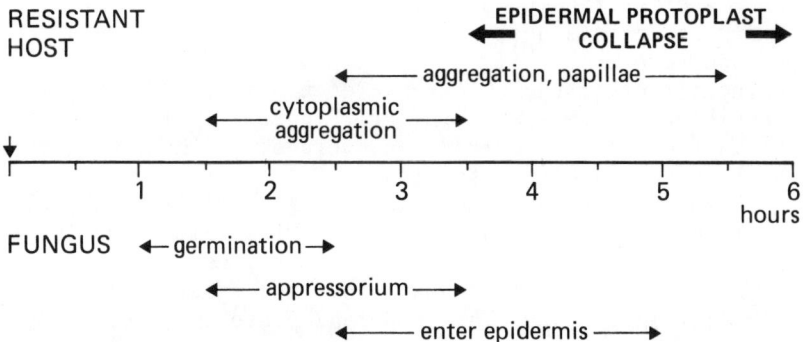

TABLE 2
Host responses observed in vivo in response to
experimental inoculation of cowpea roots by zoospore
suspensions of race 2.

Response observed	cv. Poona (Susceptible)	cv. Caloona (Resistant)
Cytoplasmic aggregation and/or papillae	58	74
Epidermal protoplast collapse	3	68
No response	41	16
Epidermal protoplast collapse without cytoplasmic aggregation or papillae	0	10

Numbers are % of germinated cysts which induce a particular
response in one or other of the two epidermal cells which
flank the penetrating hypha. A total of 60 germinated cysts
were observed for each cultivar.

This epidermal protoplast collapse is typical of an incompatible interaction. The same cultivar, cv. Caloona, inoculated with a virulent race of the pathogen (race 3), responded with only 11% of individual germinated cysts being
associated with collapse of epidermal protoplast, compared
with 68% when inoculated with race 2. This protoplast collapse was usually, but not always, associated with the other
observable changes, papilla formation and cytoplasmic aggreg-
ation. Another finding of interest is that a higher proportion of susceptible host cells than resistant host cells
showed no response.
    The timing of the observable events in the resistant and
susceptible cultivars was similar (Figure 2). The epidermal
protoplast collapse occurred, on average, two hours after the
hyphae had commenced penetration.
    By sectioning fixed material it is possible to measure
the rate of hyphal growth (Figure 4). Initially, the fungus
grows at a similar rate in both cultivars; between 4-8h post-
inoculation the rate slows down, then increases rapidly in
both cultivars to 12h. The epidermal protoplast collapse

FIGURE 3. Diagrammatic representation of the events at the epidermis of a resistant host root, in response to invasion by P. vignae (race 2). The cyst has germinated and the cytoplasm (C) has emptied into the hypha; a septum (S) has been formed, cutting off the empty cyst.
One of the epidermal cells in contact with the hypha has undergone protoplast collapse (epc); the other epidermal cell in contact does not usually collapse, but may form a papilla (P). The underlying cortical cells do not respond in an observable way.

4h – cv. CALOONA

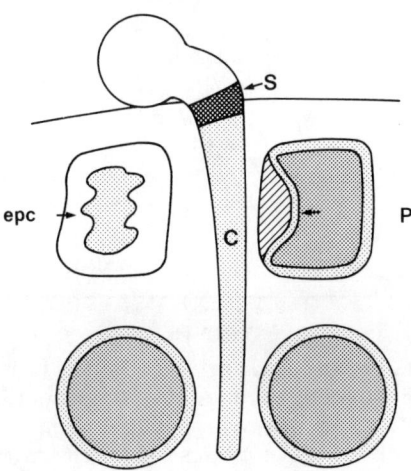

occurs on average 4.5h after inoculation, that is, about 8h before hyphal growth is arrested. This early event is specific to epidermal cells and correlates with resistance, suggesting that it may, in some way, be involved in expression of resistance. It is possible that a signal from the fungus is received by a receptor on the host plasma membrane, resulting in protoplast collapse. This may then generate signals destined for other cells distant from the epidermis, which could be involved in expression of resistance. We do not understand why only one of the two epidermal cells in contact with the hypha responds by protoplast collapse. This observation implies that adjacent cells within the epidermis are functionally distinct.

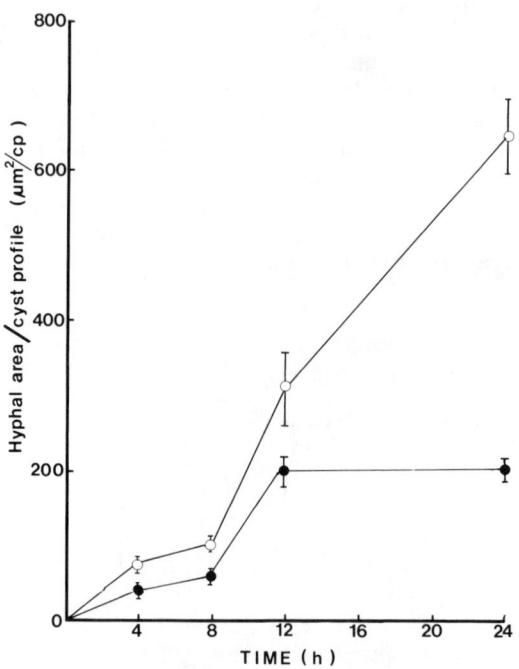

FIGURE 4. Rate of hyphal growth in sections of roots of resistant and susceptible cultivars inoculated with P. vignae race 2. The average hyphal areas, in sections taken from each of five roots of each cultivar at 4,8,12 and 24h after inoculation, were measured.

Events associated with expression of resistance in excised epicotyls of cowpea cultivars.

There are a number of difficulties associated with col-
lecting sufficient synchronously-responding, inoculated roots
for biochemical analysis, but as resistance is expressed in all
tissues examined, we have developed a system for obtaining syn-
chronously responding tissue from epicotyls. The primary leaves
of seven-day-old seedlings are excised and discarded; plugs of
agar containing fungal hyphae are placed on the cut epicotyl
surface. A soft lesion develops in the susceptible cultivar
and there is localized reddening and necrosis in resistant
plants. This method gives consistent responses to inoculation;
in control plants treated with agar plugs, cell damage from
cutting the epicotyl is restricted to the upper 150μm (three

cell layers). The rate of hyphal penetration into epicotyls
from resistant and susceptible cultivars is remarkably similar to that in intact roots. Up to 8h after inoculation the
rates of penetration are similar in the two cultivars; after
8h the differences become apparent, with the rate of growth
into the susceptible cultivar becoming rapid, while there is
essentially no further growth into the resistant cultivar.
The pattern of infection and the distribution of hyphae throughout the various tissues of the epicotyl 12h post-inoculation
are similar in both cultivars: the epidermis contains few hyphae and most are present in the pith and the cortex. The
growth is preferentially intercellular in all tissues, especially at the hyphal front (3).
    Having mapped the rate of growth of hyphae in the epi-
cotyls and established that this rate is highly reproducible,
it is possible to inoculate many plants and collect sufficient
material for biochemical studies. The material can be cut into
pieces known to include or to exclude the hyphal front.

Phenylalanine ammonia-lyase (PAL) in inoculated epicotyls.

    PAL activity is expressed at increased levels in the re-
sistant cultivar, early in the interaction, at a time when the
rates of hyphal growth in the resistant and susceptible cultivar are similar (Figure 5). By 12h, when there is a dramatic
difference in the growth rates in the two cultivars, there is
no detectable PAL activity above control levels in the susceptible cultivar although the hyphae have penetrated throughout
the segment examined. In the resistant cultivar, the hyphae
have almost reached their maximum growth into the segment and
the extractable PAL activity continues to increase. By 24h
the hyphae have penetrated throughout both segments of the
susceptible cultivar and there is a detectable increase in PAL
activity, but only in the upper segment, not in the segment
containing the hyphal front (8). In summary,
the PAL activity is detected early after inoculation in tissue
of the resistant cultivar and this increase is localized to the
segment containing the hyphal front; in the susceptible cultivar, the response is slower and is detected in tissue segments
which have borne the hyphae for some time, but is not detected
in the segment containing the hyphal front (Figure 5) (8).
    In addition to PAL, we have measured levels of expression
of mRNA encoding the hydroxyproline-rich glycoprotein (HPRGP)
and chalcone synthase by Northern blots. A genomic clone for
HPRGP was kindly provided by Professor J. Varner and a cDNA

clone for chalcone synthase was kindly provided by Professor
K. Hahlbrock. PAL activity is detected at increased levels
three hours after inoculation, which is several hours before
arrest of hyphal growth, while mRNAs for HPRGP and chalcone
synthase are not detected until several hours after arrest of
hyphal growth.

   Information on the rate of fungal growth, and the time at
which various genes (known to be associated with expression of
resistance in other systems) are expressed in relation to fungal
growth, is valuable for designing molecular studies. For example, preparing mRNA for cDNA libraries from epicotyl tissue
8h after inoculation would be likely to give cDNA clones for
PAL, but not for chalcone synthase or HPRGP. Similarly, cDNA
libraries prepare from mRNA from root tissue 3.5h after inoculation would be likely to contain cDNA clones for genes involved in the epidermal protoplast collapse but these genes may
not be present in libraries prepared from tissue collected
hours after the event.

   FIGURE 5. Diagram showing increases in phenylalanine ammonia lyase activity in excised epicotyls of cowpea cultivars
   after inoculation with <u>Phytophthora vignae</u> (race 2). The
   average distance hyphae have grown into the epicotyls after
   various times is indicated by the vertical lines. The agar
   plug containing the inoculum (stipple) was discarded and
   two segments, each 4mm in length, were examined.
   C = control - levels of PAL activity obtained in segments
   of epicotyls treated with agar plugs containing no inoculum. x2 = levels of PAL twice that obtained in controls.
   x3 = levels of PAL three times that obtained in controls
   (0.7kat/kg fresh weight, 70 kat/kg protein).

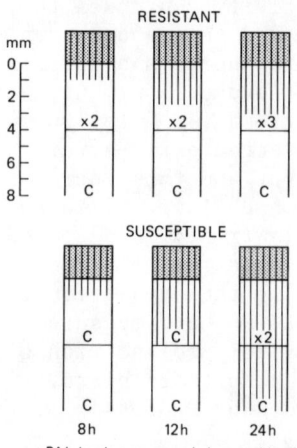

PAL levels over wounded control levels

Arabinogalactan-proteins (AGPs) in inoculated epicotyls

AGPs are proteoglycans which are widely distributed in
flowering plants and are associated, in some situations, with
wounding (9,10). We have detected an increase in the extractable AGP levels in inoculated epicotyls of the resistant cultivar (11). This increase is detected only in epicotyl segments of the resistant cultivar harvested 24h after inoculation
and is not detected in the susceptible cultivar. The increase
is localized to the segment of wounded tissue immediately below
the cut epicotyl surface. Thus the increase in AGP levels correlates with expression of resistance; but it is not detected
until well after differential growth of the fungus in the resistant and susceptible cultivars is established. This does
not rule out the possibility that there is earlier production
of increased levels of AGPs, as the sensitivity of the methods
for their detection (12) is the limiting factor. In this case
we are detecting a gene product which has undergone extensive
post-translational modification before it is measured; in contrast, PAL is measured as enzymic activity which requires extremely low levels of protein for detection.

Hybridization histochemistry as a method for detection of
gene expression at the cellular level.

Hybridization histochemistry involves taking a section
of inoculated material, incubating it with either DNA or synthetic oligonucleotide probes bearing suitable labels, and
detecting where within the section, binding of the probe to
mRNA occurs. This technique has proved valuable in studies
of animal gene expression (13,14) and we have started to adapt
the technology for studies with plant tissues.
We found increased binding of a genomic clone for hydroxyproline-rich glycoprotein (HPRGP) (kindly provided by Professor Varner) throughout the upper 8mm of the resistant, inoculated epicotyl after 24h, that is, the hybridization zone
was in advance of the hyphal front. In contrast, in the susceptible cultivar, there was a low level of hybridization
throughout the upper 8mm. This can only be regarded as a preliminary experiment, but serves to show what the problems and
the potential are. Experiments of this type can be paralleled
by Northern blots which give more specific information about
the nature of the hybridizing species, but information obtained
by Northern blots regarding the tissue origin of the hybridizing species is limited by the capacity to dissect the tissue

accurately and reproducibly for RNA preparation.

In hybridization histochemistry less background binding is often obtained using oligonucleotide probes. We have successfully used 20-mers in our studies of self-incompability of Nicotiana alata (14). Oligonucleotide probes in the reverse sense can be used as controls (15). The hybridization experiments using the genomic DNA for the HPRGP show binding throughout the cells, although these cells are highly vacuolated. We believe that this distribution is an artefact caused by rapid thawing of the frozen section before fixation of the tissue. This problem can probably be overcome by technical modifications at this step.

CONCLUSIONS

For the cowpea-P. vignae interaction:

1. The first observable event associated with resistance in roots inoculated with zoospores is epidermal protoplast collapse.

2. The pattern and rate of fungal growth in root and epicotyls is remarkably similar; this is true for both susceptible and resistant cultivars.

3. By establishing the rate of hyphal growth and the cellular responses in resistant and susceptible cultivars, it is possible:

   . to select host material either ahead, at or behind the hyphal front for molecular studies;
   . to measure biochemical responses (e.g. PAL levels) known to occur in other host-pathogen systems in relation to the fungal growth within the host;
   . to design experiments to clone genes associated with particular resistance responses defined either from microscopic or molecular studies.

4. The techniques of hybridization histochemistry can be used to follow gene expression at the cellular level in plant tissue to unite cellular and molecular studies.

ACKNOWLEDGEMENTS

We are most grateful to our colleagues, Dr David Guest, Dr Tony Bacic and Professor Bruce Stone for discussion; to Ms Anne Pottage for preparing the diagrams and to Ms Anne Poulton and Ms Stacey Malcolm for skilful editorial assistance.

REFERENCES

1. Aist JR (1983). Structural responses as resistance mechanisms. In Bailey JA, Deverall, BJ (eds): "The Dynamics of Host Defence", New York: Academic Press, p 33.
2. Ralton JE, Smart MG, Clarke AE (1986). Recognition and infection processes in plant-pathogen interactions. In Kosuge E, Nester E (eds): "Plant-microbe Interactions: Molecular and Genetic Perspectives", Vol 2, Macmillan (in press).
3. Ralton JE, Howlett BJ, Clarke AE, Irwin JAG, Imrie B (1986) Cowpea and Phytophthora vignae (in preparation).
4. Purss GS (1972). Pathogenic specialization in Phytophthora vignae. Aust J agric Res 23: 453.
5. Purss GS (1958). Studies on varietal resistance to stem rot (Phytophthora vignae Purss) in the cowpea. Qld J Ag Sci 15:1.
6. Purss GS (1963). Caloona - stem rot resistant cowpea. Qd agric J 89: 756.
7. Smart MG, Clarke AE (1986). In vivo observations of cowpea roots inoculated with Phytophthora vignae (in preparation).
8. Howlett BJ, Clarke AE (1986). Phenylalanine ammonia-lyase production as a response to infection of cowpea in the stem-rotting fungus, Phytophthora vignae (in preparation).
9. Clarke AE, Anderson RL, Stone BA (1979). Form and function of arabinogalactans and arabinogalactan-proteins. Phytochemistry 17: 521.
10. Fincher GB, Stone BA, Clarke AE (1983). Arabinogalactan-proteins - structure, biosynthesis and function. Ann Rev Plant Physiol 34: 47.
11. Kapuscinski M, Clarke AE (1986). Arabinogalactan protein levels in cowpea epicotyls inoculated with Phytophthora vignae (in preparation).
12. van Holst GJ, Clarke AE (1985). Quantification of arabino-galactan-protein in plant extracts by single radial gel diffusion. Analytical Biochemistry 148 (in press).
13. Coghlan JP, Penschow JD, Hudson PJ, Niall HD (1984). Hybridization histochemistry: use of recombinant DNA for tissue localizations of specific mRNA populations. Clin and Exper Hyper - Theory and Practice A6(1&2):63.

14. Coghlan JP, Aldred P, Haralambidis J, Niall HD, Penschow JD, Tregear GW (1985). Hybridization histochemistry. Analytical Biochemistry 149:1.
15. Anderson MA, Cornish EC, Mau S-L, Williams EG, Hoggart R, Atkinson A, Bonig I, Grego B, Simpson R, Roche P, Haley J, Penschow J, Niall H, Tregear G, Coghlan J, Crawford R, Clarke AE (1986). Molecular cloning of cDNA for a stylar glycoprotein associated with expression of self-incompatibility in Nicotiana alata Link & Otto. Nature (in press).
16. Cornish EC, Bonig I, Clarke AE (1986). Tissue-specific expression of genes associated with self-incompatibility in Nicotiana alata (Link & Otto) using hybridization histochemistry (in preparation).

Molecular Strategies for Crop Protection, pages 49–58
© 1987 Alan R. Liss, Inc.

ACTIVATION OF DEFENSE GENES IN RESPONSE TO ELICITOR AND INFECTION[1]

C.J. Lamb, J.N. Bell, D.R. Corbin, M.A. Lawton, M.C. Mehdy, T.B. Ryder, N. Sauer and M.H. Walter

Plant Biology Laboratory, Salk Institute,
P.O. Box 85800, San Diego, California 92138

ABSTRACT Induction of defense responses by fungal elicitors or infection involves transcriptional activation of defense genes, in some cases within 2 to 3 minutes of signal perception. Induction occurs in distant, hitherto uninfected tissue as well as directly infected tissue implying intercellular transmission of elicitation signals. Defense gene systems are highly polymorphic at the gene, RNA and protein levels. Implications for defense gene function and the perception and transduction of stress signals are discussed.

PLANTS EXHIBIT INDUCIBLE DEFENSE RESPONSES

Plants exhibit natural resistance to disease and while a number of static protection mechanisms have been described, disease resistance appears to be an active process which from inhibitor studies is dependent on host RNA and protein synthesis. Detailed studies of the interaction between Phytophthora infestans and potato by Muller and associates in the 1940's led to the suggestion that plants resist

---

[1] Supported by grants to C.J.L from the Samuel Roberts Noble Foundation, Tan Sri Tan and USDA (CRGO 84-CRCR-1-1464) and Fellowships from ACS (D.R.C), NSF (M.C.M), EMBO (N.S) and DFG (M.H.W)

infection by inducible mechanisms including the synthesis of low molecular weight anti-microbial compounds termed phytoalexins. Subsequent investigations have shown that accumulation of phytoalexins in response to infection occurs in a wide variety of plant species. Furthermore, there are a number of other inducible defense responses including: accumulation of hydroxyproline-rich glycoproteins (HRGP), deposition of lignin-like material and other wall-bound phenolics, stimulation of the activity of certain hydrolytic enzymes such as glucanase and chitinase, accumulation of proteinase inhibitors as well as the appearance of so-called pathogenesis-related (PR) proteins (1,2).

Induction of these defense responses associated with expression of localized hypersensitive resistance is observed in the early stages of attempted infection by a non-pathogen or by an avirulent form of a pathogen in a genetically incompatible interaction (1,2). In addition, defense responses are also often induced in the later stages of a genetically compatible interaction with a virulent form of a pathogen associated with the onset of lesion formation (3), as well as in tissue distant from the initial site of microbial attack associated with establishment and/or subsequent expression of induced systemic immunity (4). These defense responses can also be induced by elicitors present in microbial cell walls and culture filtrates as well as by a variety of structurally unrelated artificial inducers (5).

ELICITORS AND INFECTION INDUCE MARKED CHANGES IN THE PATTERN OF PROTEIN SYNTHESIS

Two-dimensional gel electrophoretic analysis of the pattern of polypeptides synthesized in vivo or in vitro by translation of isolated RNA has shown that elicitor and infection cause marked changes in the overall pattern of protein synthesis, which can often be correlated with induction of specific defense responses and expression of various forms of disease resistance (6). Likewise, over the last several years, studies in a number of systems have demonstrated stimulation by elicitor or infection of the synthesis of specific proteins functionally implicated in these defense responses. Examples include: (a) phenylpropanoid biosynthetic enzymes involved in the synthesis of lignin precursors and isoflavonoid and coumarin phytoalexins (6), (b) chitinase (unpublished), (c) HRGP apoproteins (7), (d) proteinase inhibitors (8) and (e) specific PR proteins (9). These observations focus attention on the organization and structure of the corresponding plant defense genes in relation to their activation by biological stress and specific functions in disease resistance.

INDUCTION OF RESPONSES INVOLVES TRANSCRIPTIONAL ACTIVATION OF DEFENSE GENES

In order to investigate the activation and function of specific plant defense genes, we have generated a series of cDNA libraries complementary to mRNA isolated from bean (Phaseolus vulgaris L.) cells at various intervals after elicitor treatment. We have identified clones containing sequences encoding HRGP apoproteins (unpublished) and several enzymes of phenylpropanoid biosynthesis including phenylalanine ammonia-lyase (PAL: first enzyme of overall pathway: lignin and isoflavonoid phytoalexin synthesis); chalcone synthase and chalcone isomerase (CHS, CHI: first 2 enzymes of a branch pathway specific for flavonoids and isoflavonoid phytoalexins); cinnamyl alcohol dehydrogenase (CAD: second enzyme of the branch pathway for synthesis of lignin precursors) (10,11, unpublished).

Using these cloned cDNA sequences as probes for RNA blot hybridization, we have demonstrated that elicitor treatment of bean cells causes very rapid, marked but transient accumulation of PAL, CHS, CHI and CAD mRNAs. Kinetics for induction of these mRNAs follows closely those previously observed for the synthesis of the respective proteins in vivo and changes in translatable mRNA activity in vitro. In particular, accumulation of this set of mRNAs was observed within 10 to 20 min of elicitor treatment with maximum levels at 2 hr after elicitor treatment for CAD mRNA and 3 to 4 nr after elicitor treatment for PAL, CHS and CHI mRNAs (10,11, unpublished).

Elicitor stimulation of chitinase mRNA levels coordinately with PAL, CHS and CHI mRNAs can be inferred from recent studies demonstrating stimulation of chitinase synthesis by in vivo protein labeling and in vitro translation of isolated mRNA coupled with specific immunoprecipitation (unpublished). cDNA clones containing sequences complementary to elicitor-induced chitinase mRNAs have recently been isolated by antibody screening of cDNA libraries cloned in λgt11, and are currently being used in RNA blot hybridizations to confirm rapid induction of chitinase mRNA. In contrast, elicitor caused a less rapid but more prolonged accumulation of HRGP mRNA with maximum rates of accumulation between 4 and 10 hr after elicitor treatment and maximum levels between 24 and 36 hr after elicitor treatment (12). Overall it can be concluded that accumulation of these mRNAs is the major factor governing the stimulation of enzyme or protein synthesis throughout the phase of rapid increase in the respective enzyme or protein levels.

The marked accumulation of these mRNAs from very low, almost undetectable basal levels strongly suggests that elicitor initially causes a transient increase in the transcription of the corresponding defense genes. This has been confirmed by in vivo and in vitro labeling of newly-synthesized transcripts (13,14). Newly-synthesized mRNAs were labeled in vivo with 4-thiouridine and then purified by organo-mercurial affinity chromatography. Blot hybridization of these newly-synthesized mRNAs showed that elicitor markedly stimulated the rate of synthesis of defense gene transcripts including PAL, CHS and CHI mRNAs as part of a pronounced change in the overall pattern of RNA synthesis (13).

Transcriptional activation of PAL, CHS and HRGP genes by elicitor has recently been confirmed by analysis of transcripts completed in vitro in nuclei isolated from control and elicitor treated cells (14). Maximum rates of transcription of PAL and CHS genes were observed 1 hr 20 min after elicitor treatment concomitant with the onset of the phase of rapid accumulation of the corresponding transcripts. Detailed analysis of the early stage of the response revealed a marked stimulation of PAL and CHS gene transcription within 5 min of elicitor treatment and possibly as early as 2 to 3 min after elicitor treatment. In contrast, transcriptional activation of HRGP genes was preceded by a lag of 1 to 2 hr and reached maximum levels only 7 hr after elicitor treatment concomitant with the phase of rapid accumulation of HRGP transcripts. Elicitor induction of PAL and CHS gene transcription in bean cells represents the most rapid response hitherto observed that can be causally related to expression of defense responses with kinetics comparable to the most rapid gene activation systems in animal cells. Together with auxin responsive genes of unknown function in soybean plumules (15), this represents the earliest transcriptional event yet observed for specific plant genes in response to an exogenous signal.

Similarly, we have recently observed, by RNA blot hybridization, marked accumulation of hybridizable PAL and CHS mRNAs in wounded or infected hypocotyls. Accumulation of hybridizable PAL and CHS mRNAs during race:cultivar specific interactions with <u>Colletotrichum</u> <u>lindemuthianum</u> exhibited similar kinetics to those previously observed for stimulation of enzyme synthesis (16). Thus, in an incompatible interaction (host resistant), following molecular recognition of the fungal infection peg, there was an early induction of mRNA in directly infected tissue and, less markedly, in distant, hitherto uninfected tissue. mRNA induction occurred prior to the onset of phytoalexin accumulation, lignin deposition and expression of localized hypersensitive resistance in directly infected tissue, and establishment of induced systemic immunity in distant uninfected tissue. In contrast, in a compatible interaction (host susceptible) there was no early induction of PAL and CHS mRNA but rather a delayed widespread accumulation associated with the onset of phytoalexin accumulation and lignification in response to trauma signals during attempted lesion limitation. In parallel, we have demonstrated that HRGP mRNA accumulation follows a broadly similar temporal and spatial pattern as that observed for PAL and CHS mRNAs, although, as in elicited cells, accumulation of HRGP mRNA is less rapid and more prolonged (12). Stimulation of the transcription of PAL, CHS and HRGP genes in infected hypocotyls has been confirmed by analysis of nuclear run-off transcripts (14).

These and similar observations in other systems indicate that stimulation of defense gene transcription and mRNA induction by elicitor and infection characteristically underlies the expression of specific defense responses.

POLYMORPHISM OF PLANT DEFENSE GENES AND PROTEINS

In the course of studying the mechanisms underlying the activation of plant defense responses in bean we have uncovered extensive polymorphism for the plant defense genes and proteins examined so far, viz: PAL: 4 genes, 6 elicitor-induced mRNAs, 10-11 isopolypeptides, 4-5 native tetrameric enzymes; CHS: 6-8 genes, >6 elicitor-induced mRNAs, 8-10 isopolypeptides; CHI: several distinct genes, HRGP: >5 genes, >4 elicitor-induced transcripts (12,16,17, unpublished). In addition, preliminary evidence indicates the presence of several chitinase genes within the bean genome.

This polymorphism may be of considerable biological importance and may reflect gene amplification to enhance the ability of the plant to respond effectively to biological stress. In antirhinum and petunia, where the phytoalexins are not isoflavonoid derivatives and CHS is involved only in the synthesis of flavonoid pigments and UV-protectants, there are only 1 or 2 CHS genes per haploid genome. There is extensive clustering of the CHS genes within the bean genome. We therefore propose that in bean, CHS underwent a series of gene duplication events specifically associated with the evolution in this species of isoflavonoid synthesis as a phytoalexin response to biological stress.

Furthermore, genetic polymorphism allows evolution of a family of genes encoding closely related polypeptides with similar but subtly different functional properties attuned to optimal operation in specific biological circumstances. One example already uncovered is the selective induction by elicitor of PAL native enzyme forms exhibiting a low Km for phenylalanine, which asserts metabolic priority for phenylpropanoid biosynthesis in the cellular economy of phenylalanine under conditions of biological stress (17). Likewise, sequence analysis of polymorphic HRGP cDNAs complementary to different elicitor-induced transcripts indicate that the encoded polypeptides exhibit very characteristic variations in a highy repetitive structural motif. The nature of these variants suggests functional significance (unpublished).

A corollary of this polymorphism is that sets of genes encoding identical or very similar polypeptides may be positioned in different regulatory circuits, thereby allowing a flexible response to a variety of environmental conditions. The emerging data indicate differential expression of distinct PAL, CHS and HRGP isogenes at various stages following perception of a biological stress signal and also differences in the relative induction of isogenic members of defense gene families in compatible and incompatible interactions. For example, in the case of HRGP, a 2.4 kb transcript is the major induced form in the incompatible interaction but is only relatively weakly induced in the compatible interaction. In contrast a distinct 3.6 kb HRGP transcript, which by sequence analysis is derived from a different isogene, is very weakly induced in the incompatible interaction but is markedly induced in the later stages of the compatible interaction.

FUNCTIONAL IMPLICATIONS OF DEFENSE
GENE ORGANIZATION AND ACTIVATION

The observation of extensive polymorphism of defense genes and rapid induction of multiple arrays of genes encoding proteins involved in distinct defense responses has considerable implications for interpretation of the genetics of host-pathogen interactions and the overall function of inducible defense responses in disease resistance.

Thus in the gene-for-gene hypothesis, a specific disease resistance gene in the host can be paired with a specific avirulence gene in the pathogen such that gene products of the corresponding dominant alleles form a molecular recognition complex leading to incompatibility (18). Analysis of the segregation of plant disease resistance genes has failed to reveal other plant genes which modify the effects of the disease resistance genes. This has been interpreted to indicate that the formation of the molecular complex between avirulence gene product and disease resistance gene product is not merely the recognition event, but also represents the actual resistance response (18). In this model, recognition would itself be sufficient for expression of hypersensitive resistance and ensuing defense responses including phytoalexin accumulation, lignification etc. would not have a primary role but rather some ill-defined secondary function.

However, our recent analysis of the organization and expression of defense genes in bean argues against this interpretation. Thus a mutation in a specific gene (e.g. one of the six CHS isogenes) is individually unlikely to have any significant effect on the outcome of the host:pathogen interaction in view of the multiple induction of several isogenes encoding that biochemical function (e.g. CHS activity) and furthermore the multiple induction of several complementary defense responses, only one of which involves that particular defense gene family (e.g. phytoalexin accumulation).

The magnitude and rapidity of elicitor stimulation of defense gene activation in suspension cultured bean cells also strongly argues that the resultant defense responses play a primary role in disease resistance. Likewise, induction of PAL and CHS mRNA accumulation within 39 hr after inoculation of bean hypocotyls with the spores of an incompatible race of C. lindemuthianum represents an extremely early event in relation to the ingress of the pathogen, since following spore inoculation there is a period of 30 to 40 hr during which the spores germinate and produce an infection peg before fungal hyphae come in contact with the first host epidermal cell (3). Thus, in relation to the timing of this initial cell:cell contact at which the putative recognition event occurs, induction of defense mRNAs in an incompatible interaction is extremely rapid and occurs well before the first signs of hypersensitive flecking and expression of hypersensitive resistance. These observations do not appear to be consistent with the hypopothesis that the induction of defense mRNAs is causally preceded by hypersensitive cell death and initial expression of hypersensitive resistance.

In the light of the recent analysis of the organization and activation of defense genes, our working hypothesis is that disease resistance genes encode proteins involved in molecular recognition of pathogens leading to rapid activation of defense genes and hence expression of localized hypersensitive resistance. Induction of defense mRNAs in response to general trauma signals in the later stages of a compatible interaction in hitherto uninfected cells ahead of the invading pathogen provides a plausible mechanism for the process of attempted lesion limitation, which under appropriate physiological conditions can restrict the size of the lesion and prevent complete rotting of the organ and hence plant death, even though the interaction is genetically compatible (3,16). Likewise, the preactivation of defense genes in hitherto uninfected cells distant from the initial site of infection by an incompatible pathogen may represent a molecular mechanism by which induced systemic resistance is established (16).

IMPLICATIONS FOR BIOLOGICAL STRESS SIGNALING

The rapidity of the induction of PAL and CHS genes by elicitor implies that there are few intervening steps between elicitor binding to a putative plant cell receptor and specific transcriptional activation of these genes. Hence PAL and CHS gene activation might involve direct interaction of an elicitor-receptor complex with cis-acting regulatory DNA sequences as postulated for transcriptional activation by steroid hormones in animal cells. Alternatively, elicitor binding to a receptor putatively located on the cell surface might activate a short signal transduction cascade leading to modulation of trans-acting nuclear regulatory proteins that interact with the target cis-acting regulatory DNA sequences. The different kinetics for elicitor stimulation of HRGP gene transcription relative to PAL and CHS gene activation presumably reflect two distinct stimuli or a single stimulus leading to either sequential effects or divergent pathways.

Operation of similarly complex signal systems in infected tissue can likewise be inferred from our recent analysis of the kinetics of activation of defense genes. Moreover, within the syndrome of biological stress, different regulatory circuits appear to be involved in defense gene activation in different circumstances. Perhaps the clearest example to date is the preferential activation of two HRGP genes in, respectively, the early stages of an incompatible interaction and the later stages of a compatible interaction, implying the operation of distinct signal transduction systems in the two situations. Likewise, the observation of defense gene activation in tissue distant from the site of infection implies intercellular transmission of endogenous elicitation signals in addition to localized transduction of exogenous signals in directly infected cells.

Elucidation of the molecular mechanisms underlying the operation of these regulatory networks is likely to have a considerable impact on emerging rationales for enhancement of disease resistance by gene transfer techniques.

REFERENCES

1. Bell AA (1981). Biochemical mechanisms of disease resistance. Annu. Rev. Plant Physiol. 32:21–81.
2. Sequeira L (1983). Mechanisms of induced resistance in plants. Annu. Rev. Microbiol. 37:51–79.
3. Bailey JA (1982). Physiological and biochemical events associated with the expression of resistance to disease. In Wood RKS ed: "Active Defense Mechanisms in Plants", New York, Plenum, pp 39–65.
4. Dean RA, Kuc J (1985). Induced systemic resistance protection in plants. Trends in Biotechnol. 3:125–129.
5. Darvill AG, Albersheim P (1984). Phytoalexins and their elicitors: A defense against microbial infection in plants. Annu. Rev. Plant Physiol. 35:243–275.
6. Dixon RA, Dey PM, Lamb CJ (1983). Phytoalexins: Enzymology and Molecular Biology. Adv. Enzymol. 55:1–136.

7. Roby D, Toppan A, Esquerre-Tugaye MT (1985). Cell surfaces in plant-microorganism interactions. V. Elicitors of fungal and of plant origin trigger the synthesis of ethylene and of cell wall hydroxyproline-rich glycoprotein in plants. Plant Physiol. 77:700-704.
8. Ryan, CA (1984). Systemic responses to wounding. In Kosuge T, Nester EW, eds: "Plant-Microbe Interactions: Molecular and Genetic Perspectives", New York, MacMillan, pp 307-320.
9. Van Loon, LC (1985). Pathogenesis-related proteins. Plant Mol. Biol. 4:111-116.
10. Ryder TB, Cramer CL, Bell JN, Robbins MP, Dixon RA, Lamb CJ (1984). Elicitor rapidly induces chalcone synthase mRNA in Phaseolus vulgaris cells at the onset of the phytoalexin defense response. Proc. Natl. Acad. Sci. USA 81:5724-5728.
11. Edwards K, Cramer CL, Bolwell GP, Dixon RA, Schuch W, Lamb CJ (1985). Rapid transient induction of phenylalanine ammonia-lyase mRNA in elicitor-treated bean cells. Proc. Natl. Acad. Sci. USA 82:6731-6735.
12. Showalter AM, Bell JN, Cramer CL, Bailey JA, Varner JE, Lamb CJ (1985). Accumulation of hydroxyproline-rich glycoprotein mRNAs in response to fungal elicitor and infection. Proc. Natl. Acad. Sci. USA 82:6551-6555.
13. Cramer CL, Ryder TB, Bell JN, Lamb CJ (1985). Rapid switching of plant gene expression by fungal elicitor. Science 227:1240-1243.
14. Lawton MA, Lamb CJ (1986). Transcriptional activation of defense genes by fungal elicitor, wounding and infection. Submitted.
15. Hagen G, Guilfoyle TJ (1985). Rapid induction of selective transcription by auxins. Mol. Cell. Biol. 5:1197-1203.
16. Bell JN, Ryder TB, Wingate VPM, Bailey JA, Lamb CJ (1986). Differential accumulation of plant defense gene transcripts in a compatible and an incompatible plant:pathogen interaction. Mol. Cell. Biol., in press.
17. Bolwell GP, Bell JN, Cramer CL, Schuch W, Lamb CJ, Dixon RA (1985). L-Phenylalanine ammonia-lyase from Phaseolus vulgaris: Characterization and differential induction of multiple forms from elicitor-treated cell suspension cultures. Eur. J. Biochem. 149:411-419.
18. Ellingboe AH (1981). Changing concepts in host:pathogen genetics. Annu. Rev. Phytopathol. 19:125-143.

Molecular Strategies for Crop Protection, pages 59–70
© 1987 Alan R. Liss, Inc.

ADAPTATION OF PATHOGENIC FUNGI TO TOXIC CHEMICAL BARRIERS IN PLANTS: THE PISATIN DEMETHYLASE OF *NECTRIA HAEMATOCOCCA* AS AN EXAMPLE[1]

Hans D. VanEtten, David E. Matthews, and Susan F. Mackintosh

Department of Plant Pathology, Cornell University, Ithaca, NY 14853

ABSTRACT *Nectria haematococca* is a facultative parasite with a broad range of plant hosts, though individual isolates of the fungus differ in virulence on different hosts. The mechanisms underlying this specialization within *N. haematococca* are probably complex, but one genetic trait identified as a requirement for virulence on *Pisum sativum* is the ability to detoxify the pea phytoalexin pisatin. This trait can be found not only in *N. haematococca* field isolates collected from diseased pea but also in many of the isolates obtained from other hosts, which do not produce pisatin. Isolates differ in their expression of the detoxifying enzyme, pisatin demethylase, and only those genes which confer the higher demethylation rates contribute to virulence on pea. The multiple demethylase genes probably encode a family of cytochrome P-450 isozymes, and those which are more specific for pisatin may have evolved as an adaptation to the disease interaction with pea.

INTRODUCTION

It is a well recognized phenomenon experimentally that when organisms are constantly exposed to toxic chemicals they develop tolerance to these compounds, and in nature there are many examples of organisms which can grow in toxic environments. Frequently the biochemical basis of this adaptation is the acquisition of the ability to metabolize the toxin to a nontoxic derivative. Higher plants often harbor either preformed toxic chemicals or precursors that quickly generate toxic metabolites when the plant tissue is injured. In addition many plants have the potential to synthesize toxins *de novo* (phytoalexins) in response to stresses such as microbial infection. Thus it is not unexpected that many of the specific microorganisms which are frequently exposed to a given plant's toxic chemicals, i.e. its normal

[1]This work was supported by Department of Energy contract DE-AC02-83ER13073 and USDA S&E grant 84-CRCR-1-1388.

pathogens, show tolerance to these compounds. Often the biochemical basis of this tolerance is the presence of an enzyme that will detoxify the plant constituent.

In this article we summarize some of the evidence that such an adaptation occurred in the fungus *Nectria haematococca* as it became a pathogen of garden pea (*Pisum sativum*). Other examples of this general phenomenon have been described elsewhere (1,2,3).

RESULTS AND DISCUSSION

Pathogenicity of *Nectria haematococca* on Pea

N. haematococca normally invades the hypocotyl area of pea seedlings and causes a progressive necrosis of the lower epicotyl and upper tap root regions of the plant (4). High concentrations of the phytoalexin pisatin can accumulate during infection of this tissue by *N. haematococca* (5). However early studies showed that this fungus was highly tolerant of pisatin *in vitro* and that it could demethylate pisatin to produce a less toxic product (6,7,8,9,10). These observations suggested that *N. haematococca* had adapted to the toxic environment presented by its host, by acquiring the ability to detoxify its host's phytoalexin. The results of subsequent genetic studies strongly support this hypothesis.

Genetic Diversity in *Nectria haematococca* for Pisatin Demethylating Activity (Pda) and Pathogenicity

Nectria haematococca is the sexual stage of *Fusarium solani*, a fungus with a very wide distribution in nature. Though not all *F. solani* isolates are sexually fertile, and not all of these fall into the same compatibility group (mating population VI) as the *N. haematococca* isolates which are pea pathogens, members of this smaller group can also be found in a wide variety of habitats (11). They have been isolated from lesions of twenty angiosperm species and two animals, and have been demonstrated to be pathogens of ten of these. Isolates can also be found as saprophytes in soil. When a collection of field isolates obtained from all of these habitats was surveyed for the ability to demethylate pisatin, sensitivity to pisatin, and pathogenicity on pea (12,13), natural variation for each of these traits was found and the traits were interrelated. All isolates that lacked the ability to demethylate pisatin (Pda⁻) were sensitive to pisatin and essentially non-pathogenic on pea while all the highly pathogenic isolates were tolerant of pisatin and were Pda⁺. Although Pda⁺ isolates were found that were non-pathogenic on pea, the absence of any Pda⁻ isolates that were tolerant of pisatin or pathogenic on pea suggested that pisatin demethylation was essential for both of these traits.

Since *N. haematococca* is a heterothallic ascomycete it is possible to perform controlled crosses between isolates that differ in these traits. Dozens of such crosses have been made in our laboratory and thousands of progeny analyzed for virulence (13,14,15, unpublished). No Pda⁻ progeny isolate has been found that is highly pathogenic on pea or highly tolerant of pisatin, providing strong evidence that this pisatin detoxifying activity is required by *N. haematococca* to be an aggressive pathogen on this host.

The distribution of Pda in field isolates obtained from different habitats suggests that this trait also confers a selective advantage in natural populations of *N. haematococca*. All of the 25 isolates in our collection which were originally obtained from pea lesions are Pda⁺. Interestingly, 51 of 82 isolates obtained from other habitats, none of which contain pisatin, are also Pda⁺. Indeed some isolates from non-pea habitats show significant virulence towards pea (Fig. 1). Possibly these isolates had had contact with pea recently (in evolutionary time). *N. haematococca* may be a broadly adapted organism not only as a species, but also individually, such that a single isolate has the potential to move from host to host. Some field isolates virulent on pea have been found to be virulent also on chickpea (11) or mulberry (16). And probably most isolates are capable of living at least in saprophytic habitats in addition to their host of origin.

Two other hypotheses for the existence of Pda⁺ isolates in non-pea habitats suggest themselves. One, that a gene for pisatin demethylation is disseminated via the sexual stage, seems unlikely: sexual reproduction of

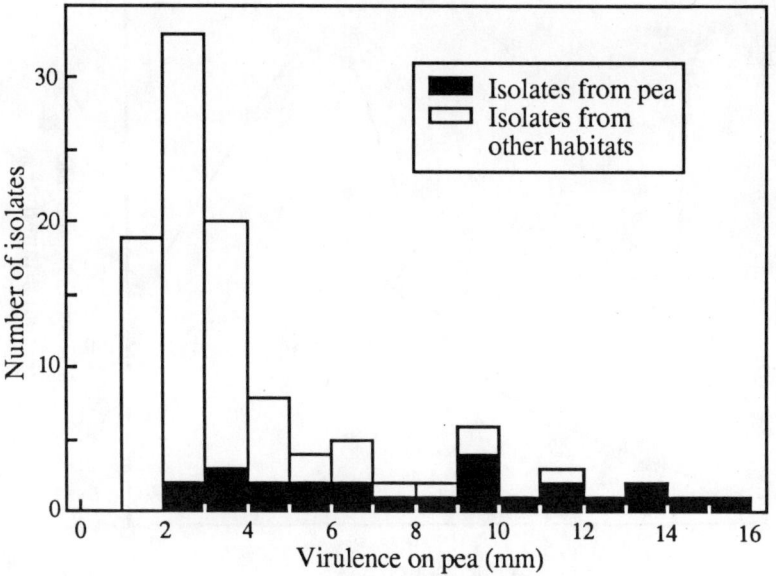

FIGURE 1. Distribution of virulence on pea in field isolates of *Nectria haematococca* collected from different habitats. Virulence measured as lesion length on epicotyls 4 days after inoculation.

N. haematococca appears to be rare in nature and has been observed in only one habitat, diseased mulberry. Alternatively, Pda might be an incidental function of a trait which adapts the fungus to a variety of habitats, such as a relatively nonspecific enzyme for degradation of toxic compounds. Further genetic and biochemical studies of pisatin demethylation have not resolved these questions, but have shed some additional light on them.

Identification of *Pda* Genes Coding for Different Levels of Enzyme Activity

Studies on the expression of Pda in culture by a highly virulent isolate of *N. haematococca* (isolate T-63) indicated that pisatin demethylase was an inducible enzyme (17). When pisatin was added to mycelium of T-63 suspended in buffer, there was a lag before pisatin disappeared from the culture, and the rate of demethylation measured for pisatin added subsequently increased from a nondetectable level to a maximum 8 hours

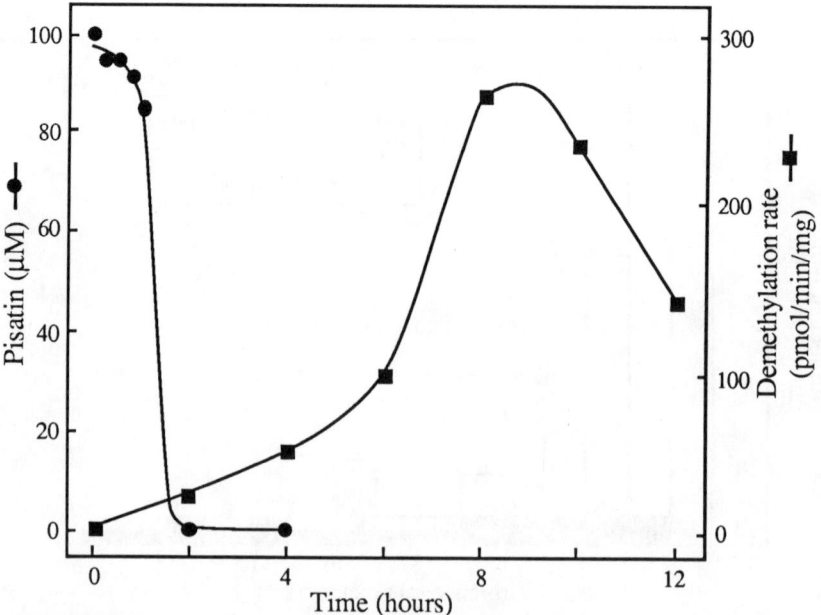

FIGURE 2. Demethylation of pisatin by *Nectria haematococca* isolate T-63. ● , Disappearance of the first addition of pisatin. ■ , Rate of demethylation of subsequent additions.

later (Fig. 2). To determine if there was natural variation in the regulation of pisatin demethylase in *N. haematococca*, 59 Pda$^+$ field isolates were surveyed for quantitative differences in demethylating activity. Large differences were observed in both the lag time before pisatin disappeared and the demethylation rate after induction (18).

Genetic analysis of a few field isolates exhibiting different levels of Pda expression revealed that there are several genes for pisatin demethylation (*Pda* genes) in this fungus. These behave like independent structural genes for demethylase enzymes, in that the positive allele of any one of them confers the Pda$^+$ phenotype. In addition however each of these genes confers a characteristic level of Pda expression (15,19), indicating that the corresponding enzymes differ in their activity towards pisatin, or their

TABLE 1
PDA GENES OF *NECTRIA HAEMATOCOCCA*[a]

| Gene | Phenotype | Source | | Pisatin demethylation[b] | | Virulence (lesion length, mm) |
		Field isolate	Habitat	T50 (h)	Maximum rate (pmol/min/mg)	
*Pda*1-1	Pda^i	T-2	pea	(2)	31 - 138	3 - 20
*Pda*2-1	Pda^n	T-219	soil	(20)	0.2 - 1^c	2 - 5
*Pda*3-1	Pda^n	T-2	pea	(20)	0.2 - 1^c	2 - 5
*Pda*3-2	Pda^n	T-161	unknown	8 - 59	0.7 - 28	1 - 5
*Pda*4-1	Pda^{si}	T-23	pea	3 - 8	43 - 92	2 - 13
None	Pda$^-$	---	---	---	0	0 - 5

[a]Values are the ranges observed for ascospore progeny possessing the gene. Parentheses indicate that the inheritance of the trait has not been examined quantitatively; the number given is a typical value for the source isolate or for one of its progeny known to contain a positive allele for only the one *Pda* gene.

[b]T50 is the time required for 50% demethylation of pisatin (3.3 nmol/mg mycelium) by a culture not previously exposed to pisatin. Maximum rate is the rate measured at this time point, except for *Pda*1-1 which gave higher rates in response to a second addition of pisatin 6 to 8 h after the initial exposure. Under the latter condition *Pda*4-1 showed rates less than 30 pmol/min/mg (cf. Fig. 2) and rates for the other genes were too low for accurate quantitation.

[c]Values given for maximum rate of *Pda*2-1 and *Pda*3-1 are underestimates because the degree of nonlinearity in the time course of demethylation by these isolates was not appreciated in our earlier work (14). When more detailed time courses were examined for isolate T-219, it showed rates similar to those of *Pda*3-2 (15).

from different field isolates, but they might also prove to be identical or to be separate, closely linked genes. A number of additional genetic determinants have also been found (unpublished results), but at present we do not know whether these are new genes or distinct alleles of the known genes.

Thus pisatin demethylation is not a unique genetic trait, and its presence in multiple habitats does not seem to be due to simple sexual or parasexual dispersal of a single gene, or to migration of individual isolates from habitat to habitat. A family of these genes exists, genes which either originated independently or have evolved independently long enough to have diverged in phenotypic expression and chromosomal location.

Different Levels of Expression of the Pisatin Detoxifying Activity are Associated with Different Levels of Disease Producing Potential.

Both the genetic analyses and the survey of field isolates indicated that regulation of Pda expression is important for pathogenicity on pea. Only the genes conferring the higher levels of expression, *Pda*1-1 and *Pda*4-1, were capable of contributing to virulence (Table 1). Progeny with the noninducible phenotype (*Pda*2-1, *Pda*3-1 and *Pda*3-2) were no more virulent than Pda⁻ progeny (14,15).

The field isolates presented a more complex picture because they showed a continuous range of Pda expression levels, rather than falling into discrete categories corresponding to the genetically defined phenotypes. Clearly, the identified *Pda* genes do not represent all of the quantitative variation that exists in *N. haematococca* for this trait; whether this variation is due to a large number of single *Pda* genes or alleles, each with its own characteristic level of expression, or due to separate modifier genes affecting Pda expression, is unknown. It was possible to classify the field isolates into two groups, based on a somewhat arbitrary statistical test. The cutoff point corresponded approximately to the difference between the Pda^n phenotype and the higher-expression phenotypes observed in the genetic studies. All of the highly virulent isolates were in the latter group (18). We conclude that low levels of pisatin demethylase are not sufficient to permit virulence on pea, perhaps because of the high rate of pisatin synthesis occurring in the infected tissue. Whatever selective advantage these genes may confer in nature, it apparently is not a specific adaptation to pea as a host.

Table 1 is a summary of data from several studies, and the effects of the *Pda* genes on virulence were not all measured in a common genetic background. Recently the effects of genes representing each of the quantitative phenotypes were examined in a family of related crosses (15). As in previous studies, the Pda^n phenotype (*Pda*3-2) did not contribute to virulence (Fig. 3). Furthermore, progeny with the highest level of demethylase expression (*Pda*1-1) were more virulent than those with intermediate demethylase levels (*Pda*4-1). This result must be regarded as preliminary until a larger number of progeny have been scored, but it

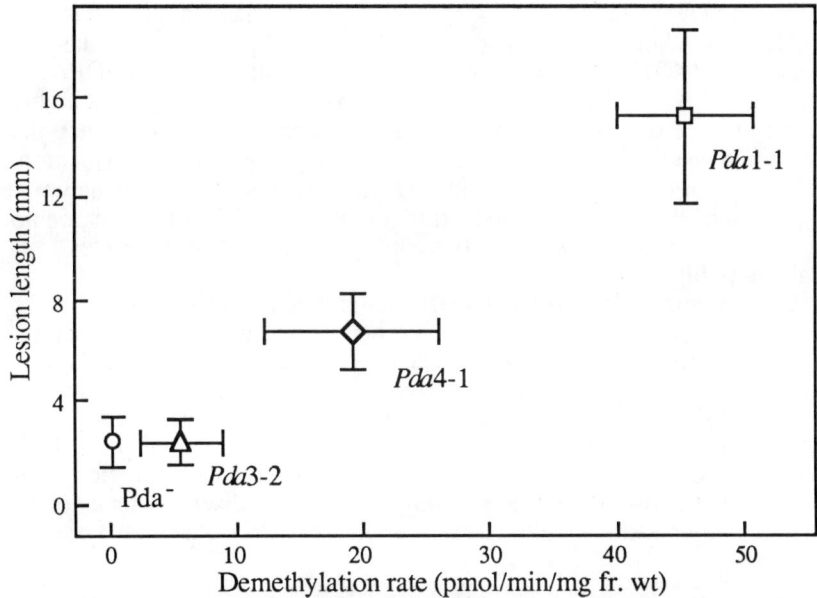

FIGURE 3. Segregation of pisatin demethylation rate and virulence on pea in tetrads from related crosses involving three *Pda* genes. Symbols indicate mean ± 1 SD. Numbers of progeny examined were 30 (Pda⁻), 15 (*Pda*3-2), 7 (*Pda*4-1) and 7 (*Pda*1-1). Demethylation rates were measured after 6 h preinduction with pisatin for both *Pda*1-1 and *Pda*4-1 progeny.

suggests that the level of pisatin demethylase activity may be a major limiting factor for virulence in some genetic backgrounds.

Pisatin Demethylase is a Cytochrome P-450 System.

The demethylation of pisatin is catalyzed by a cytochrome P-450 monooxygenase (20). This microsomal system is composed of at least two proteins: a flavoprotein (NADPH cytochrome P-450 reductase) and a hemeprotein (cytochrome P-450) where the substrate and molecular oxygen bind (21,22). Although cytochrome P-450 sytems have not been well characterized in filamentous fungi, these enzymes have been thoroughly characterized from other sources, particularly mammalian liver where cytochrome P-450 also serves a detoxification function. Mammalian NADPH cytochrome P-450 reductase is a single protein encoded at a single genetic locus, and is highly conserved among species (25). Hepatic cytochrome P-450, on the other hand, exists as multiple isozymes with different substrate specificity and differential inducibility, encoded at separate loci. Genetic variation for cytochrome P-450 isozymes has been found between lines of mice and between individual humans (23,24).

The NADPH cytochrome P-450 reductase of *N. haematococca* has been purified to near homogeneity, and a partially purified preparation of the cytochrome P-450 has been obtained (22). As with the hepatic NADPH cytochrome P-450 reductase, the *N. haematococca* reductase appears highly conserved. The reductases from both Pda⁻ and Pda⁺ isolates will function to reconstitute pisatin demethylase activity when combined with the cytochrome P-450 preparation from a Pda⁺ isolate (21,22). Comparisons with a number of other fungi showed that the reductase is immunochemically conserved not only within *N. haematococca* but also among filamentous ascomycetes (Scala, unpublished).

Cytochrome P-450 probably exists in multiple isozymic forms in *N. haematococca*, analogous to what has been found in the hepatic system, and we propose that the *Pda* genes code for different cytochrome P-450s. Since some of these genes appear to have no adaptive significance with regard to virulence on pea, the ability of the corresponding isozymes to demethylate pisatin may be only incidental to their biological function. It might be expected that such isozymes would show a broader specificity than those which are essential for detoxifying pisatin in the disease interaction. Although only pisatin demethylases from virulent field isolates have been

TABLE 2

SUBSTRATE AND INDUCER SPECIFICITY OF PISATIN DEMETHYLASE

Compound	Efficiency as substrate	Efficiency as inducer
pisatin	(100%)	(100%)
7,4'-dimethoxy-isoflavone	58	< 1
6-methoxy-1-tetralone	< 1	41

Substrates (40 µM) were compared using microsomes from isolate T-9; control rate was 6.0 nmol/min/mg protein. Inducers were compared using isolate T-63; control rate was 150 pmol/min/mg mycelium (fresh wt.). Inducers were tested at 100 µM, except for 6-methoxy-1-tetralone which was more effective at 1 mM.

characterized as yet, these have been found to be quite specific for pisatin. A variety of compounds have been tested for their ability to act as inducers, substrates or inhibitors competitive with pisatin (17,26, unpublished). Table 2 shows the most distantly related compounds found to be effective in any of these respects. The only pisatin analog which was effective as an inhibitor was (-)pisatin, the enantiomer of the pea phytoalexin.

Isolation of a *Pda* Gene by Expression in *Aspergillus nidulans*

Recently one of the *Pda* genes of *N. haematococca* has been cloned in *A. nidulans* (27). A genomic library from a Pda[+] isolate of *N. haematococca* was constructed in the cosmid vector pKBY2 containing the *trpC* gene of *A. nidulans*, and this library was used to transform a *trpC[-] A. nidulans* isolate (28). A Trp[+], Pda[+] transformant was detected which contained integrated vector and *N. haematococca* DNA. The *Pda* gene was recovered by lambda packaging of genomic DNA from the transformant, and is currently being subcloned from the 40 kb *N. haematococca* DNA insert in the cosmid.

Although wild type *A. nidulans* is Pda[-], its NADPH-cytochrome P-450 reductase can function with the cytochrome P-450 from Pda[+] *N. haematococca* to reconstitute pisatin demethylase activity (Scala, unpublished). Presumably the *Pda* gene that has been isolated by expression in *A. nidulans* codes for a specific cytochrome P-450 that can demethylate pisatin.

Role of *Trans*-Acting Genes in Pisatin Demethylase Regulation

The strong inducibility of pisatin demethylase in some isolates, and the specificity of induction, suggest the existence of a regulatory protein which recognizes pisatin and induces or derepresses the structural gene for the enzyme. The gene for such a regulator would represent an additional genetic load to *N. haematococca* with no benefit in habitats other than pea. Perhaps surprisingly, our genetic studies to date have not identified any genes of this type. As mentioned above, when *Pda* genes segregate they carry with them not only the ability to demethylate pisatin but also the same regulatory phenotype as found in the Pda[+] parent. This result could be explained if all the Pda[-] parents used in these crosses possessed the same allele of the regulator gene as the Pda[+] parents. But given the number of crosses involved, it seems more likely that the regulator gene is tightly linked to the structural gene, or that the regulatory mechanism does not conform to the simple model. The fact that pisatin demethylase expression is much lower in the *A. nidulans* transformant than in the *N. haematococca* isolate which provided the *Pda* gene (27) hints that some regulatory elements may not have been transferred, but perhaps these elements simply do not function as well in *A. nidulans*.

CONCLUSIONS

Not all plant pathogens are tolerant of the chemical toxins their hosts produce (2). Presumably some pathogens do not require such tolerance because their mode of infection allows them to avoid contact with their hosts' toxins. Though the situation that we have observed with *N. haematococca* and pisatin is not universal, we think it is common: that a pathogen evolves a means to tolerate the toxins produced by its host and that this trait, once developed, makes an important contribution to virulence on that host. Organisms like *N. haematococca* which are broadly adapted as parasites and saprophytes may be especially suited for the gradual evolution of specific detoxification enzymes, since they are not dependent on a particular host for the ability to survive and reproduce at all. It may even be significant that Pda⁻ isolates are capable of producing small lesions on pea. Although our virulence assays are performed under somewhat artificial conditions, these small lesions suggest that growth of *N. haematococca* under the selective pressure of pisatin toxicity may occur in nature as well.

ACKNOWLEDGMENTS

We are grateful to Mike Costa, Patty Matthews, Felice Scala, and Klaus-Michael Weltring for their important contributions to this work, as well as those who made earlier contributions cited below.

REFERENCES

1. Mansfield JW (1983). Antimicrobial Compounds. In Callow JA (ed): "Biochemical Plant Pathology," New York: John Wiley & Sons, p 237.
2. VanEtten HD, Matthews DE, Smith DA (1982). Metabolism of phytoalexins. In Bailey JA, Mansfield JW (eds): "Phytoalexins," Glasgow: Blackie & Son, p 181.
3. Smith DA, Wheeler HE, Banks SW, Cleveland TE (1984). Association between lowered kievitone hydratase activity and reduced virulence to bean in variants of *Fusarium solani* f. sp. *phaseoli*. Physiol. Plant Pathol. 25:135.
4. Kraft JM, Burke DW, Haglund WA (1981). Fusarium diseases of beans, peas, and lentils. In Nelson PE, Toussoun TA, Cook RJ (eds): "Fusarium: Diseases, Biology, and Taxonomy," University Park: Pennsylvania State University Press, p 142.
5. Pueppke SG, VanEtten HD (1974). Pisatin accumulation and lesion development in peas infected with *Aphanomyces euteiches*, *Fusarium solani* f. sp. *pisi*, or *Rhizoctonia solani*. Phytopathology 64:1433.
6. Cruickshank IAM (1962). Studies on phytoalexins IV. The antimicrobial spectrum of pisatin. Australian J. Biol. Sciences 15:147.

7. Nonaka F (1967). Inactivation of pisatin by pathogenic fungi. Agric. Bull. Saga Univ. 24:109.
8. De Wit-Elshove A (1969). The role of pisatin in the resistance of pea plants - some further experiments on the breakdown of pisatin. Neth. J. Plant Pathol. 75:164.
9. VanEtten HD, Pueppke SG, Kelsey TC (1975). 3,6a-Dihydroxy-8,9-methylenedioxypterocarpan as a metabolite of pisatin produced by *Fusarium solani* f. sp. *pisi*. Phytochemistry 14:1103.
10. VanEtten HD, Pueppke SG (1976). Isoflavonoid phytoalexins. In Friend J, Threlfall DR (eds): "Biochemical Aspects of Plant-Parasitic Relationships," London: Academic Press, p 239.
11. VanEtten HD, Kistler HC (1987). *Nectria haematococca* mating populations I and VI. In Sidhu GS (ed): "Genetics of Pathogenic Fungi," Advances in Plant Pathology 6: (in press).
12. VanEtten HD, Matthews PS, Tegtmeier KJ, Dietert MF, Stein JI (1980). The association of pisatin tolerance and demethylation with virulence on pea in *Nectria haematococca*. Physiol. Plant Pathol. 16:257.
13. Tegtmeier KJ, VanEtten HD (1982). The role of pisatin tolerance and degradation in the virulence of *Nectria haematococca* on peas: a genetic analysis. Phytopathology 72:608.
14. Kistler HC, VanEtten HD (1984). Regulation of pisatin demethylase in *Nectria haematococca* and its influence on pisatin tolerance and virulence. J. Gen. Microbiol. 130:2605.
15. Mackintosh SF (1986). Regulatory phenotypes of pisatin demethylase in *Nectria haematococca* in regard to pathogenicity on pea and tolerance to pisatin. M.S. Thesis, Cornell University, 65p.
16. Matuo T, Snyder WC (1972). Host virulence and the *Hypomyces* stage of *Fusarium solani* f. sp. *pisi*. Phytopathology 62:731.
17. VanEtten HD, Barz W (1981). Expression of pisatin demethylation ability in *Nectria haematococca*. Arch. Microbiol. 129:56.
18. VanEtten HD, Matthews PS (1984). Naturally-occurring variation in the induction of pisatin demethylating ability in *Nectria haematococca* mating population VI. Physiol. Plant Pathol. 24:149.
19. Kistler HC, VanEtten HD (1984). Three non-allelic genes for pisatin demethylation in the fungus *Nectria haematococca*. J. Gen. Microbiol. 130:2595.
20. Matthews DE, VanEtten HD (1983). Detoxification of the phytoalexin pisatin by a fungal cytochrome P-450. Arch. Biochem. Biophys. 221:494.
21. Desjardins AE, Matthews DE, VanEtten HD (1984). Solubilization and reconstitution of pisatin demethylase, a cytochrome P-450 from the pathogenic fungus *Nectria haematococca*. Plant Physiol. 75:611.
22. Desjardins AE, VanEtten HD (1986). Partial purification of pisatin demethylase, a cytochrome P-450 from the pathogenic fungus *Nectria haematococca*. Arch. Microbiol. 144:84.

23. Adesnik M, Atchison M (1986). Genes for cytochrome P-450 and their regulation. CRC Critical Reviews in Biochemistry 19:247.
24. Iversen PL, Hines RN, Bresnick E (1986). The molecular biology of the polycyclic aromatic hydrocarbon inducible cytochrome P-450; the past is prologue. BioEssays 4:15.
25. Simmons DL, Lalley PA, Kasper CB (1985). Chromosomal assignments of genes coding for components of the mixed-function oxidase system in mice. J. Biol. Chem. 260:515.
26. VanEtten HD (1982). Phytoalexin detoxification by monooxygenases and its importance for pathogenicity. In Asada Y, Bushnell WR, Ouchi S, Vance CP (eds): "Plant Infection, the Physiological and Biochemical Basis," Tokyo: Japan Scientific Societies Press, p 315.
27. Weltring K-M, Matthews P, Turgeon G, Yoder OC, VanEtten HD (1986). Isolation and expression of a *Nectria haematococca* gene for phytoalexin detoxification in *Aspergillus nidulans*. J. Cellular Biochem. Supplement 10c:29.
28. Yelton MM, Timberlake WF, van den Hondel CAMJJ (1985). A cosmid for selecting genes by complementation in *Aspergillus nidulans*: Selection of the developmentally regulated yA locus. Proc. Natl. Acad. Sci. USA 82:834.

Molecular Strategies for Crop Protection, pages 71–81
© 1987 Alan R. Liss, Inc.

CARBOHYDRATE-SPECIFIC MONOCLONAL ANTIBODIES
AND IMMUNOSUPPRESSION IN THE STUDY
OF PLANT DISEASE RESISTANCE[1]

Arthur R. Ayers and Keith L. Wycoff

Department of Cellular and Developmental Biology
Harvard University, Cambridge, MA 02138

ABSTRACT   The interactions between plants and
their pathogens are dominated by the expression of
pairs of genes; a resistance gene in the plant
host is paired with each complementary avirulence
gene in the pathogen.   This common genetic
relationship has suggested a corresponding
interaction between plant and pathogen molecules
in the primary recognition event that initiates
the biochemical responses observed as resistance.
Study of resistance gene function is thus reduced
to identification of the plant molecules
(presumably resistance gene products) mediating
recognition of pathogen determinants associated
with particular avirulence genes.
     Here we describe immunochemical approaches to
identify pathogen molecules with antigenic
determinants coded by avirulence genes.   We have
developed immunological suppression with
cyclophosphamide to permit direct selection of
mouse hybridoma clones producing antibodies to
antigenic determinants that differ between races
of the pathogen. This approach may permit the use
of infected plant tissue as an inject antigen for
the production of antibodies specific for antigens
of pathogens, including obligate parasites.

[1]This work was supported by grants from The National
Science Foundation (PCM 83-02789), the U.S. Dept. of
Agriculture (USDA 85-CRCR-1-1632) and the Maria Moors Cabot
Foundation.

INTRODUCTION

A Molecular Model for Pathogen Recognition

Plants have a variety of defenses that can be invoked upon infection by microorganisms. In some cases a particular plant is unable to thwart the advances of a particular pathogen. This apparent breech in an otherwise comprehensive mantle of defense reflects a lack of adequate surveillance, not a fault in the defensive capacity of the plant. Thus, genes for disease resistance may be more appropriately called "pathogen recognition genes".

The molecules that mediate pathogen recognition are not known, but the genetics of host-pathogen interactions permit some predictions. In many instances resistance genes in a plant species are paired with avirulence genes in the pathogen (a "gene-for-gene" relationship,3,6). The molecular implications of the dominance and independent inheritance of both resistance and avirulence genes have been recently reviewed (6). One model incorporating the characteristics of the genetics of the interactions hypothesizes that avirulence genes code for glycosyl-transferases involved in the synthesis of carbohydrate constituents that decorate the extracellular products of the pathogen, e.g. extracellular enzymes or structural components of fungal walls (1,2,5,6). Corresponding resistance genes code for receptors that bind to the carbohydrate determinants of the pathogen and trigger elaboration of host defenses. The exquisite specificity of the host-pathogen interaction is thus mediated by a corresponding specificity in the synthesis and binding of the carbohydrate determinants.

Immunochemical Approaches

The model presented here makes several predictions about host-pathogen interactions that may be tested experimentally. The extracellular carbohydrate determinants from races of the pathogen with different avirulence genes must be different. Thus, antibodies raised to a representative mixture of antigens from one race should be able to distinguish that race from others. An immunochemical approach is complicated by several features of the immunization process.

All antigens are not equal; a few antigenic

determinants dominate the immunological response and reduce the production of antibodies to the remaining antigens (4). Also, antigenic determinants that are also present as mouse antigens will not yield a response due to the animal's tolerance of these antigens. It is, therefore, likely that some antigens will not be represented by corresponding antibodies following a challenge with a mixture of antigens. Polyclonal sera from mice or rabbits injected with different races of Phytophthora megasperma f.sp. glycinea (Pmg), for example, cannot distinguish between antigens from different races (5).

Another complicating factor is the general absence of isogenic pathogen races. Many antigenic differences would be expected to result from the random variations in protein primary sequences, and generate antigenic differences unrelated to expression of avirulence genes. Carbohydrate antigens, on the other hand, result from the action of specific enzymes, and would be expected to be more stable to random variation.

We have applied the use of monoclonal antibodies and immunosuppression in the search for antigenic determinants associated with avirulence genes. With monoclonal antibodies it is possible to produce clones of cells that produce single antibodies starting with a mouse injected with a complex mixture of antigens. In our case we injected a mixture of either mycelial wall or glycoprotein antigens from one race of Pmg and screened for clones producing antibodies reacting differentially with equivalent antigens from either the original or other races of Pmg (1,5,9).

The impact of immunodominance and tolerance in the production of antibodies to complex antigen mixtures can be minimized by tolerization. A common difficulty in producing antibodies specific for a purified glycoprotein is the immunodominance of the carbohydrate constituents (4). Since the carbohydrate antigens are shared by many glycoproteins with similar cellular localization, antibodies raised to a pure glycoprotein enzyme my crossreact with even unrelated enzymes. One approach to this problem of immunodominance is to reduce the response of mice to the common antigens by injection with these antigens while their immune system is still immature. This process is termed tolerization, and although technically inconvenient, it has proved useful in minimizing immunodominance (4).

An alternative approach is immunosuppression (Fig. 1). Cyclophosphamide is an antineoplastic drug that eliminates B lymphocyte responses to injected antigens. Thus, a mixture

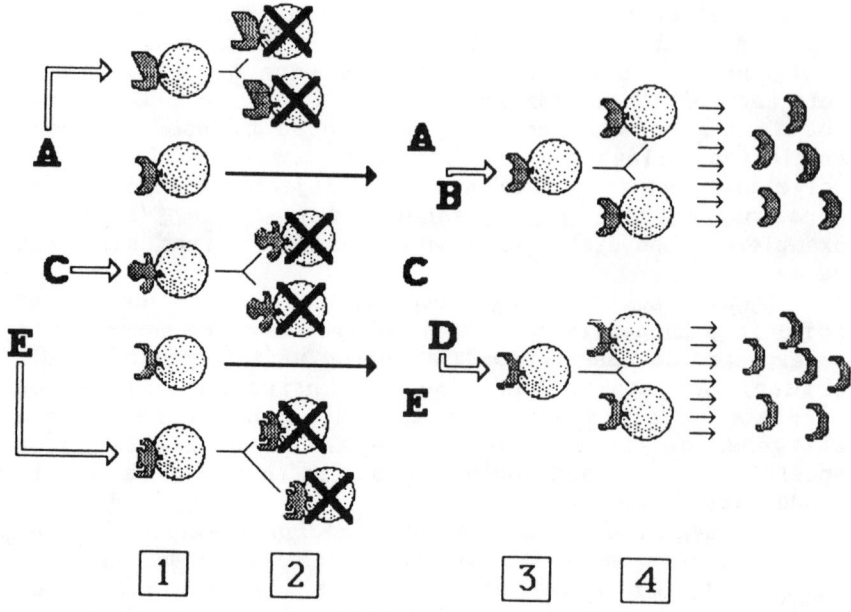

FIGURE 1.   Immunosuppression with cyclophosphamide

of antigens that would normally lead to proliferation of B lymphocytes that subsequently produce serum antibodies (Fig. 1, step 1) will, upon addition of cyclophosphamide (step 2), result in the elimination of all of the lymphocytes that respond to the antigens (7). In subsequent exposures to new mixtures of antigens (step 3), there will be no response to antigens present in the original mixture, but new antigens (step 4) will result in B lymphocyte proliferation and antibody production. In our case antigen mixtures from two different races of Pmg can be used to yield antibodies specific for avirulence gene associated antigens.

Another effect of cyclophosphamide is to eliminate the T lymphocytes that suppress B lymphocyte response to antigens to which the animal is tolerant. Thus, immunosuppression can also be used to relieve tolerance to foreign antigens of interest and result in antibodies otherwise unavailable.

METHODS

Monoclonal Antibody Production

Monoclonal antibodies were produced to mixtures of antigens from Pmg races 1, 4 and 12 following established procedures (5). Race 1 has at least 14 identified avirulence genes with only six of these present in race 12. Race 4 is intermediate in sharing more genes with the other two races. Antigen mixtures consisted of extracellular glycoproteins or mycelial wall as described (5,9).

Immmunosuppression with Cyclophosphamide

Six-week old Balb/C mice were injected with either 6 ug of extracellular glycoproteins or 12 ug of purified walls from race 12 of Pmg. After 2 min, 12hr and 24hr, the challenged mice were injected with cyclophosphamide (equivalent to 100 mg/kg body weight in PBS). Two weeks after the first injection they were injected with 120 ug of extracellular glycoproteins or 200 ug of purified mycelial walls from Pmg race 1. One week later serum samples were taken and the titer of antibodies to races 1, 4 and 12 antigens was tested by ELISA as described (9).
A second round of challenges with antigens from Pmg race 12, followed by cyclophosphamide and antigens from Pmg race 1 was initiated four weeks from the first injection. A final additional injection with race 1 walls was given one week later and a fusion for hybridoma production was performed four days later. Colonies were screened by ELISA against immobilized mycelial walls of Pmg races 1, 4 and 12 after one week of incubation (9).

RESULTS

Carbohydrate-Specific Monoclonal Antibodies

In our initial studies, we produced a library of monoclonal antibodies to the immunodominant antigens of different races of a fungal soybean pathogen, Phytophthora megasperma f.sp. glycinea (5,9). Antibodies binding to antigens associated with particular avirulence genes were tentatively identified by screening for preferential binding to antigen mixtures from races of the pathogen with

differing avirulence genes.   These antibodies are being
extensively tested against numerous isolates of different
races of the pathogen to establish an obligatory correlation
between individual avirulence genes and antigenic
determinants identified by the antibodies.

All of the antibodies examined were specific for
carbohydrate antigens (9).   Monoclonal antibodies from mice
challenged with Pmg glycoproteins were specific for one of
six classes of glycoproteins as observed by complex binding
patterns on Western blots.   The specificity for the
carbohydrate determinants of the glycoproteins was
determined by perturbation of binding of the antibodies to
glycoproteins modified by reagents (periodate, a-
mannosidase, Endo-H, etc.) known to affect selectively
carbohydrate constituents.

Several antibodies were observed to have higher
affinities for race 1 glycoproteins than for the equivalent
glycoproteins from race 12 (9).   These differences in
affinity were attributed to differences in the structures of
the carbohydrate constituents of the glycoproteins.

Immunosuppression

Two cycles consisting of injections of Pmg 12 antigens,

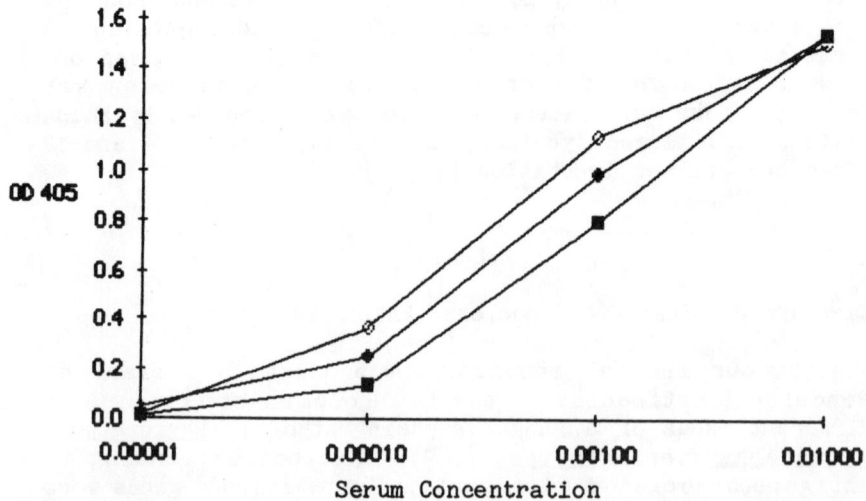

FIGURE 2.   Pmg 1 antiserum binding to 1(◆),4(◇),12(■).

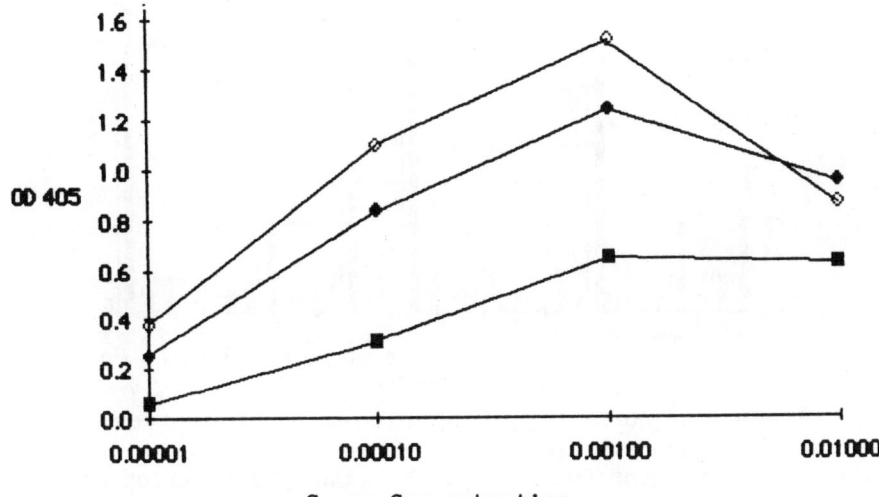

FIGURE 3.   Antiserum produced by suppression of response to Pmg 12 and challenge with Pmg 1.   Binding to antigens from Pmg 1(◆), 4(◇) or 12(■).

immunosuppression, and then Pmg 1 antigens were performed prior to hybridoma fusion and production of antibodies for screening.   Analysis of the polyclonal sera after the first cycle show a dramatic decrease in the binding of the antibodies to race 12 walls relative to races 1 or 4 (Figs. 2 and 3).   At a dilution of 1:1000, antiserum from a mouse receiving three injections of Pmg race 1 antigens shows approximately 80% as much binding to antigens from race 12 as race 1 (Fig. 2).   After the first cycle of immunosuppression, the response to race 12 has been reduced to half that of race 1 antigens (Fig. 3).   Thus, the effects of immunosuppression in reducing the production of antibodies to antigens common to different races of Pmg is observable even in polyclonal sera.

Hybridoma colonies from mice receiving two cycles of suppression show dramatic differences in the response of their antibodies to antigens from Pmg races 1 and 12 (Fig. 4).   The yield of hybridomas from the spleens of mice treated with cyclophosphamide is approximately 10% of control mice.   Some antibodies react equivalently with antigens from both races (e.g. hybridomas 8, 17 and 31, Fig. 3), whereas others react only to race 1 (e.g. hybridomas 10,

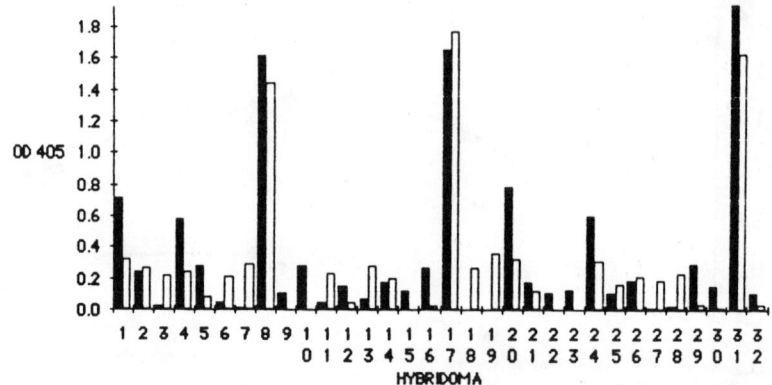

FIGURE 4. Reaction of hybridoma colonies from two cycles of suppression with Pmg 12 antigens and challenge with Pmg 1 antigens in ELISA assays with antigens from Pmg1 (■) and 12 (□).

23 and 30), or to race 12 (e.g. hybridomas 7, 18, 19 and 27) antigens. These race-specific patterns differ from those observed with monoclonal antibodies obtained by a series of injections with antigens from a single race. The reaction pattern of the antibodies in Western blots (data not shown) is also different. In some cases the pattern is consistent with binding to protein determinants rather than carbohydrate.

DISCUSSION

Multiple Classes of Glycoproteins

The structure of the carbohydrate components of the extracellular glycoproteins of Pmg and related fungi has not been extensively studied. Glycosyl linkage composition has been determined for a Pmg extracellular enzyme, invertase, and yielded a very complex pattern of mannosyl, glucosyl and glucosaminosyl residues with a large variety of linkages (10). Glycoproteins released by hot water extraction of purified mycelial walls have been similarly analyzed and with similar results (2). Work with monoclonal antibodies specific for the carbohydrate residues of Pmg glycoproteins give an indication that at least six different types of oligosaccharides are present on the full spectrum of glycoproteins found in extracellular culture filtrates (9). Thus, the differences in glycosyl linkage composition of invertases initially attributed to avirulence gene differences (10) may reflect differences in the ratio of alternate carbohydrate decorations of the same equivalent enzymes. Carbohydrate-specific antibodies may be very useful in sorting out the multiplicity of carbohydrates present on glycoproteins to determine if there are qualitative differences in structure between different races, as indicated by binding studies (9).

Glycosyltransferases as Avirulence Gene Products

Carbohydrates have been implicated as the determinants mediating race-cultivar specific plant disease resistance, because they have many of the molecular characteristics that are consistent with the genetics of host-pathogen interactions (6). It has also been suggested that resistance genes may be related to components of mammalian immune systems, the major histocompatibility antigens (6). One striking implication of the hypothesized carbohydrate-receptor interaction is that the glycosyltransferases and corresponding resistance genes may have very similar characteristics. As a corollary, avirulence genes and resistance genes may be evolutionarily related (5). In animals, glycosyltransferases have recently been suggested to be the evolutionary antecedents of some major histocompatibity complex proteins and immunoglobulins (8). This provocative hypothesis is based on similarities in several characteristics: specificity of ligand binding, membrane and soluble forms, polymorphism, involvement in intercellular recognition, and ability to modulate expression when a ligand is bound. Resistance genes share these same traits and may represent the primitive recognition system of the ancestor of plants and animals.

Immunodominance, Tolerance and Suppression

Experiments with polyclonal sera raised to mycelial walls (10) or glycoproteins (5) of Pmg could not demonstrate race-specific binding reactions with Pmg glycoproteins. Likely explanations for these observations are similarities in the immunodominant antigens in different races of the pathogen, with an attendant lack of response to the minor antigenic determinants associated with avirulence genes. Tolerance to the antigenic determinants of interest may also play a role, since the high mannose content of the Pmg carbohydrates may reflect antigenic homologies with highly conserved mannan components of glycoproteins. We are beginning to address these issues in the study of Pmg carbohydrate antigens by the use of immunosuppression.

The immunosuppression experiments described here demonstrate the effectiveness of this approach in increasing the production of monoclonal antibodies to antigens that differ between races of Pmg with different avirulence genes. Several antibodies with new specificities have been identified and are being characterized. Among these antibodies are examples that bind only to antigens from Pmg race 1 and not to equivalent antigens from race 12. We believe that these are good candidates for antibodies specific for avirulence gene associated antigens.

Potential Applications of Immunosuppression

Immunosuppression has tremendous potential for facilitating the production of antibodies specific for antigenic determinants present in one mixture but absent in another. We are now evaluating the use of suppression with infected plant tissue as the inject antigen. Our experience suggests that we will be able to produce monoclonal antibodies (or polyclonal sera) specific for the antigens that differ in soybeans infected with two different compatible races of Pmg. This approach is expected to permit immunochemical studies with obligate parasites and for any system with the appropriate antigen pairs available.

ACKNOWLEDGMENTS

We wish to acknowledge Lisa Rayder, Douglas Brugge and John Glyphis for their contributions to this manuscript.

REFERENCES

1.   Ayers AR, Goodell JJ, DeAngelis P (1985).   Plant detection of pathogens. In Conn EE, Cooper-Driver G, Swain T (eds) "The Biochemical Interactions of Plants with Other Organisms," Recent Advances in

Phytochemistry, Volume 19, New York: Plenum Press, p 1.

2. Ayers AR, Valent B, Ebel J, Albersheim P (1976). Host-
    pathogen interactions. XI. Composition and structure
    of wall-released elicitor fractions. Pl Physiol
    57:766.

3. Day PR (1974). "Genetics of Host-Pathogen
    Interactions," San Francisco: W. H. Freeman and
    Company.

4. Golumbeski GS, Dimond RL (1986). The use of
    tolerization in the production of monoclonal antibodies
    against minor antigenic determinants. Anal Biochem
    154:373.

5. Goodell JJ, DeAngelis P, Ayers AR (1985).
    Immunochemical identification of antigens involved in
    plant/pathogen interactions. In Key JL, Kosuge T (eds):
    "Cellular and Molecular Biology of Plant Stress", UCLA
    Symposia on Molecular and Cellular Biology, New Series,
    Volume 22, New York: Alan R. Liss, Inc, p. 447.

6. Keen NT (1983). Specific recognition in gene-for-gene
    host-parasite systems. Adv Pl Pathol 1:1.

7. Matthew WD, Patterson PH (1983). The production of a
    monoclonal antibody that blocks the action of a neurite
    outgrowth-promoting factor. Cold Spring Harbor Symp
    Quant Biol 48:625.

8. Roth, S (1985). Are glycosyltransferases the
    evolutionary antecedents of the immunoglobulins? Quat
    Rev Biol 60:145.

9. Wycoff KL, Goodell JJ, Ayers AR (1987). Monoclonal
    antibodies to glycoproteins of a fungal plant
    pathogen, _Phytophthora megasperma_ f.sp._glycinea_,
    submitted to Plant Physiology.

10. Ziegler E, Albersheim P (1977). Host-pathogen
    interactions. XIII. Extracellular invertases secreted
    by three races of a plant pathogen are glycoproteins
    which possess different carbohydrate structures. Pl
    Physiol 59:1104.

Molecular Strategies for Crop Protection, pages 83–93
© 1987 Alan R. Liss, Inc.

GENETIC ANALYSIS OF HOST SPECIES SPECIFICITY IN[1]
MAGNAPORTHE GRISEA

Barbara Valent and Forrest G. Chumley

E. I. Du Pont De Nemours and Co., Inc.
Central Research and Development Department
Experimental Station, E402/2208
Wilmington, DE 19898

ABSTRACT   Crosses   between   a   field   isolate   of
Magnaporthe grisea (anamorph, Pyricularia oryzae Cav.
and Pyricularia grisea) that infects weeping lovegrass
and a field isolate that infects goosegrass have
resulted in the identification of a single gene
difference that determines pathogenicity toward weeping
lovegrass and a second unlinked single gene difference
that   determines   pathogenicity   toward   goosegrass.
Fertile laboratory strains of M. grisea that infect
goosegrass and/or weeping lovegrass have been
developed. These strains may now be utilized in a
rigorous genetic analysis of host species specificity
and general pathogenicity.

INTRODUCTION

Magnaporthe grisea, (1,2,3) is a fungal plant pathogen
with unique promise as a system for the study of
host-parasite interactions. M. grisea is the perfect state
of fungi formerly described as two distinct species,
Pyricularia oryzae, the causal agent of the rice blast
disease, and Pyricularia grisea, pathogens of grasses other
than rice. The species distinction between P. oryzae and
P. grisea has been abandoned because the two are
morphologically indistinguishable, they are interfertile

[1]We gratefully acknowledge the support of the
Department   of   Energy,   (DE-AC02-76ERO-1426   and
DE-AC02-84ER13160), which allowed us to initiate this work.

(4, 5,6), and many field isolates of the fungus infect rice and other grasses as well (7; Valent and Chumley unpublished results).

The fungus is a filamentous, heterothallic Ascomycete, and it grows on defined medium, a trait that facilitates both genetic and biochemical analysis. The availability of interfertile strains of M. grisea that differ in the grass species they infect makes genetic analysis of host species specificity possible. Genetic analysis of host cultivar specificity is also of interest since hundreds of races of the rice blast fungus have been distinguished according to the spectrum of rice cultivars they can infect (8,9,10,11,12). Thus, M. grisea offers a particularly good system for studying genetic and biochemical determinants of host specificity, as well as determinants of general pathogenicity, those characteristics required for a strain of the fungus to infect any plant.

The results reported in this paper demonstrate that host species specificity differences exhibited by some field isolates of M. grisea have a simple genetic basis. In addition, the results demonstrate that limitations on genetic analysis imposed by the low fertility of field isolates may be quickly overcome by selection of highly fertile progeny. Problems encountered in crosses between field isolates include a failure to produce ascospores, or poor germination of ascospores. With the development of highly pathogenic fertile laboratory strains that can be crossed reproducibly to yield ascospores that are all viable, rigorous genetic analysis of host specificity and general pathogenicity of M. grisea is now possible.

MATERIALS AND METHODS

Strains, Media and Strain Storage.

The M. grisea field isolates used in this study were: WGG-FA40 (finger millet and goosegrass pathogen) and K76-79 (a weeping lovegrass pathogen), both generously provided by H. Yaegashi. Strains designated by a three-part number are single ascospore progeny of genetic crosses.

The media used in this study have been described (6,13). The fungus is stored in a non-metabolizing state by freezing in dried lesion tissue, in cellulose filter paper disks (13mm diameter, S and S, #597) or on silica gel (6,13).

Genetic Crosses.

Sexual crosses are performed by pairing strains of opposite mating type on oatmeal agar. The cross plates are incubated at 20C under continuous cool white fluorescent light. Asci containing viable ascospores appear 13 to 21 days after the strains are paired. For tetrad analysis, asci are released from the perithecium using a pair of fine forceps. In good crosses, the ascus walls lose their integrity about 20 minutes after the asci are released from the perithecium. The eight spores of an ascus are separated by hand on Misato-Hara medium solidified with 4% agar using a finely drawn glass needle and a 50X Wild stereomicroscope. In crosses between highly fertile strains, mature perithecia contain loose ascospores and random ascospore analysis can by easily accomplished. Germinated ascospores are transferred to complete medium wells of Falcon 3047 Tissue Culture Plates for immediate in storage as described (6,13).

Testing the Pathogenicity of Strains.

Conidia are collected from cultures growing on oatmeal agar plates by washing with a sterile 0.25% gelatin solution (Sigma, Type IV from calf skin). Five plants growing in a single five-inch plastic pot are infected at the four leaf stage. A four ml suspension containing 5×10^5 conidia per ml is sprayed onto the plants using an artist's air brush. Infected plants are then sealed inside a plastic bag and left in the dark for 16 hours at 24C. After 24 hours, the bags are removed and the plants are placed in the greenhouse at 26-28C. Infection types are scored after 5 days as described (6). Weeping lovegrass typically shows an all-or-nothing response to infection by different strains of the fungus. That is, plants are either unblemished by the fungus, or they are completely killed. Infected goosegrass plants never show the devastating symptoms observed with weeping lovegrass: a range of lesion types is detected.

Weeping lovegrass seed was obtained from Valley Seed Service in Fresno, California. Goosegrass seed was obtained from Azlin Weed Seed Service in Leland, Mississippi. Both host plants are grown in vermiculite with fourteen hours of daylight, a constant temperature of 25C, and a relative humidity of 80%. They are watered with a nutrient solution.

RESULTS

Genetic Analysis of Host Species Specificity.

Field isolates of M. grisea exhibit a range of mating characteristics (5,14,15,16,17,18). A few field isolates that infect finger millet and goosegrass are hermaphrodites. Many other strains mate only as males and others are totally infertile. Even the most fertile crosses typically produce only 5-10% viable ascospores. For a complete description of fertility in M. grisea, see reference 6.

Many field isolates of M. grisea have been mated to screen for crosses that produce a high proportion of viable ascospores. Crosses between the weeping lovegrass pathogen, K76-79, and the goosegrass pathogen, WGG-FA40, were noteworthy in consistently producing perithecia containing 45% viable ascospores. This high fertility was somewhat surprising in view of the fact that strain K76-79 shows degraded culture morphology due to having been continuously subcultured (6), and it is not able to function as a female in crosses. Nevertheless, K76-79 is an aggressive pathogen of weeping lovegrass and typically yields ascospores with good viability in crosses with hermaphrodites. Seven complete tetrads from a cross (number 4091) between K76-79 and, WGG-FA40, have been studied in detail. Three tetrads shown in Table 1 are representative of these seven.

Determinants of host-species specificity were inherited simply in the K76-79 by WGG-FA40 cross. One half of the progeny within each tetrad were pathogens of weeping lovegrass and one half were pathogens of goosegrass. Parental and recombinant classes appeared with equal frequency, indicating that a gene determining pathogenicity to weeping lovegrass had segregated independently from a gene determining pathogenicity to goosegrass. Of the seven tetrads examined, two were parental ditypes, one was a nonparental ditype and four were tetratypes. Thus, the two parental field isolates differ at two unlinked loci, one conditioning pathogenicity toward goosegrass, and the other conditioning pathogenicity toward weeping lovegrass. These genes, named Pw11 and Pgg1 respectively, are unlinked to the mating type locus (Table 2).

TABLE 1

ANALYSIS OF THREE TETRADS FROM CROSS 4091
BETWEEN THE WEEPING LOVEGRASS PATHOGEN K76-79
AND THE GOOSEGRASS PATHOGEN WGG-FA40

Ascus	Spore	Mating Type	Pathogenicity to goosegrass[a]	Pathogenicity to weeping lovegrass[a]
10	1	Mat1-2	–	–
	3	Mat1-2	–	–
	2	Mat1-2	++	–
	8	Mat1-2	++	–
	4	Mat1-1	++	++
	5	Mat1-1	++	++
	6	Mat1-1	–	++
	7	Mat1-1	–	++
22	1	Mat1-1	++	–
	3	Mat1-1	++	–
	2	Mat1-1	++	–
	4	Mat1-1	++	–
	5	Mat1-2	–	++
	6	Mat1-2	–	++
	7	Mat1-2	–	++
	8	Mat1-2	–	++
25	1	Mat1-1	–	–
	2	Mat1-1	++	++
	4	Mat1-1	++	++
	3	Mat1-2	–	–
	6	Mat1-2	–	–
	5	Mat1-2	++	++
	7	Mat1-2	++	++

(a)   ++ indicates a pathogen, – indicates a nonpathogen.

TABLE 2
ABSENCE OF LINKAGE BETWEEN Pwl, Pgg, AND Mat

Gene Pair	Tetrad Analysis			Cross
	PD	NPD	T	
Pwl/Pgg	2	1	4	4091
Pgg/Mat	1	0	6	4091
Pwl/Mat	3	2	1	4091
Pwl/Mat	2	0	4	4136

Subsequent crosses confirmed the single gene segregation of host-species specificity observed in cross 4091. A single ascospore strain from a tetratype tetrad, 4091-5-8, which is a pathogen of both weeping lovegrass and goosegrass, was back-crossed to the goosegrass-infecting parent, WGG-FA40 (cross 4016). Random ascospore analysis was conducted, and 25 of 44 progeny tested for pathogenicity infected weeping lovegrass, while 19 of 44 did not infect weeping lovegrass. All 44 progeny infected goosegrass as expected.

The results of a cross between two progeny from cross 4091 also support single gene segregation of host species specificity. Strain 4091-1-6, which infects goosegrass but not weeping lovegrass, was crossed with strain 4091-14-3, which infects both goosegrass and weeping lovegrass (cross 4136). In each of six tetrads tested, half the progeny infect weeping lovegrass and half do not. All progeny are pathogens of goosegrass as predicted. The results of this cross show that the gene that determines pathogenicity toward weeping lovegrass is unlinked to mating type (Table 2).

Development of Improved Strains for Genetic Analysis.

Although cross 4091 (WGG-FA40 x K76-79) resulted in exceptionally good ascospore viability for a cross between field isolates, only 4% of the progeny were hermaphrodites, and many progeny were difficult to test for pathogenicity due to poor conidiation. Randomly chosen progeny from cross 4091 failed to mate with each other, produced few asci, or produced asci containing many dead ascspores. Therefore, we screened the progeny for those that exhibit improved fertility in mating. Strain 4091-5-8, which infects both weeping lovegrass and goosegrass, mates very efficiently as a male and as a female. This strain mated especially well with selected progeny of cross 4136 described above between the weeping lovegrass and goosegrass pathogen, 4091-14-3, and the goosegrass pathogen, 4091-1-6. Therefore, progeny from six tetrads from cross 4136 were screened for pathogenicity and for ability to mate with 4091-5-8. The results are shown for four of these tetrads in Table 3. Strains 4136-1-2 and 4136-1-3, 4136-2-8, 4136-3-3 and 4136-3-8, and 4136-4-3 and 4136-4-6 mate with 4091-5-8 to yield 95% ascospore viability. While all progeny tested were hermaphrodites, the strains that gave 95% viability of ascospores in crosses with 4091-5-8 were also more proficient at forming perithecia (i.e., mating as a female) than the others.

The progeny strains 4136-4-3 and 4136-4-6 are the only highly fertile strains in the six tetrads examined that infect both weeping lovegrass and goosegrass. Progeny from six tetrads of a cross between 4091-5-8 and 4136-4-3 were tested to determine if all progeny were as fertile as the parents and pathogenic toward both weeping lovegrass and goosegrass. All progeny exhibited the pathogenicity and fertility characteristics of the parental strains. Thus, these strains are examples of improved laboratory strains.

DISCUSSION

Natural isolates of M. grisea differ in host range, both with regard to the ability to attack various species of grasses and with regard to the ability to infect different cultivars of rice. The poor fertility of field isolates of the fungus presents a significant obstacle to conducting a satisfactory genetic analysis of these host range differences. The results presented in this paper demonstrate that it is possible to develop M. grisea laboratory strains that show substantial improvements over field isolates in production of perithecia and viable ascospores. We have developed strains that can be crossed, yielding progeny that are all highly fertile and pathogenic toward two grass species. Mutational analysis can now be performed with these strains to develop an understanding of

TABLE 3

INHERITENCE OF PATHOGENICITY AND FERTILITY
AMONG Mat1-1 PROGENY OF FOUR TETRADS FROM CROSS 4136

Ascus	Spore	Pathogenicity GG[a]	Pathogenicity WLG[a]	Viability of Ascospores[b]
1	2	++	+	95%
	3	++	+	95%
	6	++	++	No Asci
	8	++	++	No Asci
2	1	++	++	20%
	2	++	++	20%
	8	++	-	95%
3	3	++	-	95%
	8	++	-	95%
	5	++	++	10%
	7	++	++	10%
4	1	++	-	NT
	4	++	-	20%
	3	++	++	95%
	6	++	++	95%

(a) ++ indicates a pathogen, + indicates a weak pathogen and
    - indicates a nonpathogen.   GG refers to goosegrass.
    WLG refers to weeping lovegrass.

(b) Viability of ascospores produced in crosses with strain
    4091-5-8.

genes needed for general pathogenicity and host species specificity.

The weeping lovegrass pathogen, K76-79, a non-pathogen of goosegrass, mates with WGG-FA40 to yield many viable progeny, half of which can infect goosegrass. This result indicates that a single gene difference between K76-79 and WGG-FA40 accounts for the pathogenicity difference on weeping lovegrass. The same cross shows that alleles of another, unlinked gene determine pathogenicity toward goosegrass. These results do not imply that single genes alone determine pathogenicity toward weeping lovegrass and goosegrass. Pathogenicity is a complex phenotype that most likely involves the action of many genes. Presumably, all genes required for pathogenicity on each host are shared in common by the two two strains analyzed, except for the single genes we have identified.

Results presented in this paper have revealed single gene differences that govern the ability to attack different species of host grasses. We are now engaged in similar experiments that will allow us to undertake a genetic analysis of determinants required for pathogenicity toward rice. Our goal is to identify fungal genes that determine rice pathogenicity in general and rice cultivar specificity in particular. The identification of critical genes will lead to efforts to clone these genes via M. grisea transformation (19) and to determine how they act in determining the outcome of any particular host-pathogen interaction. We believe that this understanding will lead to insights for controlling plant diseases, perhaps by aiding in the design of stable disease resistant cultivars of crop plants.

REFERENCES

1. Barr ME (1977). Magnaporthe, Telimenella and Hyponectria (Physosporellaceae). Mycologia 69: 952-966.
2. Yaegashi H and Udagawa S (1978). The taxonomical identity of the perfect state of Pyricularia grisea and its allies. Can. J. Bot. 56: 180-183.
3. Yaegashi H and Udagawa S (1976). Additional note: the perfect state of Pyricularia grisea and its allies. Can. J. Bot. 56: 2184.

4. Yaegashi H and Hebert TT (1976). Perithecial
    development and nuclear behavior in Pyricularia.
    Phytopathology 66: 122-126.
5. Tanaka Y, Murata N and Kato H (1979). Behavior of
    nuclei and chromosomes during ascus development on the
    mating between either rice-strain or weeping
    lovegrass-strain and ragi-strain of Pyricularia. Ann.
    Phytopath. Soc. Japan 45: 182-191.
6. Valent B, Crawford MS, Weaver CG and Chumley FG (1986).
    Genetic studies of fertility and pathogenicity in
    Magnaporthe grisea (Pyricularia oryzae). Iowa State
    Journal of Research 60: 569-594.
7. Mackill AO and Bonman JM (1986). New hosts of
    Pyricularia oryzae. Plant Disease 70: 125-127.
8. Latterell FM, Marchetti MA and Grove BR (1965).
    Co-ordination of effort to establish an international
    system for race identification in Pyricularia oryzae.
    In "The Rice Blast Disease. Proceedings of a Symposium
    at The International Rice Research Institute, July,
    1963," Baltimore: The Johns Hopkins Press, pp.
    257-274.
9. Atkins JG, Robert AL, Adair CR, Goto K, Kozaka T,
    Yanagida R, Yamada M and Matsumoto S (1967). An
    international set of rice varieties for differentiating
    races of Pyricularia oryzae. Phytopathology 57:
    297-301.
10. Kiyosawa S (1976). Pathogenic variations of Pyricularia
    oryzae and their use in genetic and breeding studies.
    SABRAO Journal 8: 53-67.
11. Yamada M, Kiyosawa S, Yamaguchi T, Hirano T, Kobayashi
    T, Kushibuchi K and Watanabe S (1976). Proposal of a
    new method for differentiating races of Pyricularia
    oryzae. Cavara in Japan. Ann. Phytopath. Soc. Japan 42:
    216-219.
12. Ou SH (1980). Pathogen variability and host resistance
    in rice blast disease. Ann. Rev. Phytopathol. 18:
    167-187.
13. Crawford MS, Chumley FG, Weaver CG and Valent B (1986).
    Characterization of the Heterokaryotic and Vegetative
    Diploid Phases of Magnaporthe grisea. Genetics (in
    press).
14. Hebert TT (1971). The perfect stage of Pyricularia
    grisea. Phytopathology 61: 83-87.

15. Ueyama A and Tsuda M (1975). Formation of the perfect state in culture of Pyricularia sp. from some graminaceous plants (preliminary report). Trans. mycol. Soc. Japan 16: 420-422.

16. Yaegashi H and Nishihara N (1976). Production of the Perfect Stage in Pyricularia from Cereals and Grasses. Ann. Phytopath. Soc. Japan. 42: 511-515.

17. Kato H, Yamaguchi T and Nishihara N (1976). The perfect state of Pyricularia oryzae Cav. in culture Ann. Phytopath. Soc. Japan 42: 507-510.

18. Taga M, Waki T, Tsuda M and Ueyama A (1982). Fungicide sensitivity and genetics of IBP-resistant mutants of Pyricularia oryzae. Phytopathology 72: 905-908.

19. Parsons KA, Valent B and Chumley FG (1986). Transformation of Magnaporthe grisea using the A. nidulans argB gene. Proc. Nat. Acad. Sci. (USA) (in preparation).

Molecular Strategies for Crop Protection, pages 95–106
© 1987 Alan R. Liss, Inc.

MOLECULAR STRATEGIES FOR THE ANALYSIS OF THE
INTERACTION OF USTILAGO MAYDIS AND MAIZE

Sally A Leong, Jun Wang, Allen Budde,
David Holden, Thomas Kinscherf and Timothy Smith

USDA, ARS, Department of Plant Pathology,
University of Wisconsin, Madison, WI   53706

ABSTRACT   The high affinity iron transport system of
Ustilago maydis, the causative agent of corn smut, is
being examined. The longterm goal of this work is to
elucidate the structure and regulation of the genes
involved in the process of biosynthesis and uptake of
the siderophores of U. maydis and to assess the role
of siderophore mediated iron uptake in phytopathogenicity. U. maydis was confirmed to produce two
hydroxamate siderophores ferrichrome and ferrichrome
A. Using novel screening bioassays, four classes of
mutants defective in siderophore production were
isolated after N'-methyl-N'-nitro-N-nitrosoguanidine
mutagenesis of haploid basidiospores. Of these, three
classes were deficient in siderophore biogenesis. The
fourth class was found to produce siderophore
constitutively. Genetic analysis of mutants doubly
defective in ferrichrome and ferrichrome A synthesis
has revealed that this lesion segregates as a single
gene. The defect was corrected by supplementation of
low iron media with δ-N-OH-L-ornithine suggesting
that these mutants are unable to hydroxylate
L-ornithine, a common precursor of ferrichrome and
ferrichrome A. Efforts are also being made to develop
tools for molecular genetic analysis of U. maydis.
Low frequency, integrative transformation of U. maydis
has been obtained using a plasmid vehicle containing a
chimeric selectable marker derived from a U. maydis
Hsp70 promoter linked to the coding sequence for

[1]This work was supported by the University of
Wisconsin, the USDA, USDA Grant no. 83-CRCR-1-1350 and NIH
Biomedical Research Support Grant 2-S07-RR07098-18, MOD1.

hygromycin B phosphotransferase. A molecular
karyotype has also been developed for U. maydis strain
UM001. This fungus has at least 20 distinct
chromosomes. Chromosome size variation has been
observed for different isolates of the organism.

INTRODUCTION

The mechanism by which pathogenic organisms sequester
and regulate uptake of iron is central to an understanding
of the etiology of infectious disease in mammals. The
ability of microbial pathogens to acquire this essential
element from mammalian hosts is an important component of
virulence. In contrast, the role of iron acquisition by
phytopathogens in plant infection remains unknown (1). We
are examining the high affinity (siderophore-mediated) iron
transport system of Ustilago maydis. The long term goal of
this work is to elucidate the structure and regulation of
the genes involved in the processes of biosynthesis and
uptake of the siderophores of U. maydis and to assess the
role of siderophore-mediated iron uptake in
phytopathogenicity and saprophytic growth. Iron and/or
iron chelators have been shown to have profound effects
(inhibitory or stimulatory) on spore formation
(Aspergillus), spore germination (Neurospora, Fusarium,
Colletotrichum, Botrytis), germ tube elongation (Fusarium),
zoospore encystment (Phytophthora) and formation of
appressoria (Colletotrichum). Siderophores have also been
implicated as a mechanism to explain disease suppressive
soils (1). Thus siderophores may impact in many ways on
the general survival of phytopathogenic fungi. U. maydis
offers a tractable system in which to begin to develop an
understanding of the molecular genetics of siderophore
biogenesis and transport in fungi as the chemistry of its
siderophores is known and the organism is amenable to
Mendelian genetic analysis.

RESULTS AND DISCUSSION

Characterization of the Siderophores of U. maydis.

Some thirty lab strains (gifts from R. Holliday and P.
Day) and fresh field isolates were surveyed for siderophore
production. One strain UM001 (A$_2$B$_2$) was selected for

further biochemical and genetic studies. U. maydis has
been reported ro produce the hydroxamate siderophores
ferrichrome and ferrichrome A (2) (Figure 1). Following
the benzyl alcohol extraction procedure, two siderophores
have been recovered from the supernatants of the fungus
growing in low iron medium. The two siderophores appeared
by various physical analyses, i.e., HPLC, UV-visible
specta, TLC, and paper electrophoresis, when compared to
authentic ferrichrome and ferrichrome A, to be the same
compounds. Final verification, was obtained by NMR and
amino acid analysis.

Isolation of Siderophore Biosynthetic Mutants.

    All the mutants described were isolated after exposing
log phase haploid cells to 50 μg/ml N-methyl-N'-nitro-N-
nitrosoguanidine (NTG). Using a variety of novel screening
procedures, several groups of siderophore biosynthetic
mutants have been discovered.
    Ferrichrome-minus mutants. A Salmonella typhimurium
LT-2 mutant enb-7 was employed as a biological indicator

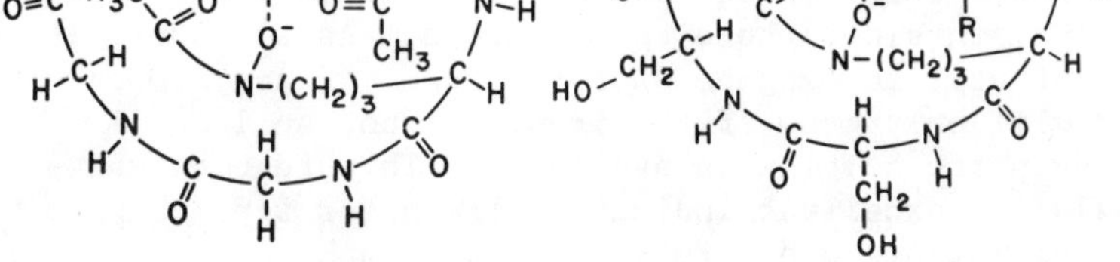

FIGURE 1. Ferrichrome (left) and ferrichrome A
(right). The R group of ferrichrome A is trans-ß-
methylglutaconic acid.

for the secretion of ferrichrome from fungal colonies (3).
This mutant has lost the ability to make its own
siderophore, enterobactin, and shows poor growth on E
medium which is a low iron medium plus citrate. Citrate is
known to inhibit iron transport in S. typhimurium. The
growth enb-7 on E medium can be restored by adding either
ferrichrome (but not ferrichrome A) or fungal culture
supernatants from low iron medium. This is due to the
presence of the Fhu A (Ton A) gene product, a ferrichrome
outer membrane receptor. NTG-treated fungi were spread
directly on a modified E medium containing 5×10^{-3}M
AlCl$_3$ and seeded with enb-7. Six ferrichrome-minus
mutants were obtained from ~ 35,000 NTG-treated cells.

Ferrichrome A-minus mutants. A similar strategy was
used to identify ferrichrome A-minus mutants. In this case
the Agrobacterium tumefaciens siderophore mutant, XVII$_{19}$
(4), was employed as a biological indicator in Luria Broth
medium containing EDDA (ethylenediamine-di-
orthohydroxyphenylacetic acid). One ferrichrome A-minus
mutant was detected after screening 7,000 colonies.

Ferrichrome and ferrichrome A double-minus mutants.
In the process of isolating the above classes of mutants, 4
strains were found to be defective in both ferrichrome and
ferrichrome A production. The high proportion of "double"
mutants obtained in the total mutants isolated suggests
that the two siderophores share a common biosynthetic or
secretion pathway. Alternatively, this could be the result
of the disruption of a common regulatory function. Since
NTG was employed as a mutagen, this could also be the
result of multiple, linked mutations.

Regulatory mutants. A novel approach has been taken
to obtain constitutive, mutants which produce high levels
of siderophore in iron replete medium. TA2701 (5) is a
S. typhimurium LT-2 mutant defective in the ability to
transport ferrichrome (Ton A-) as well as in the synthesis
of its own siderophore enterobactin. As a result, it
cannot utilize exogeneous ferrichrome an an iron carrier
and will only grow if the iron, but not an iron-
siderophore complex is available. This feature makes
TA2701 an excellent indicator strain for screening for
fungal mutants constitutive for siderophore biosynthesis.
Such mutants secrete large amounts of siderophore into iron
replete agar. These siderophores in turn form complexes
with free iron, making growth of the bacterial indicator
impossible. Seven strains showing growth inhibition zones

have been identified as constitutive mutants from a
population of 200,000 mutagenized colonies.

  Biochemical analysis of mutants. Little is currently
known about the biosynthetic pathway that leads to
ferrichrome and ferrichrome A. Based on isotope labeling
experiments in U. sphaerogena, Emery deduced that
L-ornithine is first hydroxylated to yield δ-N-OH-L-
ornithine (6) (Figure 2). This compound is then acetylated
via acetylCoA to give δ-N-acetyl-N-OH-L-ornithine, a
precursor of ferrichrome. The acetylase has been partially
purified from this organism (7).

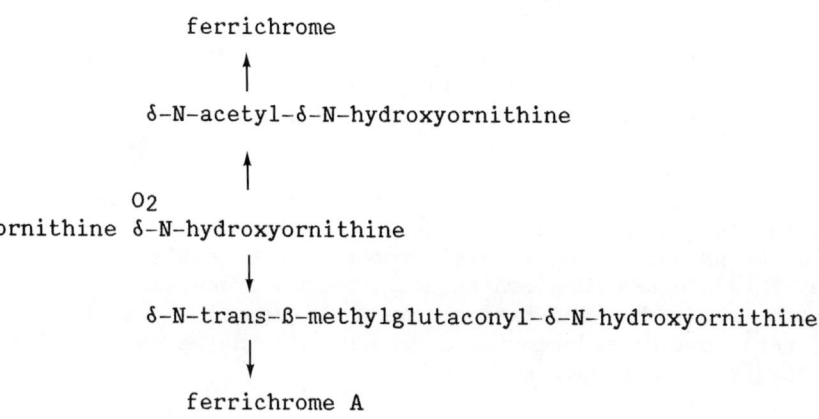

FIGURE 2. Proposed biosynthetic pathway to ferrichrome
and ferrichrome A.

   The latter steps in ferrichrome synthesis are unknown
but may proceed, as in the biosynthesis of the cyclic
peptide antibiotic gramicidin, via covalently bound
thioester intermediates on a multifunctional polypeptide.
This mechanism has been implicated in both rhodotorulic
acid and ferrichrome biosynthesis since an δ-N-acetyl-
δ-N-OH-L-ornithine-ATP-PPi exchange activity has been
observed in extracts prepared from Rhodotorula pilmaneae
(8) and Aspergillus quadricinctus (9). Ferrichrome and
ferrichrome A most likely share the first step in their
biosynthesis since both contain δ-N-OH-L-ornithine. The

pathways must diverge after this reaction since ferrichrome A contains trans-ß-methyl-glutaconic acid as the acyl group of the hydroxamate. Also the amino acid composition of the peptides differs: ferrichrome contains three glycines while ferrichrome A contains one glycine and two serines.

To date we have characterized the mutants with regard to the presence or absence of the end products ferrichrome and ferrichrome A as well as for total hydroxylamine in culture supernatants. No evidence was found for intermediates containing the latter functional group. Feeding tests using δ-N-OH-ornithine and ornithine were also conducted with the ferrichrome/ferrichrome A-deficient mutants. All 4 of the mutants were biochemically complemented to produce both siderophores in media supplemented with δ-N-OH-ornithine.

Segregation of the siderophore lesion was examined in crosses with a wild type isolate and one ferrichrome mutant, two ferrichrome/ferrichrome A mutants, and two constitutive mutants. In each case a 1:1 ratio was observed for the segregation of the siderophore biosynthetic defect (Table 1). These data, in conjunction with the feeding assays, suggest that the mutants defective in the production of both siderophores are unable to hydroxylate δ-N-OH-L-ornithine, a common precursor of both ferrichrome and ferrichrome A (Figure 2). This is the first genetic evidence for a common step in the pathways to these two siderophores.

TABLE 1

SEGREGATION OF SIDEROPHORE MUTATION

$J_2(a_2b_2$ fc$^-$/fcA$^-$) x 288 (a_1b_1, inos$^-$, pan 1-1, nar-1, recl-1)

Progeny	Cross 1	Cross 2
fc$^+$/fcA$^+$	194	111
fc$^-$/fcA$^-$	192	104

Complementation analysis of one ferrichrome mutant, two ferrichrome/ferrichrome A mutants and one constitutive

mutant was carried out by forcing diploids between the
siderophore auxotrophs and a strain which was wild-type
with respect to the siderophore lesion. In each case the
siderophore defect was found to be recessive.

Gene Transfer.

One of our goals has been to develop a transformation
system for U. maydis. A routine gene transfer system for
Ustilago will enable the molecular genetic analysis of
pathogenicity determinants. Because of its yeast-like
properties and well-defined genetics, U. maydis offers an
accessible system to develop this methodology as well as
techniques for the analysis of gene expression in
basidiomycetes. Ustilago might also serve as an alternate
host for the isolation of genes involved in host-parasite
interactions from other plant pathogenic basidiomycetes
such as the rusts, which cannot be cultivated axenically.
In support of this approach is the recent finding that a
virulent Xanthomonas campestris pv. glycines could be
converted to avirulence on soybean when harboring a cloned
dominant avirulence gene from Pseudomonas syringae pv.
glycinea (Brian Staskawicz, personal communication).
We have made considerable progress in this effort. An
efficient method for the generation and regeneration of
protoplasts has been established. After treatment with
Novozyme 234 up to 90% of the starting basidiospores can be
formed into protoplasts and completely regenerated.
Conditions for controlled fusion of protoplasts with 40%
polyethylene glycol and $CaCl_2$ have also been established.
We are following two approaches for the development of
selectable markers for transformation. Chimeric markers
are being constructed from transcription control sequences
of genes identified from U. maydis linked to the antibiotic
resistance gene for hygromycin B phosphotransferase. The
Hsp70 gene has been cloned from U. maydis by heterologous
hybridization with a cloned yeast Hsp70 gene (10). One
clone encoding an U. maydis Hsp70 gene has been extensively
characterized by restriction mapping, S1 nuclease
protection assays and DNA sequence analysis. The N-terminus
region is 70% homologous at the DNA sequence level with an
Hsp70 gene from yeast. Conserved transcription promotion
signals found in other heat shock genes have also been
identified. The gene encodes a heat shock inducible mRNA
of 2.2 Kb and was found to map on the twelfth smallest
chromosome of strain UM001. The promoter region was linked

to the hygromycin B phosphotransferase gene by
transcription fusion (Figure 3) and preliminary
evidence--both genetic and physical--for low frequency
(10 per μg DNA per 10^7 protoplasts), integrative
transformation of two strains of U. maydis has been
obtained (Figure 4). The DNA was introduced by
polyethylene-CaCl$_2$ induced fusion of protoplasts. The
copy number of the transforming plasmid appeared to vary
from transformant to transformant and the level of
hygromycin B resistance was roughly correlated with the
number of integrated copies. It should be noted however,
that we have not formally established whether or not these
transformants are monokaryotic. In the one case examined,
the transforming DNA was tentatively found to map on the
same chromosome as the heat shock gene (Figure 5). Also
consistent with the integrated state of the transformation
vector is the observation that the transformants were both
mitotically and meiotically stable. Finally, substantial
hygromycin B phosphotransferase was detected in selected
transformants but not in the recipient strain. Additional
selectable markers are currently being constructed which
include a translation fusion of the Hsp70 promoter and
N-terminus to a truncated hygromycin B phosphotransferase
gene flanked by the Trp C terminator of Aspergillus
nidulans.

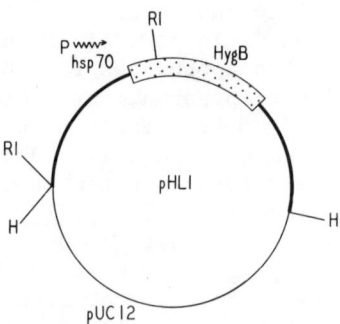

FIGURE 3.   Transformation vector pHL1 contains a
U. maydis Hsp70 promoter fused to the BamHI fragment
containing the hygromycin B phosphotransferase gene of
pLG90 (11).

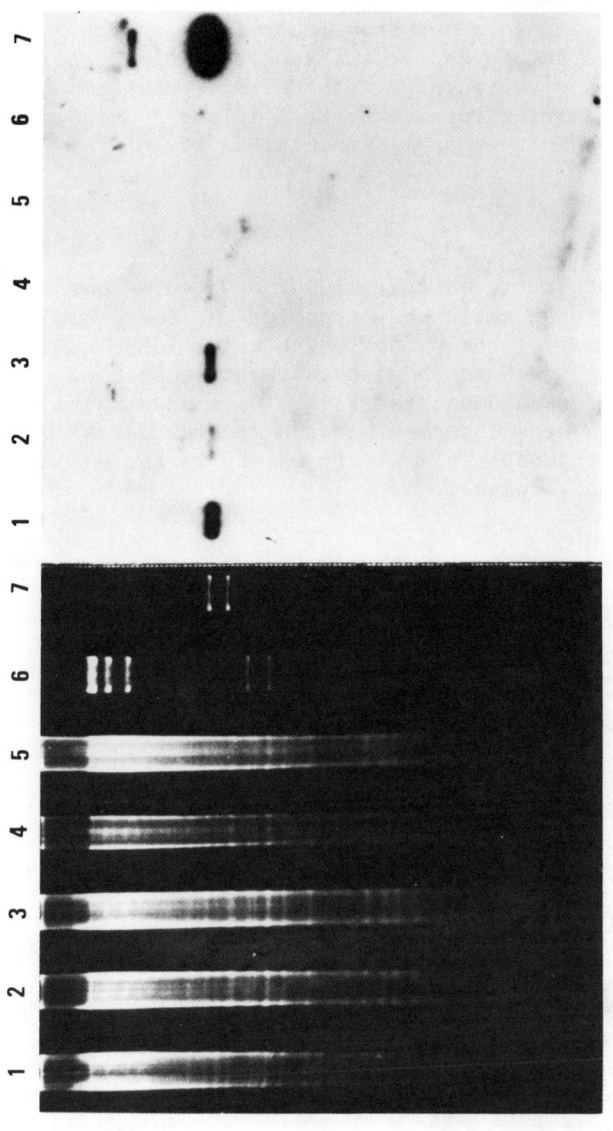

FIGURE 4. Southern hybridization analysis of transformants. HindIII-digested total DNA (7.5 µg) was electrophoresed in 1% agarose, transferred to nitrocellulose and probed with a 1.15 Kb BamHI containing the coding sequence of the hygromycin B phosphotransferase gene. The lanes contain: (1) Peking strain, transformant N-2; (2) UM001, transformant N-1, (3) UM001, transformant 7; (4) UM001, transformant 4; (5) UM001; (6) λ, HindIII-digested; and (7) pHL1, Hind-III-digested plasmid that was employed as the transformation vector.

As a second strategy for marker development, mutants defective in uracil biosynthesis have been obtained by selection for resistance to 5-fluoro-orotic acid (12). Attempts are being made to complement these mutants with the complementary cloned genes from other fungi and to use these cloned genes as probes for the isolation of uracil biosynthetic genes of U. maydis. In addition, complementation of uracil-requiring mutants of E. coli and A. nidulans with a gene library of U. maydis genomic DNA is being attempted.

Molecular Karyotype.

As an aid to genetic analysis in U. maydis, we have developed an electrophoretic karyotype for the fungus using OFAGE (13). By varying the pulse-length conditions, strain UM001 of U. maydis has been shown to have at least 20 distinct yeast-sized chromosomes (Figure 5). Chromosome size variation has been observed between strains which suggests that considerable plasticity exists in the genome organization of this fungus.

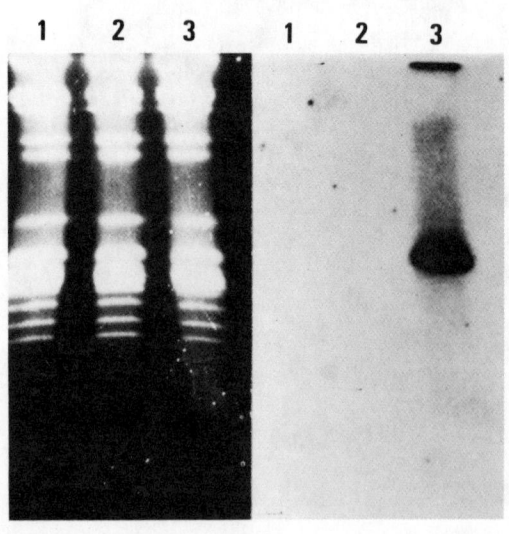

FIGURE 5. Southern hybridization analysis of the
electrophoretic karyotype of transformants . OFAGE was
carried out essentially as described by Carle and Olson
(13) using protoplasts embedded in 2.5% low melting agarose
1 M sorbitol, 0.5 M EDTA, pH 7.4. The DNA was transferred
to Zetaprobe and probed with a 1.15 Kb <u>Bam</u>HI fragment
containing the coding region of the hygromycin B
phosphotransferase gene (11). The lanes contain the
chromosome complement of: (1) UM001; (2) UM001, a putative
transformant that was resistant to low levels of hygromycin
B; and (3) UM001, transformant 7. Hybridization is
observed to the chromosome doublet containing the <u>hsp</u>70
gene.

ACKNOWLEDGEMENTS

We thank Debi Schaefer and Steve Vicen for their
expert help in preparing this manuscript.

REFERENCES

1.   Neilands JB, Leong SA (1986). Siderophores in relation
     to plant growth and disease.   Ann Rev Pl Physiol
     37:187-208.
2.   Garibaldi JA, Neilands JB (1955). Isolation and
     properties of ferrichrome A. J Amer Chem Soc
     77:2429-2431.
3.   Pollack JR, Ames BN, Neilands JB (1970). Iron
     transport in <u>Salmonella</u> <u>typhimurium</u>: mutants blocked
     in the biosynthesis of enterobactin.   J Bacteriol
     104:635-639.
4.   Leong SA, Neilands JB (1981). Relationship of
     siderophore-mediated iron assimilation to virulence in
     Crown Gall Disease. J Bacteriol 147:482-491.
5.   Luckey M, Pollack JR, Wayne R, Ames BN, Neiland JB
     (1972). Iron uptake in <u>Salmonella</u> <u>typhimurium</u>:
     utilization of exogenous siderophores as iron
     carriers. J Bacteriol 111:731-738.
6.   Emery T (1966). Initial steps in the biosynthesis of
     ferrichrome.   Incorporation of δ-N-hydroxyornithine
     and δ-N-acetyl-δ-N hydroxyornithine. Biochem
     5:3694-3701.

7.   Ong DE, Emery TF (1972). Ferrichrome biosynthesis: enzyme catalyzed formation of the hydroxamic acid group. Arch Biochem Biophys 148:77-83.

8.   Akers HA, Neilands JB (1978) Biosynthesis of rhodotorulic acid and other hydroxamate type siderophores. In Gorrod JW (ed): "Biological Oxidation of Nitrogen," Elsevier/North-Holland: pp. 429-436.

9.   Hummel W, Diekman H (1981). Preliminary characterization of ferrichrome synthetase from Aspergillus quadricinctus. Biochem Biophys Acta 657:313-320.

10.   Ingolia JD, Slater MR, Craig EA (1982). Saccharomyces cerevisiae contains a complex multigene family related to the major heat shock-inducible gene of Drosophila. Mol Cell Biol 2:1388-1398.

11.   Gritz L, Davies J (1983). Plasmid-encoded hygromycin B resistance: the sequence of hygromycin B phosphotransferase gene and its expression in Escherichia coli and Saccharomyces cerevisiae. Gene 25:179-188.

12.   Boeke JD, LaCroute F, Fink GR (1984). A positive selection for mutants lacking orotidine-5'-phosphate decarboxylase activity in yeast: 5'-fluoro-orotic acid resistance. Mol Gen Genet 197:345-346.

13.   Carle GF, Olson MV (1985). An electrophoretic karyotype for yeast. Proc Natl Acad Sci USA 82:3756-3760.

Molecular Strategies for Crop Protection, pages 107–114

A MITOCHONDRIAL OPEN READING FRAME ASSOCIATED
WITH MUTATION TO MALE FERTILITY
AND TOXIN INSENSITIVITY IN T-CYTOPLASM MAIZE

Roger P. Wise[1], Daryl R. Pring[1,2],
and Burle G. Gengenbach[3]

[1]Plant Pathology Department, University of Florida,
Gainesville, 32611, U.S.A., [2]USDA-ARS, and
[3]Department of Agronomy and Plant Genetics,
University of Minnesota, St. Paul, 55108, U.S.A.

ABSTRACT   Tissue culture derived mutations to male fertility and disease toxin insensitivity in T-cytoplasm
maize were used to identify a T-specific, 345 base pair,
mitochondrial open reading frame (T ORF13) associated
with these traits. T ORF13 is deleted in all mutants
except one. The one exceptional mutant, T-4, contains a
5 base pair insertion internal to T ORF13.   The
insertion generates a frameshift resulting in a premature stop codon and would truncate the predicted
polypeptide from 12,961 Mr to 8,305 Mr.

INTRODUCTION

The production of hybrid maize has been greatly facilitated by the discovery and use of cytoplasmic male sterility
(cms). The use of cms can eliminate the need for manual
detasseling in hybrid production programs. There are three
major groups of male-sterile cytoplasms: S (USDA), C
(Charrua), and T (Texas). These groups are distinguished by
the genetics of fertility restoration (1) and by mitochondrial DNA restriction profiles (2).
T-type cytoplasmic male sterility (Tcms) was widely used
in the U.S. until the 1970 epiphytotic of southern corn leaf
blight (3).   At that time approximately 85% of the U.S.
maize crop was produced using Tcms. Tcms maize is sensitive
to the host-selective toxin (T toxin) produced by race T of
Helminthosporium maydis Nisik. and Miy., the causal organism

of southern corn leaf blight, and to the host-selective toxin produced by Phyllosticta maydis Arny and Nelson, which causes yellow leaf blight. Male-fertile N-cytoplasm maize plants are insensitive to these toxins. The primary site of toxin action appears to be the mitochondrion (4). An important step in the identification of the molecular basis for male sterility and disease toxin sensitivity in T-cytoplasm maize was obtained by the regeneration of plants through tissue culture (5). These plants displayed mutation to male fertility and insensitivity to the two pathotoxins. The mutations were stable and maternally inherited.

Previous work demonstrated that 15 of 16 male fertile, toxin insensitive regenerants carried a rearranged 6.7 kb XhoI mitochondrial DNA restriction fragment (6). Alternatively, the 6.7 kb fragment was retained in all of 34 regenerated male sterile, toxin sensitive control plants, correlating presence of the fragment with the traits. One interesting regenerant, T-4, retained the 6.7 kb XhoI restriction fragment, yet it was male fertile and insensitive to the toxin. We were interested in any alterations within and flanking the 6.7 kb XhoI fragment. In this paper we will describe the alteration in the T-4 mutant and how it was used to identify a T-specific 345 base pair open reading frame associated with mutation to male fertility and disease toxin insensitivity in T-cytoplasm maize.

RESULTS

Organization of the Region Containing the 6.7 kb XhoI Fragment

Cosmid libraries of parental T and T-4 mutant mitochondrial DNA were constructed in the vector pHC79. A 6.5 kb BamHI fragment from Wf9(N) mtDNA, which shares homology with the 6.7 kb XhoI, was used to select cosmids from both libraries, which were isolated and mapped (7). We found that the 6.7 kb XhoI fragment extends into a 5 kb repeat through which recombination occurs in T and T-4 mitochondrial DNA (Fig. 1). Analogous cosmids were recovered from the T and T-4 libraries demonstrating all four possible configurations through the repeat (AB, CD, AD, and CB in Fig. 1). T and T-4 did not differ at this level of analysis.

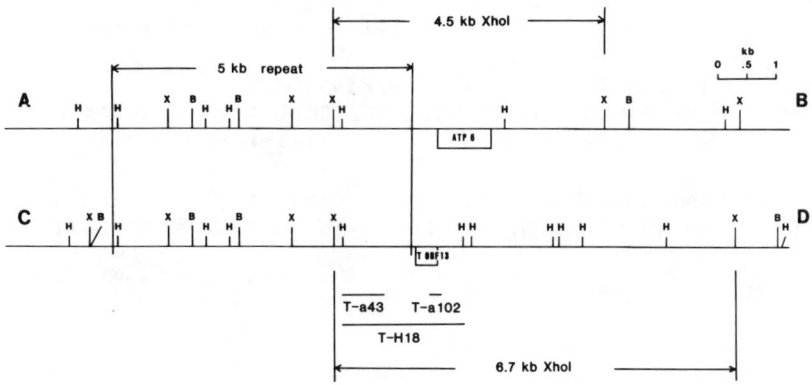

FIGURE 1. Map of the 6.7 kb XhoI fragment and flanking regions. The central portion of the map shows that there is an approximate 5 kb repeat extending into the 6.7 kb XhoI fragment. The upper sequences represent a portion colinear with a region of the master circle in the Wf9(N) (male-fertile) maize mitochondrial genome. The lower sequences represent another copy of the 5 kb repeat and flanking regions. We have recovered analogous cosmids from both the T and T-4 libraries demonstrating all four products of recombination through the repeat. There appears to be no difference between T and T-4 at this level of analysis. T-a43, T-a102, and T-H18 pUC8 subclones were mapped as indicated and are cited in the text. Restriction sites are indicated by vertical lines: B, BamHI; H, HindIII; X, XhoI.

Organization of the 6.7 kb XhoI Fragment

Further dissection of T and the T-4 mutant was accomplished by the following experiment: 6.7 and 4.5 kb XhoI fragments from T and T-4 (Fig. 1) were isolated from cosmids and digested with a series of tetradeoxynucleotide recognizing restriction enzymes followed by electrophoretic separation through 2% agarose gels. By electrophoresing the products of these digestions in adjacent lanes, we could assign changes to the single copy regions of these fragments. Using AluI, it was possible to identify a small shift in one fragment derived from the 6.7 kb XhoI in the T-4 mutant. This 188 base pair fragment (T-a102) was not within the 5

```
3' 26S:  CTGCATGAGCTATCCTTCTCATCTCATGGTTGAGGGGGGTTAAAATGAGG
         ++++++++++++++++++++++++++++++++++++++++++++++++ ++++ +++

              LeuHisGluLeuSerPheSerSerHisGly@@@
T-4:     CTGCATGAGCTATCCTTCTCATCTCATGGTTGAGGGGGGTTCAAATTAGG
         +++++++++++++++++++++    +++++++++++++++++++++++++

              LeuHisGluLeuSerPheSer     TrpLeuArgGlyValGlnIleArg
T:       CTGCATGAGCTATCCTTCTCG-----TGGTTGAGGGGGGTTCAAATTAGG
             +           +           +           +
            200         210         220         230
```

FIGURE 2. The nucleotide and predicted amino acid sequences of a portion of T ORF13. The T-4 mutant has a G to A transition adjacent to a TCTCA insertion, resulting in a sequence which matches a portion of a sequence 3' to 26S rDNA. The 5 base pair insertion generates a frameshift, placing a TGA stop codon in frame. Numbers are base pairs from ATG codon of T ORF13.

kb repeat. Further mapping and sequencing revealed that the alteration in the T-4 mutant was due to a 5 base pair insertion located internal to T ORF13 (Fig. 2), a T-cytoplasm specific 345 base pair open reading frame (8). The insertion is located 214 nucleotides 3' to the ATG initiation codon. Adjacent to the 5 base pair insertion is a G to A transition. These changes in the T-4 mutant result in 86 base pairs of perfect homology with a region 3' to 26S rDNA (9). This suggests that the altered sequences in the T-4 mutant arose by homologous recombination or gene conversion with a region 3' to 26S rDNA. The most important consequence of the insertion, however, is that it generates a frameshift placing a TGA stop codon in frame four nucleotides after the insertion. This would truncate the predicted polypeptide from 12,961 Mr to 8,305 Mr.

T ORF13 was deleted in four mutants (T-6, T-7, R-2, and R-5) which had lost the 6.7 kb XhoI fragment (Fig. 3). This deletion extended from a region just 3' to T ORF13 to a region located between the two BamHI sites within the 5 kb repeat. The upper sequences (AB in Fig. 1) were completely conserved, however.

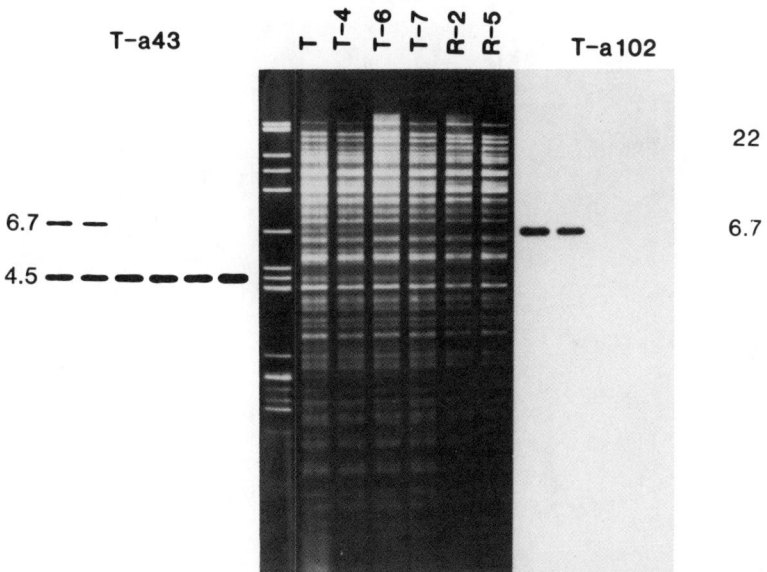

FIGURE 3. Electrophoresis of mitochondrial DNA digested with XhoI. Mitochondrial DNAs from parental T and T-mutant (T-4, T-6, T-7, R-2 and R-5) were digested with XhoI, electrophoresed on 0.8% agarose gels, and bidirectionally transferred to nitrocellulose. The membranes were probed with nick translated pUC8 clones T-a43 and T-a102. T-a43 (left) detected the absence of the 6.7 kb XhoI junction fragment although the 4.5 kb XhoI fragment was conserved in both parental and mutant lines. T-a102 specific sequences were deleted in T-6, T-7, R-2, and R-5. The 6.7 kb junction fragment and T-a102 specific sequences are also not present in N.

Expression of T ORF13

The T ORF13 region displays a complex transcriptional pattern (Fig. 4). T-H18 (Fig. 1) contains T ORF13 and sequences in the 5 kb repeat. T-H18 hybridizes to 9 transcripts, two of which (2.9 and 2.7 kb) appear to be transcribed from the upper sequence organization in Figure 1.

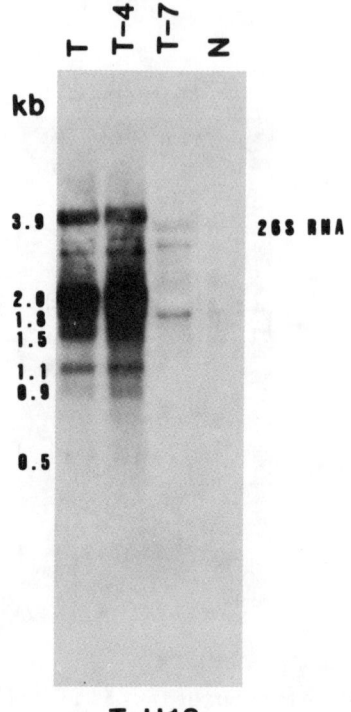

T-H18

FIGURE 4. Northern blot of transcripts from T ORF13 and the 5 kb repeat. Mitochondrial RNAs from T, T-4, T-7, and N were denatured with glyoxal, electrophoresed on 1.0% agarose gels, and transferred to nitrocellulose. The membrane was probed with nick translated T-H18 (Fig. 1).

This sequence includes the ATPase subunit 6 gene (10). T-a102, which carries the 5 base pair insertion in T-4, hybridizes to six transcripts of 3.9, 2.0, 1.8, 1.5, 1.1, and 0.9 kb in T and T-4 (data not shown). There were no differences between T and T-4 observed at the transcript level. This complex pattern is common among other plant mitochondrial genes examined. Only one of the transcripts (1.8 kb) hybridizes to T-a102 in T-7 or N.

DISCUSSION

Plants regenerated from tissue culture of T-cytoplasm maize exhibit a high degree of mutation to male fertility and resistance to H. maydis race T and to P. maydis. The changes in phenotype occur with or without selection with toxin in the media. The use of these mutations has enabled us to identify a T-cytoplasm specific mitochondrial open reading frame (T ORF13) associated with these traits. T ORF13 is deleted in all mutants except T-4. In the T-4 mutant this region has undergone a G to A transition adjacent to a five base pair insertion, duplicating a region 3' to 26S rDNA. This alteration generates a frameshift resulting in a premature stop codon. No alterations were detected in transcription through this region, suggesting that a truncated protein may be translated that is biologically inactive. A polypeptide of 12,961 Mr has been predicted from the T ORF13 sequence (8). This correlates with a 13,000 Mr mitochondrial translation product seen in T but not N-cytoplasm maize (11). The molecular events reported here are also consistent with the loss of this 13,000 Mr mitochondrial translation product in most mutants (12).

It has been established that the maize mitochondrial genome recombines prolifically (13). T ORF13 may have been generated through a series of recombination events with other parts of the genome including 26S ribosomal DNA and a region 3' to 26S rDNA, as described by Dewey et. al. (8). They detected at least seven recombination points by sequencing a 3.5 kb region including T ORF13. The 6.7 kb XhoI fragment in T-4 appears to be intact, other than the transition-insertion event, as evidenced by comparison with T using AluI, HaeIII, HincII, TaqI, Sau3A, and HinfI. The conversion of T to T-4, presumably resulting from recombination with the region 3' to 26S rDNA, seemingly occurs only in tissue culture, since the insertion event has not been detected in mtDNA from intact maize plants. The mechanism can be postulated, but we cannot explain why the event is prone to occur only in culture.

LITERATURE CITED

1. Laughnan, JR, and Gabay-Laughnan, S (1983). Cytoplasmic male sterility in maize. Annu Rev Genet 17:27.

2. Pring, DR and Levings, CS, III (1978). Heterogeneity of maize cytoplasmic genomes among male-sterile cytoplasms. Genetics 89:121.

3. Levings, CS, III and Pring, DR (1979). Molecular bases of cytoplasmic male sterility in maize. In Scandalios, JG, ed., "Physiological Genetics," Academic Press, New York: p 171.

4. Miller, RJ and Koeppe, DE (1971). Southern corn leaf blight: Susceptible and resistant mitochondria. Science 173:67.

5. Gengenbach, BG, Connelly, JA, Pring, DR, and Conde, MF (1981). Mitochondrial DNA variation in maize plants regenerated during tissue culture selection. Theor Appl Genet 59:161.

6. Umbeck, PF and Gengenbach, BG (1983). Reversion of male-sterile T-cytoplasm maize to male fertility in tissue culture. Crop Science 23:584.

7. Wise, RP, Pring, DR, and Gengenbach, BG (1985). Mitochondrial DNA rearrangements associated with reversion of T cytoplasm to male fertility and disease resistance. Maize Genetics Cooperation Newsletter 59:50.

8. Dewey, RE, Levings, CS, III, and Timothy, DH (1986). Novel recombinations in the maize mitochondrial genome produce a unique transcriptional unit in the Texas male-sterile cytoplasm. Cell 44:439.

9. Dale, RMK, Mendu, N, Ginsburg, H, and Kridl, JC (1984). Sequence analysis of the maize mitochondrial 26S rRNA gene and flanking regions. Plasmid 11:141.

10. Dewey, RE, Levings, CS, III, and Timothy, DH (1985). Nucleotide sequences of ATPase subunit 6 gene of maize mitochondria. Plant Phys 79:914.

11. Forde, BG, Oliver, RJC, and Leaver, CJ (1978). Variation in mitochondrial translation products associated with male-sterile cytoplasms in maize. Proc Natl Acad Sci USA 75:3841.

12. Dixon, LK, Leaver, CJ, Brettell, RIS, and Gengenbach, BG (1982). Mitochondrial sensitivity to Drechslera maydis T-toxin and the synthesis of a variant mitochondrial polypeptide in plants derived from maize tissue cultures with Texas male-sterile cytoplasm. Theor Appl Genet 63:75.

13. Lonsdale, DM, Hodge, TP, and Fauron, C M-R (1984). The physical map and organization of the mitochondrial genome from the fertile cytoplasm of maize. Nucl Acids Res 12:9249.

Molecular Strategies for Crop Protection, pages 115–124
© 1987 Alan R. Liss, Inc.

MOLECULAR BIOLOGY IN BIOLOGICAL
CONTROL OF SOILBORNE PATHOGENS

Ralph Baker

Department of Plant Pathology and Weed Science
Colorado State University
Fort Collins, Colorado   80523

INTRODUCTION

Studies in molecular biology, particularly directed
toward elaboration of phenomena associated with biological
control of soilborne plant pathogens, have not been extensive. Our hoard of knowledge is little but in a mind-
stretching symposium such as this, we can be courageously
speculative--hopefully, from a safe distance.

There are two areas in biological control where studies
in molecular biology have progressed. These involve the
mechanisms of exploitation and competition.

MOLECULAR INTERACTIONS BETWEEN MYCOPARASITES
AND THEIR HOSTS

Fungal parasites attacking higher plants can be exploited in turn by mycoparasites. Descriptions of parasitic
interactions and the ecological and pathogenic consequences
of such activity on fungal hosts have been reviewed (1, 2).

Interactions involving recognition are associated with
pollen on pistils, fungi, bacteria and rhizobia on plants
and cells of slime molds (3). Recognition may also be a
feature of fungal-fungal host parasitic interactions (4).

Lectins are defined as sugar-binding proteins (glycoproteins) of nonimmune origin, which agglutinate cells
and/or precipitate glucoconjugates (5). They are involved
in interactions between a cell's surface components and its
extracellular biochemical environment (6). Their property
of initiating development of cohesiveness between cells
suggests that they could be involved in recognitions (4),

coiling responses (1, 2) and attachment (7, 8) in mycoparasitic interactions.

A hemagglutinin was detected in hyphae and their
extract of Rhizoctonia solani (4), a host of Trichoderma
spp. and had a high specific activity in agglutination of
type O erythrocytes. Yet, the sugar specificity, hemagglut-
ination activity and stability of the agglutinins of R.
solani were quite different from those characteristics
associated with Sclerotium rolfsii which is also a fungal
host of Trichoderma spp. (9). The lectin produced by S.
rolfsii agglutinated certain Gram-negative bacteria and
yeasts but not human blood cells. In this case for example,
D-glucose, D-mannose and several of their derivatives specifically inhibited the agglutination of cells of Escherichia
coli. Agglutinin activity was associated with an extracellular polysaccharide produced by S. rolfsii. Evidence for
specific recognitions as a factor in mycoparasitism was
obtained in that agglutinin of S. rolfsii was capable of
agglutinating conidia only of T. hamatum strains capable of
attacking the fungal host.

Molecular changes also may occur in the fungal host
prior to penetration by mycoparasites. Observations suggest
that there is an increase of a mucilaginous substance, presumably a polysaccharide, originating from either the host
or the attacking fungus (10, 11). Such morphological
changes resulting from mycoparasitic interactions may play a
role in recognition or be a type of defense response against
invasion (10).

The cell walls of Oomycetes are composed of β-glucans,
cellulose, and less than 1.5% chitin (12). Basidiomycetes
and Ascomycetes contain relatively more chitin as well as β-
glucans but no cellulose. Obviously, a mycoparasite must
penetrate through such cell wall components to infect
hyphae. Indeed, visual observations referenced by Elad and
Misaghi (10) indicate enzymatic degradation of cell walls of
host fungi by a number of mycoparasites. Several of these
antagonists produced hydrolytic enzymes such as β-1,3- glucanase, cellulase and chitinase. Again, the fluorescence
(13) localized about cell wall lyses at points of interaction stained with Calcofluor White M2R New also suggested
enzymatic activity because this stain binds to oligomers in
regions of incomplete cell wall polymers.

A mycoparasite, Pythium nunn (14), produced extracellular β-1,3- glucanase, cellulase and chitinase, respectively,
when grown on appropriate substrates as sole carbon sources
(13). Large amounts of β-1,3- glucanase and chitinase were

produced in liquid cultures containing cell walls of two
pathogenic fungi belonging to the class Basidiomycetes.
Cellulase was produced in the presence of cell walls of two
pathogens in the Oomycetes but not chitinase. This confirms
theory related to the initiation of enzymatic activity which
depends on substrates in the cell walls of such fungi.

An interesting circumstance related to the host range
of P. nunn was described (13). Hydrolytic enzymes, appropriate for initiation of cell wall degradation, were produced in dual cultures of P. nunn with six host fungi but
were not detected in 10 non-host deuteromycete fungi. Trypsin or KOH-treated hyphae of the non-host F. oxysporum f.
sp. cucumerinum released N-acetyl-D-glucosamine in the
presence of crude enzyme preparations from P. nunn. Such
treatment of cell walls removes an outer layer of mucilaginous material associated with non-host fungi (15). This
suggests that the potential host range of P. nunn is limited
by components on the outer layer of fungal cell walls.

CHELATION OF IRON BY SIDEROPHORES

Competition is a mechanism that operates in a number of
biological control systems (16). It is "active demand in
excess of immediate supply of material or condition on the
part of two or more organisms" (17). Obviously, some microorganisms compete successfully in the contentious milieu of
soil because they produce molecules that are more efficient
in the utilization of limiting substrates. Less efficient
microbes are deprived of elements and/or substrates that are
essential for their activities -- in the case of pathogens,
for initiation of infection. In the soil, such interactions
occur in microenvironments exterior to the host and affect
penetration processes leading to infection by or survival of
pathogens.

The factors involved in competitive saprophytic ability
were advanced by Garrett (18) and include enzyme efficien-
cies, production of and/or tolerance to antibiotics and
relative growth rates. All of these factors are based on
metabolic processes but strategies for exploitation of such
processes involving molecular biology in biological control
are in their infancy.

The mechanism of competition on a molecular basis most
extensively studied is that related to management of available iron (Fe) in the rhizosphere of hosts susceptible to
the Fusarium wilt pathogens. This involves the production

of siderophores by fluorescent pseudomonads. The molecular biology involved in siderophore production and rhizosphere activity of fluorescent pseudomonads is reviewed by Neilands elsewhere in this volume. Here, aspects related to management of Fe for biological control will be treated.

The addition of fluorescent pseudomonds to soil induces suppressiveness to Fusarium wilt diseases (19). Extensive experimentation (20-28) suggests that at least a substantial proportion of suppressiveness relates to reduction in availability of Fe which limits the inoculum potential (29) of the pathogen.

The interactions and competitive relationships in Fe competition leading to suppressiveness are presented in Fig. 1. The Fe available in soil for plants and/or microorganisms is defined by the equilibrium equation: $Fe(OH)_3 + H^+ \leftrightarrows Fe^{3+} + 3H_2O$ (30). The principal factor mediating the equilibrium of this equation is the pH of the soil. Therefore, Fe^{3+}, the only form of Fe in the equation available to plant and/or microorganisms, is in low supply in alkaline soils. A clue to the importance of Fe in a suppressive soil system was obtained by adjusting soil pH to various values. The Fusarium-wilt suppressive soil in the Salinas Valley, where antagonistic fluorescent pseudomonads were found, has a native pH of ~8.0. When this was reduced to pH 7 about half of the suppressiveness was lost. At pH 6.0 the soil was conducive (19).

When **Pseudomonas putida**, ethylenediaminedi-0-hydroxyphenyl-acetic acid (EDDHA) or the ferrated form (FeEDDHA) was added to conducive soil, the soil became suppressive to Fusarium wilt of flax, cucumber or radish (21, 23, 27). Suppressiveness was not induced by Fe ethylenediaminetetra acetic acid (FeEDTA) or Fe-diethylene diaminetetra acetic acid (FeDTPA). The three ligands bind Fe^{3+} at various stability constants: EDDHA, $\log_{10} K = 33.97$ DTPA, $\log_{10} K = 27.37$ EDTA, $\log_{10} K = 25$ (30). The biological constituents of the system also chelate Fe^{3+}. Indeed, the Fusarium wilt pathogens require this element for successful germination of chlamydospores leading to infection and produce siderophores of the hydroxamate class (stability constant $\log_{10} = 29$) in Fe-deficient conditions (23). This constant is many powers of ten less than most bacterial catechol siderophores. P. putida produces a mixed hydroxamate catechol siderophore (pseudobactin) as reported by Teintze et al (31). The critical Fe^{3+} level associated with biocontrol was measured in an analog system and was below $10^{-22} \underline{M}$ (26).

Theory expansion suggests that adequate Fe^{3+} is avail-

BIOLOGICAL CONTROL BY IRON COMPETITION ON THE RHIZOPLANE

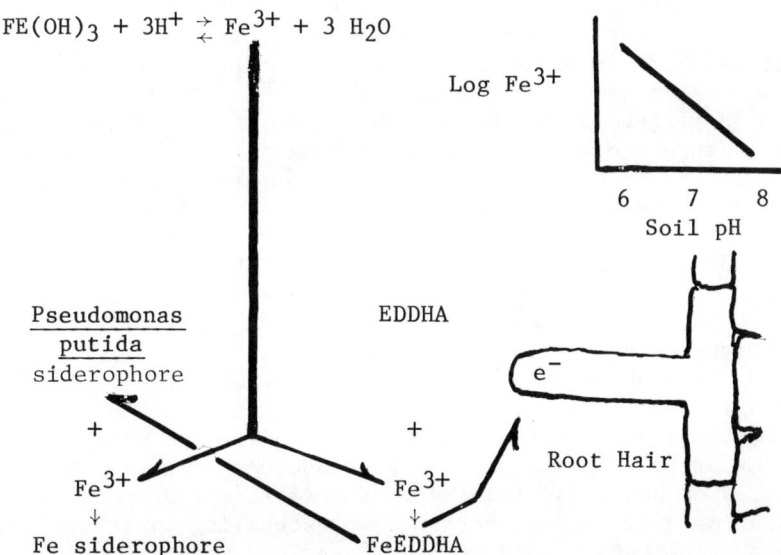

$$FE(OH)_3 + 3H^+ \rightleftharpoons Fe^{3+} + 3 H_2O$$

Log Fe^{3+}

6   7   8
Soil pH

<u>Pseudomonas</u>
<u>putida</u>
siderophore

EDDHA

e^-

+

+

Root Hair

Fe^{3+}

Fe^{3+}

Fe siderophore

FeEDDHA

Stability constants:

<u>Pseudomonas</u> <u>putida</u> (catechol hydroximate) $> 10^{34}$

FeEDDHA $10^{33.9}$

<u>Fusarium</u> <u>oxysporum</u> (hydroximate) 10^{29}

Figure 1. Mechanisms and pathways involved in soil suppressive to Fusarium wilt pathogens through competition for iron (Fe).

able to Fusarium wilt pathogens for germination and penetration through root tips of the host in conducive soil. Since many pseudomonads have high rhizosphere competence, there should be intense competition for Fe^{3+} at the rhizoplane (infection court) when P. putida produces siderophores, especially in alkaline soils. Therefore, Fe^{3+} is bound in such a way that it is unavailable to F. oxysporum. The critical Fe^{3+} level associated with biocontrol was measured in an analog system and was below $10^{-22}\underline{M}$ (26).

The ligand EDDHA has a higher stability constant that the siderophore produced by the pathogen. Fe^{3+} complexed by this ligand in the rhizosphere would be unavailable to the pathogen; however, EDDHA-bound Fe can be utilized by the host. This frees EDDHA to bind more Fe^{3+} at the root surface (30) and the soil is suppressive. This is a dynamic system in which the siderophores produced by pseudomonads, the EDDHA, and the root combine to render Fe^{3+} limiting for pathogenesis by Fusarium wilt pathogens.

Management of Fe by use of such manipulations to enhance biological control or suppression of plant pathogens, therefore, is possible. When FeEDDHA was applied to soil, activity of antagonistic fluorescent pseudomonads in the rhizospheres of hosts was favored to the detriment of pathogens producing siderophores with a lower stability constant (25). Again, siderophore production was reduced proportionately more than the rate of bacterial cell division when fluorescent pseudomonds were grown in dilution series of synthetic medium containing low Fe (21). Therefore, plants producing higher amounts of exudates in the rhizosphere should support greater siderophore production by pseudomonads than plants exuding lower nutrient levels.

Evidence for this hypothesis was obtained by observing the level of chlamydospore germination of many isolates of Fusarium spp. germinating in rhizospheres of their respective hosts in soil infested or not infested with a siderophore-producing strain of P. putida. There was a significant, direct correlation between germination of chlamydospores in soil to which the pseudomonad was not added and the degree of inhibition induced when the strain was introduced. This suggests that the more nutrients exuded in the rhizosphere (as reflected by higher levels of chlamydospore germination) the more inhibition induced by the pseudomonad. Therefore, increased inhibition results in the rhizosphere when more nutrients (except Fe) are available for the production of siderophores.

Recently Magazin et al (31) identified a structural

gene from a fluorescent pseudomonad encoding an outer
membrane receptor protein for (siderophore) ferric pseudobactin. Biosynthetic genes flanked this receptor gene on
both sides. The production of the siderphore and membrane
proteins in the pseudomonad was coordinately regulated by
the level of intracellular iron. In the same laboratory
Byer and Leong (32) investigated the mode of Fe deprivation
involved when beneficial growth-enhancing fluorescent
pseudomonds were inhibited by certain deleterious bacteria
in the same group. Some of the beneficial strains were
unable to utilize siderophores from certain deleterious
florescent pseudomonades and were inhibited. Conversely,
deleterious strains able to utilize siderophores from
beneficial pseudomonads were not inhibited. The ability of
a given pseudomonad to utilize an exogenous siderophore from
another pseudomonad, therefore, may depend upon the production of a specific outer membrane receptor protein (31) for
that pseudomonads siderophore.

IMPLICATIONS

The examples of molecular biology describing mechanisms
associated with biological control, elaborated above, are
interesting and satisfying from a basic academic standpoint.
Such ivory-tower adventures into the realm of the unseeable
often are criticised as not representative of the "real"
world, but "the microscope cannot find the animalcule which
is less perfect for being little" (Emerson).
A knowledge of mechanisms enables intelligent manipulation of the complex elements associated with biological
control systems so that antagonism can be enhanced. This
can be accomplished by management of the molecular environment in the infection court as, for example, the selective
enhancement of activity of fluorescent pseudomonads together
with deprivation of Fe^{3+} for the Fusarium wilt pathogens.
Ultimately, molecular biology can contribute to the
improvement of the efficiency activity of biocontrol agents.
The limiting factor in application of the new biotechnology
to biological control is the lack of information on what
attributes should be inserted into an antagonist's genome.
Once identified, these attributes can be added to the
arsenal of a biocontrol agent capable of activity in the
infection court under a variety of environmental conditions.

REFERENCES

1. Barnett HL, Binder FL (1973). The fungal-host parasite relationship. Annu Rev Phytopathol 11:273.
2. Martin SB, Abavi GS, Hoch HC (1985). Biological control of soilborne pathogens with antagonists. In Hoy M, Herzog DC: "Biological Control and Integrated Pest Management Systems," New York: Academic Press, p 433.
3. Sequiera L (1980). Defense triggered by the invader: recognition and compatibility phenomena. In Horsfall, J. G., Cowling, E. B: "Plant Disease, An Advanced Treatise, Vol. 5," New York: Academic Press, p 179.
4. Elad Y, Barak I, Chet I (1983). The possible role of lectins in mycoparasitism. J Bacteriol 154:1431.
5. Goldstein IJ, Hughes RC, Monsigny M, Osawa T, Sharon N (1980). What should be called a lectin? Nature (Lond) 285:66.
6. Barondes S H (1981). Lectins: their multliple endogenous cellular functions. Annv Rev Biochem 50:1259.
7. Elad Y, Barak R, Chet I, Henis Y (1982). Ultrastructural studies of interactions between **Trichoderma** spp. and pathogenic fungi. Phytopath Z 107:168.
8. Lifshitz R, Dupler M, Elad Y, Baker R (1984). Hyphal interactions between a mycoparasite, **Pythium nunn**, and several soil fungi. Can J Microbiol 30:1482.
9. Barak R, Elad Y, Mirelman D, Chet I (1985). Lectins: a possible basis for specific recognition in the interaction of **Trichoderma** and **Sclerotium rolfsii**. Phytopathology 75:458.
10. Elad Y, Misaghi IJ (1985). Biochemical aspects of plant-microbe and microbe-microbe interactions in soil. In Cooper-Driver GA Swain T Conn EE: "Chemically Mediated Interactions Between Plants and Other Organisms", New York: Plenum Press, p. 21.
11. Hoch, HC (1978). Mycoparasitic relationships. IV. **Stephanoma phaeospora** parasitic in a species of **Fusarium**. Mycologia 70:370.
12. Bartnicki-Garcia S, Lippman E (1973). Fungal cell wall composition. In Laskin AL Lechevalier HL: "Handbook of Microbiology", Vol V, 2nd Edition, Cleveland OH: p 229.
13. Elad Y, Lifshitz R, Baker R (1985). Enzymatic activity of the mycoparasite **Pythium** nunn during interaction with host and non-host fungi. Physiol Plant Pathol 27:131.

14. Lifshitz R, Stanghellini ME, Baker R (1984). A new species of Pythium isolated from soil in Colorado. Mycotaxon 20:373.

15. Kleinschuster SJ, Baker R (1974). Lectin-detectable differences in carbohydrate-containing surface moieties of macroconidia of Fusarium roseum 'avenaceum' and Fusarium solani. Phytopathology 64:396.

16. Baker R (1985). Biological control of plant pathogens: definition. In Hoy M, Herzog DC: "Biological Control in IPM Systems," New York: Academic Press, p 25.

17. Clark FE (1963). The concept of competition in microbial ecology. In Baker KF Snyder WC: "Ecology of Soil-borne Pathogens," Berkeley: Univ of Calif Press, p 339.

18. Garrett SD (1970). "Pathogenic Root-Infecting Fungi," Cambridge: Cambridge Univ Press, p 116.

19. Scher FM, Baker R (1980). Mechanism of biological control in a Fusarium – suppressive soil. Phytopathology 70:412.

20. Elad Y, Baker, R (1985). Influence of trace amounts of cations and siderophore – producing pseudomonds on germination of Fusarium oxysporum chlamydospores. Phytopathology 75:1047.

21. Elad Y, Baker R (1985). The role of competition for iron and carbon in suppression of chlamydospore germination of Fusarium spp. Phytopathology 75:1053.

22. Kloepper JW, Leong J, Teintze M, Schroth MN. (1980). Pseudomonas siderophores: a mechanism explaining disease-suppressive soils. Current Microbiol 4:317.

23. Scher FM, Baker R (1982). Effect of Pseudomonas putida and a synthetic iron chelator on induction of soil suppressiveness to Fusarium wilt pathogens. Phytopathology 72:1567.

24. Scher FM, Baker R (1984). A fluorescent microscopic technique for viewing fungi in soil and its application to studies of a Fusarium-suppressive soil. Soil Biol Biochem 15:715.

25. Scher FM, Dupler M, Baker R (1984). Effect of synthetic iron chelates on population densities of Fusarium oxysporum and the biological control agent Pseudomonas putida in soil. Can J Microbiol 30:1271.

26. Simeoni LA (1986). Critical iron level associated with biological control of Fusarium wilt. PhD Thesis, Colorado State Univ.

27. Sneh B, Dupler M, Elad Y, Baker R (1984). Chlamydospore germination of Fusarium oxysporum f. sp. cucumerinum as affected by fluorescent and lytic bacteria from a Fusarium-suppressive soil. Phytopathology 74:1115.

28. Wong PTW, Baker R (1984). Suppression of wheat take-all and Ophiobolus patch by fluorescent pseudomonads from a Fusarium-suppressive soil. Soil Biol Biochem 16:397.

29. Baker R (1977). Inoculum potential. In Horsfall JG Cowling EB: "Plant Disease an Advanced Treatise, Vol II," New York: Academic Press, p 137.

30. Lindsey WL (1974). Role of chelation in microenvironment availability. In Carson RW: "The Plant Root and it's Environment," Charlottesville: Univ Press Va, p 507.

31. Teintze M, Hassain MB, Baines CL, Leong J, van der Helm D. Structure of ferric pseudobactin a siderophore from a plant growth promoting Pseudomonas. Biochem 20: 6446.

32. Magazin MD, Moores JC, Leong J (1986). Cloning of the gene coding for the outer membrane receptor protein for ferric pseudobactin, a siderophore from a plant-growth promoting Pseudomonas strain. J Biol Chem 261:795.

33. Buyer JS, Leong J (1986). Iron transport-medicated antagonism between plant growth-promoting and plant-deleterious Pseudomonas strains. J Biol Chem 261:791.

Molecular Strategies for Crop Protection, pages 125–134
© 1987 Alan R. Liss, Inc.

ENHANCEMENT OF ROOT HEALTH
AND PLANT GROWTH BY RHIZOBACTERIA

R. James Cook, D. M. Weller, and L. S. Thomashow

USDA, Agricultural Research Service
Washington State University
Pullman, Washington 99164-6430

ABSTRACT   Fluorescent Pseudomonas spp., obtained from
roots of healthy wheat plants grown in soil naturally
infested with Pythium spp., Gaeumannomyces graminis,
or both pathogens, and selected for their ability to
inhibit one or both of these pathogens in vitro, have
improved yield of wheat by 10-25% when applied as
seed treatments in field trials conducted since 1979.
Plant growth was not improved by the bacteria in
pathogen-free soil. Introduced Pseudomonas spp.
colonize roots in a two-phase process involving a)
passive carriage of the bacteria with the advancing
root tip and b) multiplication limited by the
carrying capacity of the rhizosphere and competition
with indigenous strains. The higher the ratio of
introduced to indigenous bacteria, the better the
protection. Protection also depends on ability of
the introduced strain to produce one or more siderophores and/or antibiotics. Mutants that have lost
the ability to produce these compounds colonize but
provide less protection of wheat roots. The
suppressiveness of one strain effective against both
pathogens is probably due to the production of a
dimer of phenazine-1-carboxylate. Of 43 species of
fungi tested, G. graminis and Pythium species were
the most sensitive in vitro to this antibiotic.
Research is now focused on the genetics and ecology
of the beneficial Pseudomonas strains, on finding or
developing superior strains, and on identifying wheat
germplasm more supportive of the bacteria.

INTRODUCTION

Plants support large populations of nonpathogenic microorganisms as epiphytes and endophytes, both on the phylloplane and in the rhizosphere. These microorganisms exist at little or no expense to the plant since they live on leaf or root exudates, deposits of pollen grains and aphid honey dew, and sloughed materials such as epidermal cells and mucilage from roots. Epiphytes live on the plant surface, whereas endophytes maintain a nonpathogenic relationship within the leaf, root or stem tissues. Both types of microorganisms are ideally positioned in advance of their competitors to colonize plant parts upon senescence or death of that tissue (or plant). However, while the plant lives, this community of plant inhabitants presents a living barrier against pathogens. The protection results from competition for nutrients, from inhibitory substances produced by the colonists, and from hyperparasitism or other forms of direct exploitation (1). Moreover, antagonism may continue in lesions caused by pathogens and thereby limit the rate of lesion development or shorten the survival time for the pathogen in the infected tissue. Our challenge is to improve or expand the effectiveness of these agents of plant protection.

Successful Use of Root-colonizing Fluorescent Pseudomonas spp. for Protection of Wheat Roots in the Field.

We have isolated strains of fluorescent Pseudomonas species from roots of wheat that, when applied to wheat seeds, protect roots against take-all caused by Gaeumannomyces graminis var. tritici (2) and Pythium root rot caused by several Pythium species (3,4). In 20 field tests conducted since 1979, where either take-all, Pythium root rot, or both diseases have been the yield-limiting factors, wheat yields have been increased by 10-25% in about two-thirds of the tests. The absence of a significant response about one-third of the time may reflect a failure of the introduced strains to compete with the indigenous rhizosphere colonists. Consider, however, a cultivar of wheat that outyielded all others by 10-25% two-thirds of the time wherever take-all or Pythium root rot were problems: it would be considered genetically superior or resistant and probably would be released for

commercial use. Control of Pythium root rot in areas of
the Pacific Northwest receiving 40-45 cm or more of annual
precipitation would increase yields by 1-2 t/ha (10-25%)
on about 500,000 ha, without requiring more water or
fertilizer. Complete protection against take-all in the
irrigation districts, where this disease is most
important, would increase wheat yields by 3-4 t/ha
(15-30%) on about 50,000 ha in Washington, Oregon, and
Idaho, again without requiring more water or fertilizer.
Equally important is that both diseases are favored by
conservation tillage and their control would therefore
make this method of soil conservation more acceptable to
farmers.

Obtaining Effective Strains from the Natural Population.

     Two rules are followed when isolating candidate
strains for testing on seeds and roots of wheat. The
first is to isolate from roots of wheat growing in soil
infested with the target pathogen(s). This helps insure
that the candidates will be able to compete in the
environment created by the host root-target pathogen
interaction. Microorganisms adapted to healthy roots may
not be adapted to roots infected by the target
pathogen(s). The second rule is to isolate candidate
strains from roots of the healthiest plants growing in
pathogen-infested soil, or use soil from fields where the
pathogen occurred but causes little or no disease in spite
of an apparently favorable physical environment for
disease development. Soils of this type are known as
suppressive soils (5,6) and are proving to be an excellent
source of effective antagonists (1).
     Our candidate antagonists selected for use against
take-all are isolated from soil where take-all decline, a
state of natural suppressiveness, which may develop after
years of wheat monoculture has occurred (1,6).
Effective isolates are primarily of the Pseudomonas
fluorescens-putida group. Indeed, our evidence continues
to indicate that these bacteria as colonists of the roots
or lesions are responsible for take-all decline (7). We
have no known suppressive soil for Pythium root rot of
wheat; we therefore isolate from roots of the healthiest
plants growing in Pythium-infested soil.

In Vitro and In Vivo Screening of Candidate Antagonists.

Strains inhibitory to G. graminis var. tritici or Pythium spp. in vitro are selected on media that favor the production of antibiotics (potato dextrose agar, PDA) and/or siderophores (King's medium B, KMB). Production of these substances has been correlated with disease suppression by fluorescent pseudomonads in vivo in several instances (8,9). Our best strains for suppression of take-all and Pythium root rot produce both antibiotics and siderophores. Moreover, loss of either trait, achieved experimentally by mutagenesis, results in less protection (more disease) in vivo, although mutant strains still colonize the roots (3,7,10, Thomashow, unpublished).

The first in vivo screen is carried out in the greenhouse using plastic tubes (2.5 cm in diameter by 16.5-cm-deep) filled with rooting medium and seeded with 2 bacteria-treated seeds (11). For G. graminis var. tritici, the rooting medium includes a 2-cm-thick layer of soil infested with the pathogen contained in particles of oat grains. For Pythium, the tubes are filled entirely with soil naturally infested with oospores of a mixture of species. Wheat chaff added at 1% (w/w) to the soil increases the inoculum potential of Pythium species and is therefore used in tests where high disease pressure is desired. Pasteurized soil (60 C/30 min) is used as a check in all greenhouse tests.

The final tests are conducted in the field on sites where take-all, Pythium root rot, or both diseases are the main yield-limiting factors. For small field plots of wheat with take-all the soil is fumigated to eliminate indigenous pathogens and G. graminis var. tritici is then reintroduced into the seed furrows at the time of planting. Checks include nontreated seed sown in fumigated soil with or without inoculum of the pathogen. In tests for Pythium control, plots are established at sites where the soil contains 400-600 propagules of Pythium per gram and check plots are fumigated. Large-scale plots (up to 0.2 ha each) are conducted in naturally infested fields in cooperation with growers.

Root Colonization by Beneficial Bacteria.

Root colonization by fluorescent Pseudomonas species introduced on wheat seeds occurs in two phases (12). During phase I, bacteria are carried passively on roots elongating downward into soil. During phase II, which may occur both during and after phase I, bacteria spread locally on the roots, multiply, and survive or avoid displacement by indigenous microorganisms in the rhizosphere.

Phase I colonization. The conclusion that bacteria are carried passively downward with the elongating root is based mainly on data that indicate how the bacteria do not spread downward. Movement downward by percolating water has been ruled out since long-distance movement (7-9 cm below the seed) of the bacteria occurred in soils at constant rhizosphere soil matric potentials where the only water movement would have been toward the root in response to gradients produced during water uptake. Movement by flagellar motility was ruled out on the basis that three different nonflagellated, nonmotile mutants were present at the same populations 7-9 cm below the seed as were their respective flagellated, motile parents. The nonmotile mutants were as suppressive to take-all in vivo as were their parents. Flagellar motility and movement with mass flow of water may contribute to phase II of colonization but are not essential for phase I (12).

Passive carriage downward with the root is also indicated by the population distribution along the root, being progressively smaller with increasing distance away from the seed (source) (13). For example, starting with 10^6-10^7 cfu per seed, the population along the radicle after 10 days at -0.3 bars matric potential is typically 10^4-10^5/cm root at 1-3 cm below the seed and 10^2-10^3/cm root at 7-9 cm below the seed. Without multiplication (phase II), and perhaps even with multiplication, the number of cfu would be nil given sufficient distance from the source.

Accepting that phase I of colonization is passive, then adhesion of cells to roots may be necessary for root colonization. Attachment could be mediated by a specific mechanism such as described for Rhizobium on roots of legumes (14) but more likely involves a nonspecific mechanism such as proposed for P. fluorescens on the roots of radish (15). Preliminary work indicates that the

initial populations of pseudomonads introduced on wheat
seeds are significantly higher with some strains than
others and for some wheat cultivars than others (16).
Binding may thus be under genetic control of both the
bacterium and the plant, opening the way for genetic
studies of the bacterium-root association.

Phase II colonization. Multiplication, local spread,
and survival probably begin on the seed and especially in
the embryo region within the first few hours after the
seed is planted, with the introduced bacteria multiplying
to the limits of the niche available. Multiplication of
the introduced strain on the seeds and roots is affected
by its ability to compete for the limited supply of carbon
and energy available from the root and by physical
constraints of the environment.. Maximum populations of
our introduced strains on wheat roots occur at -0.3 to
-0.7 bars matric potential and rhizosphere pH
(pH_r)=6.0-6.5 (12).

The effectiveness of select strains of fluorescent
pseudomonads as agents of root protection is directly
related to their percentage of the total rhizosphere
population. Effectiveness may also depend on specific
sites of colonization on the seed or root. Strains
effective against G. graminis var. tritici occur at 1-2
log units higher population in the presence of the fungus
than in its absence and are thought to multiply in
lesions, thereby limiting lesion development. In this
case, a small fraction of the total population could
represent a significant source of inoculum for
colonization of lesions and suppression of take-all. With
infections by Pythium species, a major cause of plant
stunting can be traced to embryo infections during the
first 2-7 days after planting, and much of the increased
growth response to inoculation of seeds with bacteria
inhibitory to Pythium probably results from bacterial
protection of the embryos.

Significance of Antibiosis and Iron Competition.

The presence of a fast-growing highly competitive
strain in the embryo region, on the root tip, in soil
around root hairs, in young lesions, or in other infection
courts can be a deterrent in itself to colonization of
these regions by root pathogens. However, competition for

nutrients cannot totally account for the protection
afforded by the fluorescent pseudomonads. The production
of specific inhibitory substances such as antibiotics and
siderophores may both contribute to the competitive
advantage of some root-colonizing strains and account for
their effectiveness in disease suppression.

We have concentrated on the role of these substances
in suppression of take-all. Initially, nine fluorescent
Pseudomonas strains isolated from roots grown in
suppressive soil and highly inhibitory to G. graminis var.
tritici on both KMB (siderophore produced) and PDA
(antibiotics produced) were compared for ability to
suppress take-all in vivo with 18 strains (all either
noninhibitory or weakly inhibitory on either KMB or PDA)
from roots in conducive soils. The strains from the
suppressive soil, as a group, were significantly more
suppressive of take-all than those from the conducive
soils (7,10). In further work, P. fluorescens strains
Rla-80 and 2-79 (suppressive in vivo and in vitro) were
treated with NTG and 11 mutants were selected on the basis
of reduced or no ability to inhibit G. graminis var.
tritici on PDA and/or KMB. The mutants colonized wheat
roots to the same extent as their respective parents, but
all were significantly less suppressive of take-all than
their parents (10).

The relative roles of siderophores and antibiotics in
suppression of take-all were then evaluated by testing
strains Rla-80, 2-79, and L30b-80 (inhibits G. graminis
var. tritici only by a siderophore) in fumigated soil
reinfested with the take-all pathogen and amended with
either FeEDTA or EDDHA. Elimination of disease suppression
by the addition of iron, and induction of suppression with
iron chelators, have been used as evidence for a role of
siderophores in biological control by fluorescent
pseudomonads (8,9,17). Take-all was highly suppressed
when EDDHA was added to soil, confirming that the pathogen
is sensitive to iron deprivation. The suppressiveness of
L30b-80 was totally nullified, that of Rla-80 was
unchanged, and that of 2-79 was reduced to approximately
half by the addition of FeEDTA to the soil. The results
indicate that Rla-80 suppresses take-all mainly by
antibiotic production, strain L30b-80 by siderophore
production, and strain 2-79 by both antibiotic and
siderophore production (10).

A greenish-yellow antibiotic has been isolated from strain 2-79 (suppressive to both take-all and Pythium root rot) grown in potato dextrose broth. The compound is a dimer of phenazine-1-carboxylic acid and is active against a wide spectrum of fungi (43 species), with G. graminis var. tritici and several Pythium species being the most sensitive fungi tested. The antibiotic completely inhibits growth of G. graminis var. tritici in vitro at less than 1μg/ml and also suppresses take-all when applied as a wheat seed treatment. This compound is thought to play a significant role in the suppression of take-all and Pythium root rot (18).

Identification of Antibiotic Genes.

Our experimental approach to identification of genes for antibiotic biosynthesis has been to generate Tn5 transposition mutants in strain 2-79 and to screen in vitro for changes in antibiotic production. Tn5 was introduced via the suicide vehicle pGS9, which can be conjugally transferred but not stably maintained in Pseudomonas (19). Strain 2-79 transconjugants selected for the kanamycin resistance (Km^r) marker conferred by Tn5 are recovered at a frequency of 1-5 x 10^{-5} and have been shown to result from random Tn5 insertion (Thomashow, unpublished).

A number of 2-79 mutants with altered phenazine production have been identified, including isolates with changes in colony pigmentation or morphology of the phenazine crystals produced. A total of 13 nonpigmented prototrophic mutants and 2 nonpigmented auxotrophs have been recovered from 8500 transconjugants screened in 25 separate mating experiments. The low frequency of recovery suggests that these mutants are specifically altered in phenazine structural genes rather than generally impaired in secondary metabolism. This conclusion is further supported by the fact that mutant colonies produce pigments when crossfed with diffusable substances secreted by the parental strain. Southern blot analyses of DNA from nonpigmented prototrophic mutants indicate that at least three loci on separate Eco R1 restriction fragments are involved in phenazine biosynthesis. So far there is a direct correlation between pigment production on PDA or minimal medium, and antifungal activity. Thus, nonpigmented mutants are noninhibitory to G. graminis var. tritici, and some mutants with altered pigmentation exhibit reduced biological activity in the plate bioassay. Eight nonpigmented, noninhibitory mutants have also been tested in vivo, and all show reduced ability to suppress take-all.

REFERENCES

1. Cook RJ and Baker KF (1983). "Nature and Practice of Biological Control of Plant Pathogens." American Phytopathological Society, St Paul, MN.
2. Weller DM and Cook RJ (1983). Suppression of take-all of wheat by seed treatments with fluorescent pseudomonads. Phytopathology 73:463-469.
3. Becker JO and Cook RJ (1984). Pythium control by siderophore-producing bacteria on roots of wheat. (Abstr) Phytopathology 74:806.
4. Weller DM and Graham MC (1984). Application of fluorescent pseudomonads to improve the growth of wheat. Phytopathology 74:806. (Abstr).
5. Schroth MN and Hancock JG (1982). Disease-suppressive soil and root-colonizing bacteria. Science 216:1376-1381.
6. Cook RJ and Rovira AD (1976). The role of bacteria in the biological control of Gaeumannomyces graminis by suppressive soils. Soil Biol Biochem 8:267-273.
7. Cook RJ and Weller DM (1985). Management of take-all in consecutive crops of wheat or barley. In Chet I (ed): "Nonconventional Methods of Disease Control," Wiley. (In press).
8. Kloepper JW, Leong J, Teintze M, and Schroth MN (1980). Enhanced plant growth by siderophores produced by plant growth-promoting rhizobacteria. Nature 286:885-886.
9. Scher FM and Baker R (1982). Effect of Pseudomonas putida and a synthetic iron chelator on induction of soil suppressiveness to Fusarium wilt pathogens. Phytopathology 72:1567-1573.
10. Weller DM, Howie WJ, and Cook RJ (1985). Relationship of in vitro inhibition of Gaeumannomyces graminis var. tritici and in vivo suppression of

take-all by fluorescent pseudomonads. Phytopathology 75:1301.

11. Weller DM, Zhang BX, and Cook RJ (1985). Application of a rapid screening test for selection of bacteria suppressive to take-all of wheat. Plant Disease 69:710-713.

12. Howie WJ (1985). Factors affecting colonization of wheat roots and suppression of take-all by pseudomonads antagonistic to Gaeumannomyces graminis var. tritici. Ph.D. Thesis, Washington State University, Pullman, p 82.

13. Weller DM (1984). Distribution of a take-all suppressive strain of Pseudomonas fluorescens on seminal roots of winter wheat. Applied and Environmental Microbiology 48:897-899.

14. Dazzo FB (1984). Attachment of nitrogen-fixing bacteria to plant roots. Pages 130-135 in Klug MJ and Reddy CA (eds): "Current Perspectives in Microbial Ecology," Washington, DC: American Society for Microbiology.

15. James Jr DW, Suslow TV, and Steinback KE (1985). Relationship between firm adhesion and long-term colonization of roots by bacteria. Appl Environ Microbiol 50:392-397.

16. Weller DM (1986). Effects of wheat genotype on root colonization by a take-all suppressive strain of Pseudomonas fluorescens. Phytopathology 76: (In press) (Abstr).

17. Wong PTW and Baker R (1984). Suppression of wheat take-all and Ophiobolus patch by fluorescent pseudomonads from a Fusarium-suppressive soil. Soil Biol Biochem 16:397-403.

18. Gurusiddaiah S, Weller DM, Sakar A, and Cook RJ (1986). Characterization of an antibiotic produced by a strain of Pseudomonas fluorescens inhibitory to Gaeumannomyces graminis var. tritici and Pythium spp. Antimicrobial Agents and Chemotherapy 29:(In press).

19. Selvaraj G and Iyer VN (1983). Suicide plasmid vehicles for insertion mutagenesis in Rhizobium meliloti and related bacteria. J Bacteriol 156:1291-1300.

Molecular Strategies for Crop Protection, pages 135–144
© 1987 Alan R. Liss, Inc.

PLASMID-CHROMOSOME RECOMBINATION IN PSEUDOMONAS SYRINGAE PV. PHASEOLICOLA[1]

Dallice Mills, Marilyn Ehrenshaft[2], Janet Williams, Christopher Small and Alan Poplawsky[3]

Department of Botany and Plant Pathology, Oregon State University, Corvallis, Oregon 97331

ABSTRACT   Strain LR700 of Pseudomonas syringae pv. phaseolicola, the causal agent of halo blight of bean, harbors an indigenous 150 Kb cryptic plasmid, pMMC7105. This plasmid has been observed to integrate into and imprecisely excise from the host chromosome by a process that leads to the formation of a variety of new plasmids. Repetitive sequences (RS) found on pMMC7105 and the chromosome, as well as a repetitive sequence that is present only on pMMC7105, act as sites for recombination. One of these sequences, RSII, present on both replicons, has been demonstrated to be involved in both integration and excision of plasmid sequences. Restriction endonuclease mapping and Southern blot hybridization analysis show that it is highly conserved and between ca. 800 to 1500 bp in size. Two other repetitive sequences, designated RSI and RSIII, were shown to be involved only in recombination events that lead to the formation of deletion derivatives of pMMC7105. RSI appears to be less than 350 bp in size, whereas RSIII appears to be less than 450 bp in size.

[1]This work supported by NSF grant no. PCM-8315689 and by Science and Education Administration (USDA) grant no. 84-CRCR-1-1463 from the Competitive Research Grants Office.
[2]Present address: Department of Plant Pathology, University of Minnesota, St. Paul, MN 55108.
[3]Present address: Department of Plant Pathology, University of Wisconsin, Madison, WI, 53706.

INTRODUCTION

The occurrence of plasmids in phytopathogenic pseudomonads has stimulated interest in understanding their role in pathogenicity (2,3), resistance to antimicrobial compounds (1,7) and their evolution and distribution (8,11,12). The discovery that genes encoding virulence-determining enzymes for indole acetic acid production are linked to plasmid pIAA1 in strains of Pseudomonas syringae pv. savastanoi that attack oleander, but linked to the chromosome of strains that attack olive trees, is suggestive that some resident plasmids and the host chromosome undergo recombination. Indeed, strain LR700 of P. syringae pv. phaseolicola, the etiologic agent of halo blight disease of bean, Phaseolus vulgaris, harbors a cryptic 150 kilobase pair (kb) plasmid pMMC7105, that integrates into the bacterial chromosome (4). Imprecise excision of pMMC7105 has produced stable, F-prime-like plasmids that contain varying amounts of chromosomal and pMMC7105 DNA, as well as plasmids that contain sequences only from pMMC7105 (12). The occurrence of plasmids with varying amounts of homology to pMMC7105 in other strains of P. syringae pv. phaseolicola, and in strains of a related pathovar, glycinea (8), has provided the impetus for our efforts to develop an understanding of the genetic mechanisms underlying the evolution of plasmids among this diverse group of bacteria. It is generally acknowledged that pathovars of P. syringae exhibit substantial genetic diversity in terms of host range, toxin production, symptomatology and response to a variety of biochemical analyses (9). The ability of plasmids to integrate into the host chromosome and subsequently excise forming F-prime-like plasmids, could provide a mechanism for genetic exchange among a population of phytopathogenic pseudomonads if the plasmids are self-transmissible or mobilizable by other plasmids. Efforts aimed at devising molecular strategies for crop protection from these microorganisms would profit from a better understanding of the mechanisms that lead to their genetic diversity.

RESULTS

The Role of Repetitive Sequences in Plasmid-Chromosome Recombination.

Three repetitive sequences, RSI, RSII and RSIII on pMMC7105, have been implicated in recombination events that

produce excision plasmids from the integrated form. Two of
these sequences, RSII and RSIII, have been shown by Southern
hybridization (10) to also be present in the chromosome (11).
The plasmid site of integration has been mapped to BamHI
fragment 8 (BamHI-8). This fragment contains a copy each of
RSI and RSII (11). One of two junction fragments formed by
integration of pMMC7105 has been cloned and shown to contain
at least two subfragments that carry a copy of RSII. These
observations suggested that the genetic mechanism for both
integration and excision of pMMC7105 may be analogous to that
observed for F-plasmid of Escherichia coli (5,6). Insertion
sequence (IS) elements present on F-plasmid and in the
chromosome act as sites for general homologous recombination.

Integration of pMMC7105.

To understand the mechanism leading to pMMC7105
integration, the plasmid fragment involved in integration
(BamHI-8), as well as the chromosomal fragment (BamHI-R) and
one of the plasmid/chromosome junction fragments (BamHI-α)
were cloned and an EcoRI restriction endonuclease map was
developed of each (Fig. 1). The maps clearly show that BamHI-α
was derived by recombination at a site within the right
terminal EcoRI-BamHI fragment of BamHI-R and the left terminal
EcoRI-BamHI fragment of BamHI-8. The juncture of chromosomal
and plasmid sequences resides within a 2.8 Kb EcoRI
subfragment of BamHI-α.

Involvement of RSII in plasmid integration.

If pMMC7105 integrated by simple homologous
recombination, the subfragments from BamHI-8 and BamHI-R must
contain a common repetitive sequence. It was previously shown
that BamHI-8 carries two repetitive sequences, RSI and RSII,
in the 3.2 Kb BamHI-EcoRI subfragment (11). RSI was shown to
be present only in pMMC7105, whereas RSII was known to be
present in multiple copies in the chromosome. To test whether
RSII was at the site of recombination on BamHI-R, the
subfragments from each of these fragments were subjected to
fine structure restriction endonuclease mapping and Southern
hybridization with a probe made of BamHI-12 (3.8Kb) from
pMMC7105, which contains a copy of RSII. It can be seen upon
inspection of these maps and the results of probing that a
common sequence, RSII, is present in each of the fragments and

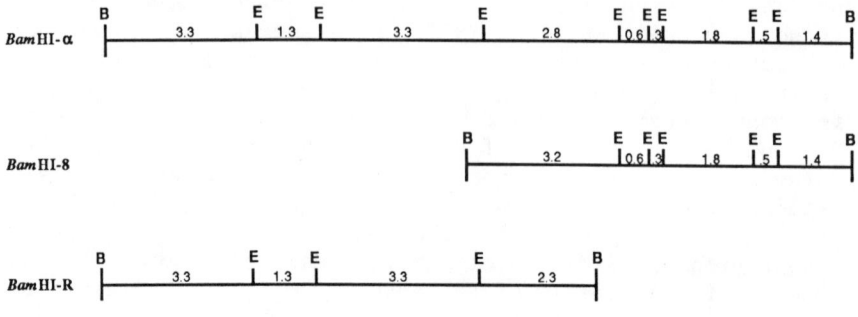

FIGURE 1. Plasmid and chromosomal BamHI fragments involved in integration of pMMC7105 and the resulting BamHI junction fragment.

it resides at the juncture of plasmid and chromosomal sequences in the BamHI-α junction fragment (Fig. 2). The copy of RSII from BamHI-8 has recently been partially sequenced and shown to have some properties expected of an insertion sequence element (Unpublished data).

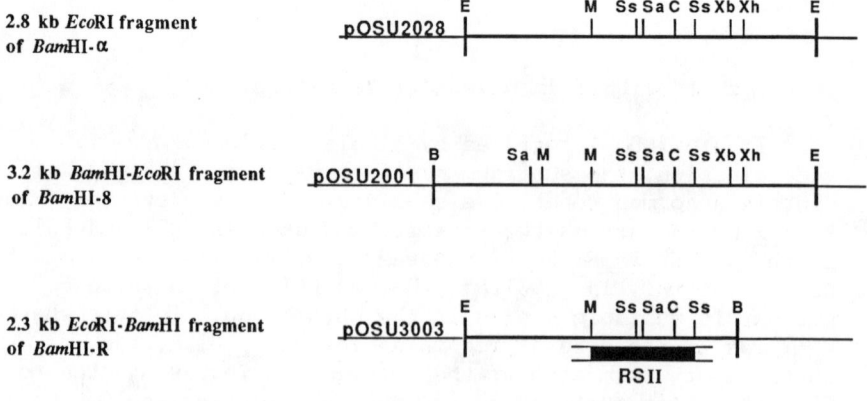

FIGURE 2. Fine structure restriction endonuclease map of pMMC7105 and chromosomal sites for integration. The position of RSII was determined by Southern blot hybridization.

The Role of RSI, RSII and RSIII in Plasmid Excision.

Excision of plasmids by a mechanism involving simple homologous recombination would occur only between fragments that carry a common repetitive element. Using fragments containing RSI, RSII and RSIII as probes, we have shown that fragments which participate in recombination and concomitant excision of a plasmid always contain a common sequence (11). Although some of the fragments carry two different repetitive sequences, as for example BamHI-12, which contains RSII and RSIII (Table 1), the second fragment frequently contains only one of the two repeated sequences. These results substantiate the hypothesis that excisive recombination of the plasmids from the chromosome occurs by simple homologous recombination.

TABLE 1

REPETITIVE SEQUENCES PRESENT IN BAMHI FRAGMENTS OF pMMC7105
AND THE CHROMOSOME OF LR700 THAT RECOMBINED TO FORM pEXC8080,
pMMC7115 AND pEXC8140

	Excision plasmid and recombined BamHI fragments[a]					
	pEXC8080		pMMC7115		pEXC8140	
	1	4	α	12	10	12
Repetitive sequence	RSI RSII	RSI	RSII	RSII RSIII	RSIII	RSII RSIII

[a]BamHI fragments are from pMMC7105 except BamHI-α which is one of two junction fragments formed when pMMC7105 integrated into the chromosome of LR700.

Three excision plasmids that were formed by recombination between fragments that carry a common repetitive sequence, either RSI, RSII or RSIII, are presented in Fig. 3. The formation of pMMC7115 involved recombination between BamHI-α and BamHI-12. Recent fine structure restriction mapping of the BamHI-α-12 junction fragment from pMMC7115, as well as

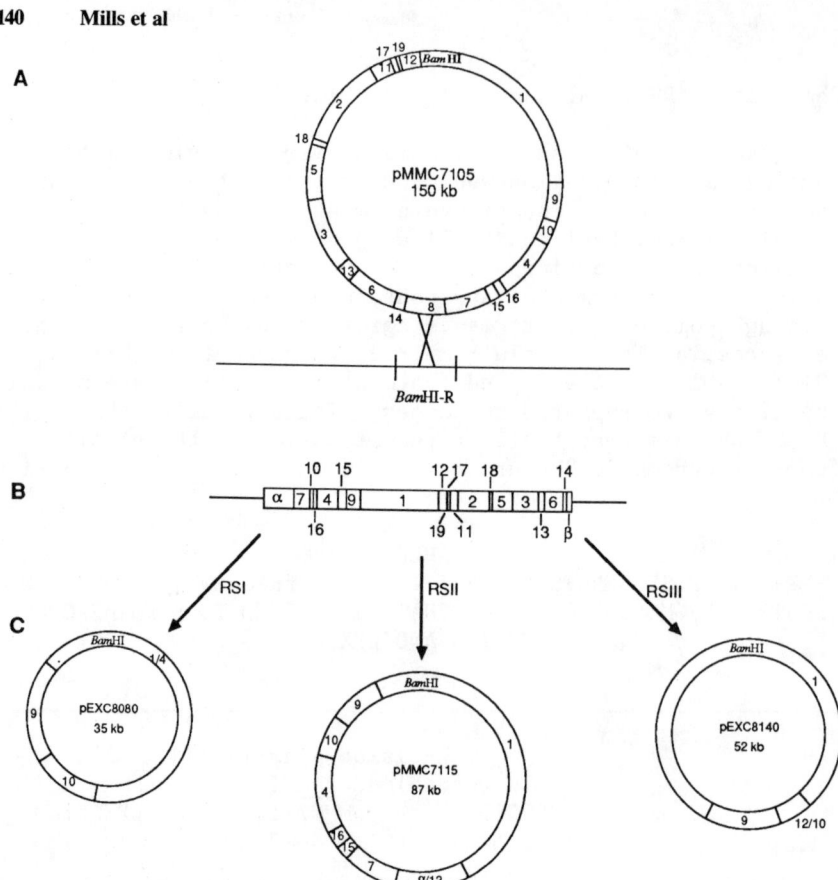

Figure 3.   Evidence for the involvement of RSI, RSII and RSIII in excision of three plasmids from the chromosome A. Integration of pMMC7105 at RSII sites on BamHI fragment 8 and BamHI-R B.   The integrated form of pMMC7105 showing junction fragments BamHI- α and BamHI- β C.   Three excision plasmids mediated by recombination at RSI, RSII and RSIII.

BamHI-α, confirmed a role of RSII in the formation of pMMC7115 (Unpublished data). Of interest, however, was the demonstration that a second copy of RSII present within chromosomal sequences of BamHI- α was the site of recombination. This site is approximately 1.5 Kb from the site at which pMMC7105 integrated into the chromosome.

The presence of a single element in one of the fragments that was involved in the recombination event that produced each of the excision plasmids (Fig. 3) provides a convincing argument for the probable site at which recombination occurred.   It can be argued from data presented in Table 1 that RSI will reside at the juncture of <u>Bam</u>HI-1 and <u>Bam</u>HI-4 sequences in the <u>Bam</u>HI-1-4 junction fragment of pEXC8080 (also see Fig. 3).   Indeed the fine structure map of these fragments revealed that the site of recombination lies within a 1.2 Kb <u>Xho</u>I-<u>Cla</u>I subfragment of <u>Bam</u>HI-1-4 (Fig. 4B).   This region hybridizes to a probe of RSI that was obtained from <u>Bam</u>HI-8 of pMMC7105.   Analysis of <u>Bam</u>HI-10 (Table 1 and Fig. 5) revealed that it contains two copies of RSIII.   The <u>Bam</u>HI-10-12 junction fragment from pEXC8140 is presently being analyzed to ascertain which of the copies of RSIII was involved.

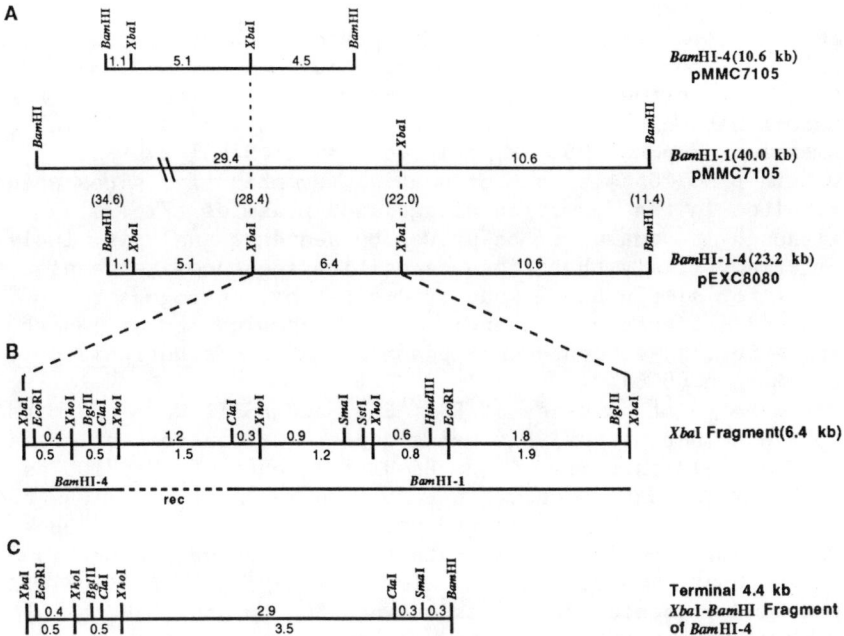

FIGURE 4.   Fine structure restriction endonuclease map of the region where recombination occurred and produced pEXC8080. A.   <u>Bam</u>HI fragments involved in recombination   B.   Fine structure map showing the site of recombination (rec)   C. Detailed map of the right end of <u>Bam</u>HI-4 showing restriction sites common to the <u>Xba</u>I fragment shown in B.

142   Mills et al

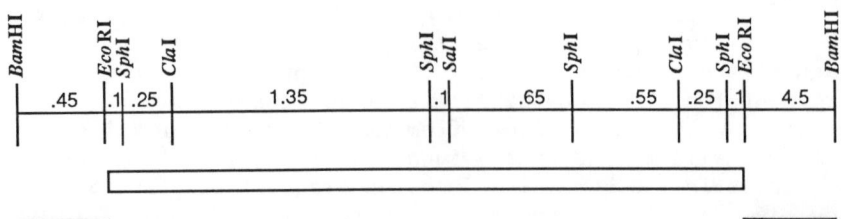

FIGURE 5. Partial restriction map of BamHI fragment 10 of pMMC7105 showing the position of RSIII at its termini (■).

DISCUSSION

It is apparent from results presented here and elsewhere (11) that homologous copies of repetitive sequence, RSII, on pMMC7105 and the chromosome of LR700, act as sites for recombination which result in both integration and partial excision of pMMC7105. Repetitive sequences RSI and RSII within pMMC7105 also are present at recombination sites which resulted in the formation of excision plasmids (Fig.3). Although it remains to be proven by sequence analysis, it is extremely likely that these repetitive sequences serve as sites for simple homologous recombination, analogous to F-plasmid integration and excision by recombination at insertion sequence (IS) elements on F-plasmid and the Escherichia coli chromosome (5,6).

The relative sizes of RSI, RSII and RSIII were estimated by Southern blot analysis using copies, or portions of copies, of these elements from other BamHI fragments of pMMC7105 as probes. The data suggest that RSI and RSIII will be less than 450 base pairs (bp) in length unless, of course, the probes were constructed from fragments that carried only a portion of these repetitive sequences. Similar analysis of RSII (Fig.2) provided evidence that it is between 800 bp and 1500 bp in size. Whether these repeated sequences are IS elements remains to be determined. Preliminary sequencing data of RSII indicates that it has some properties expected of IS elements including apparent inverted repeated sequences at its termini (unpublished data). Whether these elements are capable of transposing remains to be determined, but there is evidence that changes in hybridization banding profiles of fragments containing RSII have occurred in laboratory strains derived from LR700 (11). Plasmids that can integrate into the host

chromosome provide the opportunity for new prime-like plasmids
to be formed through imprecise excision. pMMC7115 (Fig.3) is
an example of a prime-like plasmid that contains a very small
(ca. 2Kb) segment of the chromosome. However, an excision
plasmid has been identified that contains approximately 135 Kb
of chromosomal sequences (12). The potential to recombine
genes within a pathovar population or among different
pathovars of P. syringae, may be greatly enhanced in strains
that carry self-conjugal or mobilizable episomes. These
plasmids may prove bothersome in terms of man's ability to
control bacterial populations that harbor them and adversely
affect plant health and productivity. However, their genetic
properties may also prove useful in developing new plasmids by
genetic engineering that are more suitable to phytopathogenic
pseudomonads.

ACKNOWLEDGEMENTS

    M.E. was supported in part by an ARCO Plant Cell Research
Fellowship awarded to the Center for Gene Research and
Biotechnology, Oregon State University; A.P. was supported by
NIH Molecular Biology Training Grant #GM07774 traineeship.

REFERENCES

1. Bender CL, Cooksey DA (1986). Indigenous plasmids in
    Pseudomonas syringae pv. tomato: Conjugative transfer
    and role in copper resistance. J Bacteriol 165:534.
2. Comai L, Kosuge T (1980). Involvement of plasmid
    deoxyribonucleic acid in indoleacetic acid synthesis in
    Pseudomonas savastanoi. J Bacteriol 143:950.
3. Comai L, Kosuge T (1983). Transposable element that
    causes mutations in a plant pathogenic Pseudomonas sp. J
    Bacteriol 154:1162.
4. Curiale MS, Mills D (1982). Integration and partial
    excision of a cryptic plasmid in Pseudomonas syringae
    pv. phaseolicola. J Bacteriol 152:797.
5. Deonier RC, Hadley RG (1980). IS2-IS2 and IS3-IS3
    relative recombination frequencies in F integration.
    Plasmid 3:48.
6. Deonier RC, Mirels L (1977). Excision of F plasmid
    sequences by recombination at directly repeated
    insertion sequence 2 elements: Involvement of recA.
    Proc Natl Acad Sci USA 74:3965.

7. Leary JV, Trollinger DB (1985). Identification of an indigenous plasmid carrying a gene for trimethoprim resistance in Pseudomonas syringae pv. glycinea. Mol Gen Genet 201:485.

8. Quant RL, Mills D (1984). An integrative plasmid and multiple-sized plasmids of Pseudomonas syringae pv. phaseolicola have extensive homology. Mol Gen Genet 193:459.

9. Shaad NW (1980). "Laboratory Guide for Indentification of Plant Pathogenic Bacteria". The American Phytopathological Society, St. Paul, MN.

10. Southern EM (1975). Detection of specific sequences among DNA fragments separated by gel electrophoresis. J Mol Biol 98:503.

11. Szabo LS, Mills D (1984). Integration and excision of pMC7105 in Pseudomonas syringae pv. phaseolicola: Involvement of repetitive sequences. J Bacteriol 157:821.

12. Szabo LS, Mills D (1984). Characterization of eight excision plasmids of Pseudomonas syringae pv. phaseolicola. Mol Gen Genet 196:90.

Molecular Strategies for Crop Protection, pages 145–156
© **1987 Alan R. Liss, Inc.**

OVERDRIVE, A T-DNA TRANSMISSION ENHANCER ON THE
A. TUMEFACIENS TUMOR-INDUCING PLASMID[1]

Ernest G. Peralta[2], Renate Hellmiss, and Walt Ream

Institute for Molecular Biology, Dept. of Biology,
Indiana University, Bloomington, IN 47405, USA

ABSTRACT   During crown gall tumorigenesis a
specific segment of the Agrobacterium
tumefaciens tumor-inducing (Ti) plasmid, the
T-DNA, integrates into plant nuclear DNA.
Similar 23-base-pair (bp) direct repeats at
each end of the T region signal T-DNA
borders, and T-DNA transmission (transfer and
integration) requires the right-hand direct
repeat. A chemically synthesized right
border repeat in its wild type orientation
promotes T-DNA transmission at a low
frequency; Ti plasmid sequences which
normally flank the right repeat greatly
stimulate the process. To identify flanking
sequences required for full right border
activity, we tested the activity of a border
repeat surrounded by different amounts of
normal flanking sequences. Efficient T-DNA
transmission required a conserved sequence
(5´ TAAPuTPy-CTGTPuT-TGTTTGTTTG 3´) which
lies to the right of two known right border
repeats. In either orientation, a synthetic

[1] A grant from the National Science
Foundation (PCM8316006) supported this
research. E. Peralta received a Graduate and
Professional Opportunity Program Fellowship
(USDEG-83-453) and a Bayard Franklin Floyd
Memorial Fellowship.
[2] Present address: Genentech Inc., 460 Point
San Bruno Blvd., South San Francisco, CA
94080.

oligonucleotide containing this conserved
sequence greatly stimulated the activity of a
right border repeat, and a deletion removing
15 bp from the right end of this sequence
destroyed its stimulatory effect. Thus, wild
type T-DNA transmission required both the 23
bp right border repeat and a conserved
flanking sequence which we call overdrive.

INTRODUCTION

Agrobacterium tumefaciens incites crown gall
tumors on many dicotyledonous plants when viable
bacteria infect wounded plant tissue (2, 22).
Virulent strains contain a 190 kilobase (kb)
tumor-inducing (Ti) plasmid that carries genes
essential for tumorigenesis (8, 38). During
tumorigenesis a specific segment of the Ti
plasmid, the T-DNA, integrates into plant nuclear
DNA (5, 19). Plant tumor cells express T-DNA
genes responsible for tumorous growth, but T-DNA
transmission does not require tumorigenesis or T-
DNA encoded proteins (20, 29). Similar 23 bp
direct repeats lie at both ends of four different
T regions (1, 33, 40), and these repeats signal
the T-DNA borders since T-DNA ends occur in or
near these repeats in several different tumors
(11, 32, 34, 41). Deletions that remove the T
region right border severely attenuate virulence
on most plants even though the tumor maintenance
genes remain intact (11, 15, 24, 26-27, 31, 39).
Such deletion mutants provide an assay for right
border function: we can reintroduce T-DNA borders
to the right of the T region and measure their
ability to restore virulence. T-DNA transmission
requires only a right border repeat in cis (26-27,
39), but Ti plasmid sequences that lie to the
right of the right border repeat greatly stimulate
its function (13, 25-27).
    In this study we identified a specific
sequence flanking a right border repeat that
stimulated T-DNA transmission. This flanking
sequence, which we call overdrive, lies to the
right of the TL right border repeat in the
octopine-type plasmid pTiA6NC. Related sequences

occur to the right of three other right border
repeats (1, 6, 33). We constructed a series of
deletions that extended into the overdrive
to the right of the pTiA6NC right border repeat
and demonstrated that efficient T-DNA transmission
required this flanking sequence. To determine
whether overdrive functioned as a separate
element, we synthesized this sequence and cloned
it to the right of two different border repeats.
Although right border repeats function best in one
orientation (26-27, 39), overdrive functioned in
either orientation. overdrive also retained its
activity when moved either 10 bp closer to or 7 bp
farther from the border repeat than it normally
occurs. Thus, overdrive formed a discrete element
distinct from the border repeat. These results
indicate overdrive may distinguish right and left
T-DNA borders, and they suggest a model for the
early events in T-DNA transmission. The
relationship of T-DNA right borders to overdrive
resembles that of inverted repeats to
recombinational enhancers in several site-specific
inversion systems (12, 14, 16, 28).

MATERIALS AND METHODS

We used methods and strains described earlier
(25-27).

RESULTS

Border Assay System.

To identify sequences required for T-DNA
transmission, we developed an assay system for
border sequence function (26). The octopine-type
Ti plasmid pTiA6NC contains two adjacent but
noncontiguous T regions designated TL and TR; TL,
the left T-DNA, contains all the genes required
for tumor maintenance (24, 37). A deletion mutant
(pWR113 in WR3095) of pTiA6NC lacks the three
border repeats that lie to the right of TL (the TL
right border and both TR borders) without
affecting the tumor maintenance genes (26). This

right border deletion renders WR3095 essentially
avirulent on all plants tested. A wide host range
shuttle vector, pWR64, contains sequences from
EcoRI fragment 1 (7) to the right of TL in pWR113
(26). To assay the border function of restriction
fragments containing border repeats, we insert
them into the shuttle vector at the single HindIII
site, introduce the border fragments into pWR113
by homologous recombination in A. tumefaciens
(30), and test their abililty to restore
virulence. Since deletions that remove EcoRI
fragment 1 do not affect virulence (24), our
reintroduced constructs will not interfere with
other sequences important for T-DNA transmission.
The new location of the fragments does not
adversely affect efficient right border function.
The right border functions poorly when we
reduce the normal righthand flanking sequences to
4 bp (26). Thus, our previous work indicates that
sequences normally lying to the right of the
border repeat greatly stimulate its function.

Deletion Analysis of the T-DNA Transmission Enhancer.

     To identify the Ti plasmid sequences that
stimulate border repeat activity, we constructed a
series of deletions that extended into the Ti
plasmid sequences present to the right of the TL
right repeat (Figure 1). Strains which
contained the TL right border repeat flanked on
the right by at least 40 bp of Ti sequences
exhibited full virulence (strains WR1101, WR1103;
Figures 1). An otherwise identical strain
(WR1102) that contained the TL right repeat
flanked on the right by only 25 bp of Ti sequences
exhibited greatly reduced virulence (Figure 1).
Strains (WR1105) that contained a synthetic TL
right repeat (without flanking sequences) showed
similar weak virulence (26). Therefore, efficient
T-DNA transmission required sequences located
within 40 bp to the right of the TL right repeat;
the sequences within 25 bp to the right of the
repeat did not include all of the stimulatory
sequences. A conserved sequence, 5´ TAAPuTPy-
CTGTPuT-TGTTTGTTTG 3´, begins 17 bp to the right

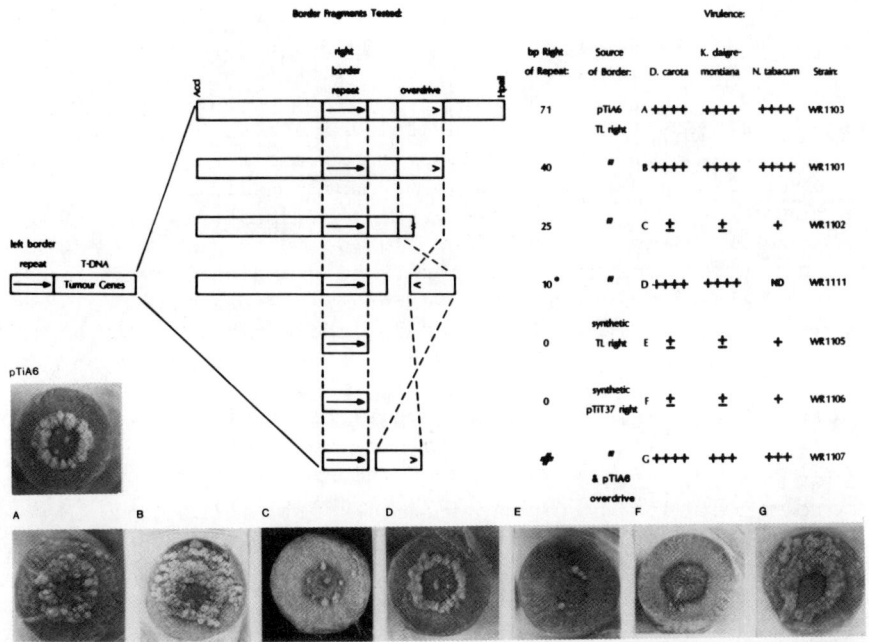

FIGURE 1.   Border Fragment Assays.

We inoculated the carrot slices shown as
follows: "pTiA6" with wild type strain A348
[A136(pTiA6NC)], "A" with WR1103, "B" with WR1101,
"C" with WR1102, "D" with WR1111, "E" with WR1105,
"F" with WR1106, and "G" with WR11007. Symbols
indicate: ++++, fully virulent; + to +++,
partially virulent; +/-, very weakly virulent; *,
WR1111 contains the pTiA6NC TL right border repeat
flanked on the right by 10 bp of normal Ti plasmid
sequences and a 24 bp synthetic <u>overdrive</u>
oligonucleotide (from the pTiA6 TL right border)
in an inverted orientation; #, WR1107 contains a
24 bp synthetic <u>overdrive</u> oligonucleotide (from
the pTiA6 TL right border) in its wild type
orientation; ➤ , T-DNA border repeat; ☐ ,
Ti plasmid sequences; ☐> , <u>overdrive</u>;
<☐ , inverted <u>overdrive</u>.

of the pTiA6NC TL right border repeat and 16 bp to
the right of the TR right repeat (1). Sequences
sharing 75 to 100 % homology with an 8 bp core

sequence (5´ TGTTTGTT 3´; underlined above) lie to
the right of T-DNA right border repeats
in the nopaline-type Ti plasmid pTiT37 (6) and
the A. rhizogenes Ri plasmid pRiA4 (33). The
deletion which removed 15 bp (including the core
sequence) from the right end of the conserved
sequence also eliminated the stimulatory effect of
flanking sequences on TL right border repeat
function (Figure 1).

Efficiency of Other Border Repeats.

     To determine whether other border repeats
with slightly different sequences function
efficiently without the conserved flanking
sequence, we tested the ability of two other T-DNA
border repeats to promote T-DNA transmission. We
tested a synthetic pTiT37 right border repeat that
lacked Ti plasmid flanking sequences (WR1106) and
a 767 bp restriction fragment containing the
pTiA6NC TL left border repeat (not shown). The
sequences that flank left border repeats do not
contain the conserved sequence found near right
border repeats (1). Each border repeat, without
the specific flanking sequence, promoted T-DNA
transmission poorly (Figure 1).

Synthetic T-DNA Transmission Enhancer: overdrive.

     To determine whether the conserved flanking
region identified by deletion analysis contained
all the sequences required to fully stimulate T-
DNA border repeat function, we synthesized an
oligonucleotide comprising the conserved flanking
sequence found to the right of the pTiA6NC TL
right border repeat (Figure 1). We inserted this
sequence in its normal orientation only 6 bp to
the right of a synthetic pTiT37 right border
repeat (Figure 1). The resulting strain
(WR1107) exhibited much greater virulence than the
parental strain (WR1106) which contained only the
synthetic pTiT37 right border repeat (Figure 1).
This synthetic oligonucleotide enhanced the
activity of a heterologous border repeat when

positioned 10 bp closer than normal to a right
border repeat. Thus, this conserved region
contained the sequences needed to stimulate border
repeat function, and the sequence retained its
function even when moved 10 bp closer than normal
to the border repeat. We call this conserved
flanking sequence overdrive (25).

Orientation Independence of overdrive.

To determine whether its orientation affected
the ability of overdrive to stimulate T-DNA
transmission, we inserted the synthetic overdrive
oligonucleotide in an inverted orientation 23 bp
to the right of the pTiA6NC TL right border repeat
(Figure 1). The resulting strain (WR1111)
contains a right border repeat in its normal
orientation (relative to the T-DNA) flanked on the
right by an inverted overdrive sequence. This
strain (WR1111) exhibited much greater virulence
than the parental strain which lacked overdrive
(Figure 1). Thus, the overdrive sequence
stimulated T-DNA transmission regardless of its
orientation relative to the right border repeat,
and overdrive retained its function when
positioned 7 bp farther than normal from a right
border repeat.

DISCUSSION

Our experiments demonstrate that efficient T-
DNA transmission requires two discrete sequences,
the 23 bp T-DNA right border repeat and a second
sequence (overdrive) lying to the right of right
border repeats. Border repeats and overdrive
apparently constitute separate elements that
interact to promote efficient T-DNA transmission.
These elements presumably do not form a single
contiguous site because: (1) a border repeat can
function inefficiently without overdrive, (2) the
exact spacing and sequence between a border repeat
and overdrive can vary slightly without reducing
border function, and (3) overdrive functions fully
in either orientation whereas the right border

repeat promotes T-DNA transmission best in one direction.

The specific sequence of a border repeat did not restrict it to function as either a "left" or "right" border repeat. The three border repeats tested (the pTiA6NC TL left and right border repeats and the pTiT37 right border repeat) all promoted T-DNA transmission at low efficiency when placed in the active orientation (26, 39) to the right of the T region in our assay system. Thus, a border repeat promoted T-DNA transmission as a function of its location and orientation with respect to the T-DNA; the flanking overdrive sequence strongly influenced the efficiency with which a repeat promoted T-DNA transmission.

Our 24 bp overdrive oligonucleotide contains all the sequences required for wild-type overdrive activity in our assay system, but full activity may require only a portion of this 24 bp sequence. We compared the sequences that flank the four known T-DNA right border repeats and found the 8 bp overdrive core sequence (5´ TGTTTGTT 3´) at slightly different locations to the right of each border repeat. Two mismatches occur within the core sequence that flanks the pTiT37 right border repeat, but the other three right borders (TL and TR of pTiA6NC and TL of pRiA4) each contain the exact core sequence. Comparisons between any two putative overdrive sequences reveal regions of homology longer than the 8 bp core sequence. For example, the pTiA6NC TL and TR right borders share homology in 19 of 24 positions in the overdrive region (5´ TAA-T--CTGT-T-TGTTTGTTTG 3´; core sequence underlined), the TL right borders of pTiA6NC and pRiA4 share homology in 15 of 17 positions in the overdrive region (5´ ATGTTTGTT--ATTGTT 3´), and the right borders of pTiT37 and TR in pTiA6NC share homology in 9 of 10 positions in the overdrive region (5´ ATTTGT-TGT 3´). However, the 8 bp core sequence constitutes the region most highly conserved in all four potential overdrive regions. overdrive activity may require only the 8 bp core sequence, or larger overdrives with somewhat different sequences may function in each plasmid due to variations in vir-encoded proteins. The two pTiA6NC overdrives share the greatest

homology possibly because both interact with
identical vir-encoded proteins. Alternatively,
some vir proteins may promote efficient T-DNA
transmission without an overdrive.

Our results differ from those of Wang et al.
(39): in the nopaline-type plasmid pTiC58, the
nopaline-type right border repeat alone promotes
efficient T-DNA transmission, but in the octopine-
type Ti plasmid pTiA6NC, we observe only partial
activity with either octopine or nopaline-type
right border repeats alone (25-26). Jen and
Chilton (13) identified a region to the right of
the nopaline-type (pTiT37) right border repeat
which exhibited overdrive activity when an
octopine-type Ti plasmid supplied the vir
functions. The region with overdrive activity
includes the core sequence identified in our
study, and the nopaline-type repeat exhibits only
weak activity in their system. Thus, efficient T-
DNA transmission may require overdrive when an
octopine-type Ti plasmid provides the vir genes
but not when a nopaline-type Ti plasmid supplies
the vir functions. Further mutagenesis will
define the sequences important for overdrive
activity in pTiA6NC, and complementation tests
with vir genes from pTiC58 (a nopaline-type Ti
plasmid) will determine whether these vir genes
can render the pTiA6NC right border repeat fully
active without overdrive.

Our experiments, together with data from
other studies, suggest a possible function for
overdrive. Efficient T-DNA transmission requires
only the right border repeat in its wild-type
orientation (26, 39) and the overdrive sequence in
cis (in pTiA6NC; 25-26). overdrive itself does
not promote T-DNA transmission in the absence of a
right border repeat (31), but overdrive greatly
stimulates T-DNA transmission. Apparently, a
directional T-DNA transmission process begins at
the right border and moves leftward through the T
region. Presumably, some of the Ti plasmid-
carried vir (virulence) genes (18), induced by
exudates from wounded plant cells (23, 35-36),
encode proteins which act at the right border
repeat to initiate T-DNA transmission. Since both
left and right border repeats (without overdrive)

exhibit weak unidirectional right border activity in our assay, we propose that <u>overdrive</u> greatly enhances interaction between the right border and the appropriate <u>vir</u> proteins. Therefore, most T-DNA transmission events will initiate at the right border and move leftward through the T-DNA.

Other site-specific recombination systems resemble the T-DNA border repeat-<u>overdrive</u> system. Several genetic recombination pathways, for example phage lambda integration (21) and Tn<u>3</u> resolution (17), require specific flanking sequences in addition to repeat sequences directly involved in a crossover site. Site-specific inversions that mediate tail fiber switching in bacteriophages (e.g. Mu, P1, P7) and variation of <u>Salmonella typhimurium</u> flagellar antigens require inverted repeat sequences and specific flanking sequences in <u>cis</u> for efficient inversion (4, 12, 14, 16, 28). These flanking sequences, called recombinational enhancers, stimulate site-specific inversion independent of their position and orientation relative to the inverted repeats. We will continue to test whether the <u>overdrive</u> sequence can also enhance T-DNA transmission in a similar fashion.

ACKNOWLEDGEMENTS

We thank Drs. M.-D. Chilton, E. Bach, and A. Montoya for providing the pTiT37 right border repeat, Drs. M.- D. Chilton and G. Jen for communicating similar results prior to publication, T. Lynch for expert technical assistance, Joon Ji for help cloning the synthetic <u>overdrive</u> sequence, and Dr. T. Alton for comments on the manuscript.

REFERENCES

1.   Barker R, Idler K, Thompson D, Kemp J (1983). Plant Mol Biol 2:335.
2.   Bevan MW, Chilton M-D (1982). Annu Rev Genet 16:357.
3.   Caplan AB, Van Montagu M, Schell J (1985). J

      Bacteriol 161:655.
4.   Craig NL (1985). Cell 4:649.
5.   DeBeuckeleer M, Lemmers M, DeVos M, Willmitzer
      L, Van Montagu M, Schell J (1981). Mol Gen
      Genet 183:283.
6.   Depicker A, Stachel S, Dhaese P, Zambryski P,
      Goodman H (1982). J Mol Appl Genet 1:561.
7.   DeVos G, DeBeuckeleer M, Van Montagu M, Schell
      J (1981). Plasmid 6:249.
8.   Garfinkel D, Simpson R, Ream L W, White F,
      Gordon M, Nester E (1981). Cell 27:143.
9.   Hepburn A, White J (1985). Plant Mol Biol
      5:3.
10.   Holsters M, Silva B, Van Vliet F, Genetello C,
      DeBlock M, Dhaese P, Depicker A, Inze D,
      Engler G, Villarroel R, Van Montagu M, Schell
      J (1980). Plasmid 3:212.
11.   Holsters M, Villarroel R, Gielen J, Seurinck
      J, DeGreve H, Van Montagu M, Schell J (1983).
      Mol Gen Genet 190:35.
12.   Huber H, Iida S, Arber W, Bickle T (1985).
      Proc Natl Acad Sci USA 82:3776.
13.   Jen G, Chilton, M-D (1986). in preparation.
14.   Johnson RC, Simon MI (1985). Cell 4:781.
15.   Joos H, Inze D, Caplan A, Sormann M, Van
      Montagu M, Schell J (1983). Cell 32:1057.
16.   Kahmann R, Rudt F, Koch C, Mertens G (1985).
      Cell 41:771.
17.   Kitts P, Symington L, Dyson P, Sherratt D
      (1983). EMBO J 2:1055.
18.   Klee HJ, White F, Iyer V, Gordon M, Nester E
      (1983). J Bacteriol 153:878.
19.   Lemmers M, DeBeuckeleer M, Holsters M,
      Zambryski P, Depicker A, Hernalsteens J, Van
      Montagu M, Schell J   (1980). J Mol Biol
      144:353.
20.   Leemans J, Deblaere R, Willmitzer L, DeGreve
      H, Hernalsteens J, Van Montagu M, Schell J
      (1982). EMBO J 1:147.
21.   Nash HA (1981). Annu Rev Genet 15:143.   31.
22.   Nester E, Gordon M, Amasino R, Yanofsky M (1984).
      Annu Rev Plant Physiol 35:387.
23.   Okker R, Spaink H, Hille J, van Brussel T,
      Lugtenberg B, Schilperoort RA (1984). Nature
      312:564.
24   Ooms G, Hooykaas P, Van Veen R, Van Beelen P,

Regensburg-Tunik T, Schilperoort RA (1982). Plasmid 7:15.

25 Peralta E, Hellmiss R, Ream W (1986). EMBO J in press.

26. Peralta EG, Ream LW (1985). Proc Natl Acad Sci USA 82:5112.

27. Peralta E, Ream LW (1985). In Szalay A, Legocki R (eds.): "Advances in the Molecular Genetics of the Bacteria-Plant Interaction," Ithaca: Cornell Univ. Publishers,p 124.

28. Plasterk R, van de Putte P (1985). EMBO J 4:237.

29. Ream LW, Gordon M, Nester E. (1983). Proc Natl Acad Sci USA 80:1660.

30. Ruvkun GB, Ausubel FM (1981). Nature 289:85.

31. Shaw CH, Watson M, Carter G, Shaw C (1984). Nucleic Acids Res 12:6031.

32. Simpson RB, O´Hara P, Kwok W, Montoya A, Lichtenstein C, Gordon M, Nester E (1982). Cell 29:1005.

33. Slightom JL, Durand-Tardif M, Jouanin L, Tepfer D (1986). J Biol Chem 261:108.

34. Slightom JL, Jouanin L, Leach F, Drong RF, Tepfer D (1985). EMBO J 4:3069.

35. Stachel S, An G, Flores C, Nester E (1985). EMBO J 4:891.

36. Stachel S, Messens E, Van Montagu M, Zambryski P (1985). Nature 318:624.

37. Thomashow MF, Nutter R, Montoya A, Gordon M, Nester E (1980). Cell 19:729.

38. Van Larebeke N, Engler G, Holsters M, Van den Elsacker S, Zaenen I, Schilperoort RA, Schell J (1974). Nature 252:169.

39. Wang K, Herrera-Estrella L, Van Montagu M, Zambryski P (1984). Cell 38:455.

40. Yadav N, Vanderleyden J, Bennet D, Barnes W, Chilton M-D (1982). Proc Natl Acad Sci USA 79:6322.

41. Zambryski P, Depicker A, Kruger K, Goodman H (1982). J Mol Appl Genet 1:361.

Molecular Strategies for Crop Protection, pages 157–168
© 1987 Alan R. Liss, Inc.

RHIZOBACTIN, A STRUCTURALLY NOVEL SIDEROPHORE
BIOCHEMICALLY RELATED TO THE OPINES[1]

M. J. Smith[2] and J. B. Neilands

Department of Biochemistry, University of California
Berkeley, California 94720

ABSTRACT The nitrogen-fixing Gram negative bacterium
Rhizobium meliloti DM4 when cultured on low iron media
forms a siderophore which corrects iron starvation in
the microorganism. The compound, N^2-[2-[(1-carboxyeth-
yl)amino]ethyl]-N^6-(3-carboxy-3-hydroxyl-1-oxopropyl)
lysine, is designated rhizobactin. The chemical structure of rhizobactin provides a new biochemical activity
for an N^2-substituted amino acid; namely, iron assimilation. The relatively large quantity of iron required
for microbial metabolism in general and nitrogen fixation in particular suggests that a siderophore-based
iron transport system is essential to the symbiosis between Rhizobium and leguminous plants. This report presents data which reconciles the finding that a rhizobactin-based iron transport system is nonfunctional in
several other 'wild-type' Rh. meliloti strains. To wit,
rhizobactin is biochemically related to certain other
unusual N^2-substituted amino acids termed opines, the
genes for the biosynthesis and utilization of which are
encoded on strain-specific virulence plasmids of Agrobacterium spp. Based on the biochemical homology a
novel biological rationale underlying opine production
is proposed. The fundamental implication of our interpretation is discussed in regard to rhizobacterial-plant
interactions.

[1]We thank Anne Bagg for her critical reading of the man-
uscript. Financial support was provided by NSF Grant No.
PCMB1-16882, NIH Grant No. 5 RO1 AI04156-22, and USDA Grant
No. 83CRCR-1-1303.
[2]Present address: Department of Chemistry, Columbia
University, New York, NY 10027.

INTRODUCTION

At biological pH, environmental iron occurs in an oxidized and polymerized state due to the fact that atmospheric
oxygen rapidly oxidizes iron(II) to iron(III), thereby forming insoluble ferric oxyhydroxide precipitates (1,2), -log
$K_s \doteq 38$. Consequently, many organisms are pressed to obtain
adequate amounts of iron even though it is the second most
abundant metal and fourth most abundant element in the
earth's crust. As a result, when the iron concentration is
less than ca. 1.0 µM or when the iron is bound in a form unavailable to the cell, a high affinity assimilation system
has been found to be expressed in virtually all aerobic and
facultative anaerobic microorganisms carefully examined for
its presence (3). Certain Lactobacilli represent one notable
exception (4).

The system consists of two parts, namely, soluble, relatively low molecular weight, virtually ferric specific ligands, generically termed siderophores (Gr. sidero = iron;
phore = bearer) and the cognate membrane receptor and transport system for the iron laden form of the siderophore (5)
(Figure 1). Based on studies in the Gram negative bacterium
Escherichia coli, both components of the high affinity iron
transport system are coordinately induced and repressed, according to the iron concentration (6,7). Such systems have
evidently been retained through evolutionary time as a device
for assuring survival of microorganisms under conditions
where the iron supply is limiting, such as within host tissue, in certain soils, and in aqueous environments where the
pH approaches, or exceeds, neutrality. The following report
stems from studies on Rhizobium meliloti, a Gram negative bacterium in the family Rhizobiaceae.

RHIZOBACTERIA AND IRON

Both agronomically important genera of rhizobacteria induce hypertrophic cell reactions on certain plants. For
example, plant infections by Agrobacterium tumefaciens or by
A. rhizogenes result in the formation of crown gall tumors
or adventitious (hairy) roots, respectively. Such overgrowths
result from the transfer and expression of a portion of a
bacterial tumor(Ti)- or root(Ri)-inducing virulence plasmid
to the plant cell nucleus (8). The transferred (T-DNA) sequence contains genes which code for enzymes involved in the
biosynthesis of opines (9,10,11), together with others which

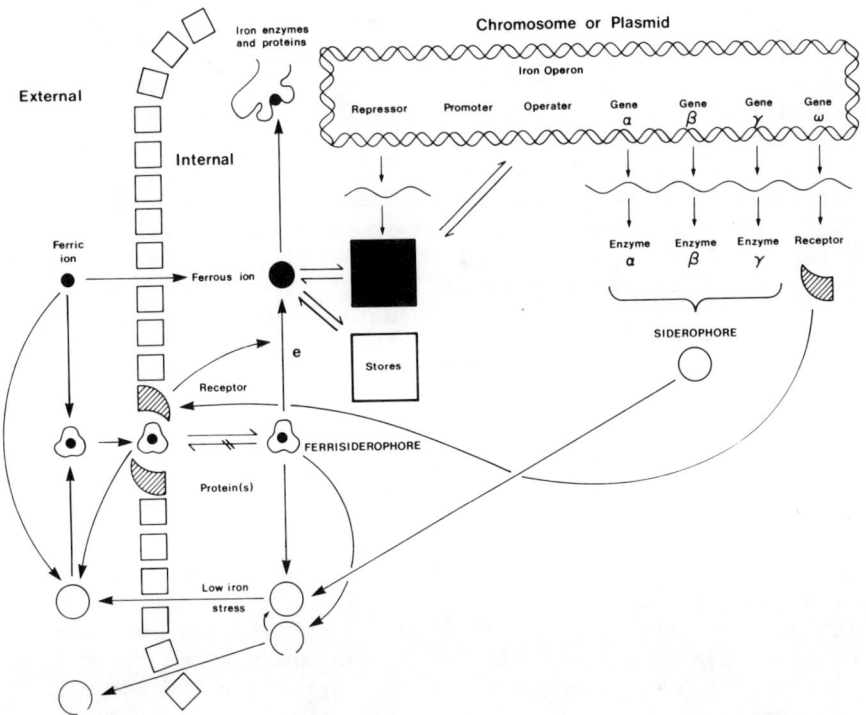

FIGURE 1.   Schematic model of low and high affinity iron assimilation pathways in aerobic and facultative anaerobic microorganisms.   Reproduced from ref. (5).

affect the phytohormone balance (i.e., auxins (12) and cyto-kinins).   In comparison, diverse Rhizobium species contain DNA sequences homologous to the T-DNA region (13) and to the virulence gene region (14) of A. tumefaciens.   By inference, the homologous DNA sequences may include certain genes re-quired for root nodulation (15).   However, it is unlikely that the homologous DNA includes the octopine synthase gene of A. tumefaciens (13).

In addition, Rhizobium spp. are generally capable of re-ducing atmospheric dinitrogen in root nodules.   The bacterial symbionts are thereby forced to acquire iron from the plant for growth and nitrogen fixation.   Ferredoxin, an electron carrier for nitrogen fixation, contains several percent iron. Nitrogenase and hydrogenase each contain more than 30 atoms of iron.   The leghaemoglobin present in nitrogen-fixing root nodules contains a haem moiety and constitutes up to 40% of the total soluble protein (16).

Developing root nodules have been observed to contain dense arrays of the plant iron storage protein phytoferritin. Even though the iron content of the plant tissue may not be limiting per se, the availability of this element influences root nodulation significantly (17). In regard to the iron nutrition of legumes, the phytosiderophore nicotianamine, N-[N-(3-amino-3-carboxypropyl)-3-amino-3-carboxypropyl]-acetidine-2-carboxylic acid, has been isolated from alfalfa (Medicago sativa L.), amongst several other diverse vascular plants (18). It is notable that alfalfa is one symbiotic host of the bacterium which secretes rhizobactin, although nicotianamine itself does not stimulate the growth of the iron-deprived bacterium (19).

RHIZOBACTIN

The large quantities of iron utilized in the Rhizobium-legume symbiosis prompted us to investigate the high affinity iron transport system of Rh. meliloti. The detection and isolation of the atypical siderophore rhizobactin was achieved after replacing the colorimetric assays used to detect phenolate and hydroxamate types of siderophore ligands with an equally sensitive yet less restrictive bacterial iron nutrition bioassay (19). In the bioassay, bacterial growth inhibition resulting from iron sequestration by ethylenediamine-di-(o-hydroxyphenylacetic acid) [log K_f iron(III) $\cong$ 30] (20) is alleviated by strain-specific siderophores and by excess iron (21).

Based on the supposition that a functional high affinity iron transport system would be essential to the aforementioned symbiosis, bacterial iron nutrition bioassays were performed on several additional isolates of Rh. meliloti (19). However, only six of the thirteen strains tested were stimulated by rhizobactin. This result was unanticipated for it indicated that a rhizobactin-based iron transport system was nonfunctional in several other Rh. meliloti strains. We shall return to this point shortly.

The chemical structure of rhizobactin has been characterized using a variety of spectroscopic techniques (22). The compound, N^2-[2-[(1-carboxyethyl)amino]ethyl]-N^6-(3-carboxy-3-hydroxy-1-oxopropyl)lysine, contains ethylenediaminedicarboxyl and α-hydroxycarboxyl moieties as metal-coordinating groups (Figure 2). Rhizobactin forms a stable, six-

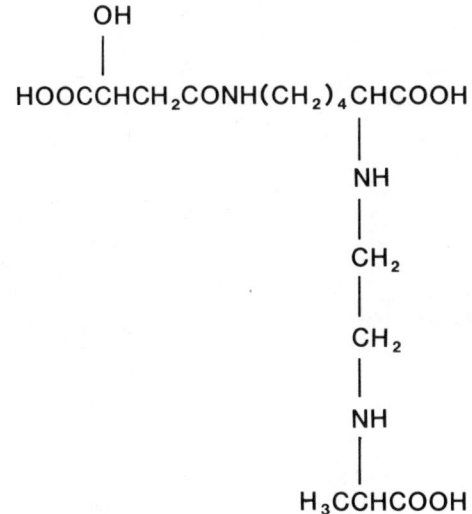

OH
|
HOOCCHCH₂CONH(CH₂)₄CHCOOH
|
NH
|
CH₂
|
CH₂
|
NH
|
H₃CCHCOOH

FIGURE 2.   Rhizobactin (19,22).

coordinate complex with iron(III) the properties of which will be published separately (B. Schwyn and J. B. Neilands, in preparation). The compound's ethylenediamine group, although well known in coordination chemistry, is novel as a natural product (23) and is unprecedented as a ligand in the siderophore series (24). Elucidating rhizobactin's chemical structure unveiled a new biochemical activity for a microbial N^2-substituted amino acid: namely, iron (metal) assimilation. This report compares the basic properties of rhizobactin with those exhibited by the general class of N^2-substituted amino acids described below.

Rhizobactin as Opine.

   N^2-substituted amino acids such as octopine, alanopine, and strombine (Figure 3) are present in the muscle tissue of many marine invertebrates, including species from four or more phyla (25). In addition, compounds falling within the octopine, nopaline, agropine, succinamopine, or leucinopine families have been isolated from axenic plant overgrowths incited by Agrobacterium spp. (26,27,28). In addition, "opine-like" compounds have been isolated from alfalfa root nodules infected by Rhizobium meliloti L 530. A related compound, N-hydroxyiminodipropionic acid occurs in fly agaric (29).

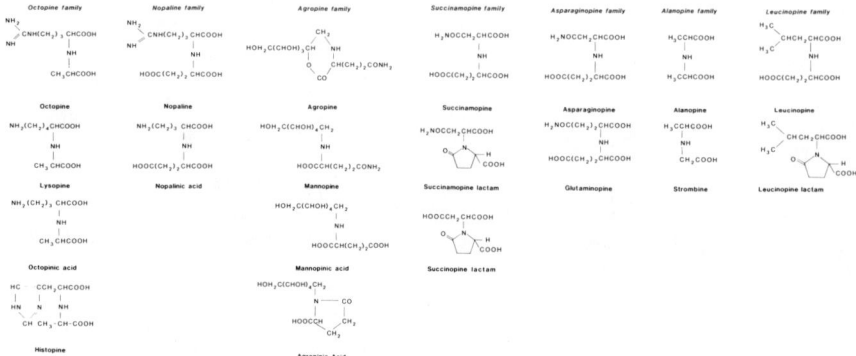

FIGURE 3.   Chemical structures of several N^2-substituted amino acids.   See refs. 25, 26, 27, 28 for reports on the individual compounds.

It is pertinent to the present discussion that the iminodiacetate functional group in the typical opine permits stable, five-membered chelate rings (30) to form with a variety of metal ions.   In fact, Chilton et al. (27) have observed that agropine chelates silver ion.

A general class of imino acid dehydrogenases (e.g., opine synthases), produce many such imino acids (25,31,32).   The enzymes catalyze the reductive imination between an α-ketoacid (e.g., pyruvic acid) and an amino acid.   In view of the chemical homology between rhizobactin and these imino acids, particularly lysopine and strombine, it is likely that rhizobactin biosynthesis requires one or more enzymes with analogous activities.   It has been proposed that marine invertebrate imino acid dehydrogenases, hence imino acid levels, contribute to the modulation of a muscle cell's redox status (33).   The available evidence indicates that the imino acids accumulate during periodic muscle activity as a result of cellular anoxia (i.e., anaerobiosis) and can reach up to 0.1% of the wet weight (25).   In effect, the enzymes appear to substitute for, or compete with, lactate dehydrogenase and thereby modify the NAD^+/NADH ratio (33).   A similar anoxic condition

might arise for rhizobacteria since parasitic infection generally results in a marked respiratory increase in plant tissue (34). Correspondingly, the catabolism of opines is believed to supply the inciting bacterium with an exclusive
source of organic substrates and energy (28,35).

Further, the "opine-like" compounds of alfalfa root nodules are specific growth substrates for Rh. meliloti L 530,
but not for Rh. meliloti RM 41. Correspondingly, rhizobactin
is "opine-like" in structure and in function since it serves
as a specific growth substrate for Rh. meliloti DM4, but not
for Rh. meliloti L 530 (Table I). This analogy reconciles

TABLE 1
GROWTH OF Rh. meliloti STRAINS DM4 AND L 530
ON VARIOUS SUBSTRATES[a]

AT medium supplements	Cell culture absorbancy	
	DM4	L 530
1. None	0.09	0.03
2. 11 mM glucose	0.13	0.02
3. 11 mM glucose 2 mM rhizobactin	0.29	0.06
4. 11 mM glucose 1 mM ammonium sulfate	1.70	0.54
5. 11 mM glucose 15 mM ammonium sulfate	2.22	1.71

[a]Cell cultures (2 ml) were grown overnight in Luria
broth at 30°C and harvested by centrifugation. The cells
were washed with C- and N-free AT medium (36) twice, and shak-
en overnight in 2 ml of AT medium containing 11 mM glucose.
AT medium was amended with filter-sterilized solutions of
thiamine hydrochloride (to 1 mg/l), biotin (to 0.2 mg/l), and
ferrous sulfate (to 5 mg/l; freshly prepared). The nitrogen-
deprived cells were used as inocula for AT media with and
without glucose, ammonium sulfate, and rhizobactin supple-
ments (ca. 10^7 cells per ml). Analytically pure rhizobactin
specimens were prepared as reported previously (19,22). Cell
culture absorbances were determined at 650 nm using a Bausch
& Lomb 710 spectrophotometer. The absorbancies listed represent the average of triplicate samples after 72 hr. The procedure outlined is a modified version of an opine utilization
assay (36).

the finding that rhizobactin was not utilized for iron transport by certain strains of Rh. meliloti. We thus anticipate
that different strains of Rhizobium (meliloti) synthesize/
utilize chemically distinct siderophores, thereby creating an
exclusive source of chelated metals for a particular bacterial population, regardless of the siderophore's catabolic
potential.

Parenthetically, a recent formulation by Buck and Ames
(37) could prove to be germane to our thesis. They propose
that one regulator of electron-acceptor pathways for certain
bacteria during anaerobic versus aerobic growth is the state
of modification of the adenosine A-37 residue in particular
U-type tRNA species. The modification is strictly dependent
upon the availability of oxygen (37) and iron (38), and appears to alter aromatic amino acid transport (39), as well
as the fidelity of the translation apparatus (40).

Concerning opine biosynthesis, even though the essential
structural gene is situated on a virulence plasmid of Agrobacterium spp., it is believed that such compounds are not
elaborated by the free-living bacterium (41,42). The nucleotide sequences of the octopine synthase (43) and nopaline
synthase (44) genes do not appear to contain typical prokaryotic ribosome-binding ('Shine-Dalgarno') sequences (45) upstream from the presumptive AUG translation initiation codons.
This finding has been interpreted as proof that the genes
only have eukaryotic regulatory sequences for expression.
However, Depicker et al. (44) note that it is not known whether agrobacteria utilize a typical (= E. coli) 'Shine-
Dalgarno' sequence or some equivalent sequence. Therefore,
the belief that such compounds are not elaborated by the
free-living bacterium is contestable. In fact, the production of rhizobactin by iron-deprived Rh. meliloti might constitute evidence to the contrary.

One should also note that A. tumefaciens strains B6 and
a B6 derivative cured of pTiB6806 (i.e., A217) produce a
phenolate-type siderophore designated agrobactin, N-[3-(2,3-
dihydroxybenzamido)propyl]-N-[4-(2,3-dihydroxybenzamido)butyl]-
2-(2,3-dihydroxyphenyl)-trans-5-methyl-oxazoline-4-carboxa-
mide (46). Moreover, a survey of 27 different strains of A.
tumefaciens and A. radiobacter plant pathogens indicated that
all but 2 produce phenolate-containing compounds during low-
iron growth (47). The non-phenolate-producing A. radiobacter
84 makes a hydroxamate-like compound, whereas no putative
iron transport growth factor has been detected in A. tumefaciens M3. In the latter instance however, the bioassay
depended upon phenolate utilization by A. tumefaciens XVII[19],

a different agrobactin-deficient strain. Further, virtually all mutants (17/18) of A. tumefaciens A217 which were defective in the synthesis of agrobactin retained their virulent phenotype on sunflower plants and on carrot root discs after pTiB6806 was reintroduced (48). Chromatographic analyses and bioassays have failed to demonstrate the presence of agrobactin in plant tissue infected with A. tumefaciens B6 (48). Given the importance of iron in microbial metabolism, it is likely that agrobactin-producing strains of Agrobacterium synthesize/utilize additional siderophores under particular growth regimes.

SUMMARY

In sum, the structural relatedness of rhizobactin and the opines, the strain-specific patterns of rhizobactin and opine synthesis/utilization, and the evident chelating ability of these N^2-substituted amino acids, lead to the conclusion that rhizobactin is biochemically related to the opines. We propose that some opines possess a hitherto unrecognized biochemical function; namely, iron (metal) assimilation. Further work in this field would refine the opine concept, as defined by Tempé and co-workers; i.e., opines are "substances synthesized by the cells of the host plant in response to a stimulus of the pathogen. Their presence creates favorable environmental conditions for the pathogen and contributes to its dissemination" (41).

REFERENCES

1. Biedermann G, Schindler P (1957). On the solubility product of precipitated iron(III) hydroxide. Acta Chem Scand 11:731.
2. Schwyn B (1983). "The Hydrolysis of Iron(III)." Dissertation, ETH, Zurich, Switzerland, #7404.
3. Neilands JB (1981). Microbial iron compounds. Ann Rev Biochem. 50:715.
4. Archibald F (1983). Lactobacillus plantarum, an organism not requiring iron. FEMS Microbiol Lett 19:29.
5. Neilands JB (1984). Siderophores of bacteria and fungi. Microbiol Sci 1:9.
6. McIntosh MA, Earhart CF (1977). Coordinate regulation by iron of the synthesis of phenolate compounds and three outer membrane proteins in E. coli. J Bacteriol. 131:331.

7. Hantke K (1981). Regulation of ferric ion transport in Escherichia coli K12. Isolation of a constitutive mutant. Mol. Gen Genet 182:288.

8. Chilton M-D, Tepfer DA, Petit A, David C, Casse-Delbart F, Tempé J (1982). Agrobacterium rhizogenes inserts T-DNA into the genomes of the host plant cells. Nature 294:432.

9. Schröder J, Schröder G, Huisman H, Schilperoot RA, Schell J (1981). The mRNA for lysopine dehydrogenase in plant tumor cells is complementary to a Ti-plasmid fragment. FEBS Lett 129:166.

10. Murai N, Kemp JD (1982). Octopine synthase mRNA isolated from sunflower crown gall callus is homologous to the Ti plasmid of Agrobacterium tumefaciens. Proc Natl Acad Sci USA 79:86.

11. Gafni Y, Chilton M-D (1985). Expression of the nopaline synthase gene in Escherichia coli. Gene 39:141.

12. Schröder G, Waffenschmidt S, Weiler EW, Schröder J (1984). The T-region of Ti plasmids codes for an enzyme synthesizing indole-3-acetic acid. Eur.J Biochem 138:387.

13. Hadley RG, Szalay AA (1982). DNA sequences homologous to the T DNA region of Agrobacterium tumefaciens are present in diverse Rhizobium species. Mol Gen Genet 188:361.

14. Prakash RK, Schilperoot RA (1982). Relationship between nif plasmids of fast-growing Rhizobium species and Ti plasmids of Agrobacterium tumefaciens. J Bacteriol 149:1129.

15. Truchet G, Rosenberg C, Vasse J, Julliot J-S, Camut S, Denarie J (1984). Transfer of Rhizobium meliloti pSym genes into Agrobacterium tumefaciens: host specific nodulation by atypical infection. J Bacteriol 157:134.

16. Nash DT, Schulman HM (1976). Leghemoglobins and nitrogenase activity during root nodule development. Can J Bot 54:2790.

17. Lie TA (1974). Environmental effects on nodulation and symbiotic nitrogen fixation. In Quispel A (ed): "The Biology of Nitrogen Fixation," Amsterdam:North-Holland, p. 555.

18. Budesinsky, M, Budzikiewicz H, Prochazka Z, Ripperger H, Romer A, Scholz G, Schreiber K (1980). Nicotianamine, a possible phytosiderophore of general occurrence. Phytochemistry 19:2295.

19. Smith MJ, Neilands JB (1984). Rhizobactin, a siderophore from Rhizobium meliloti. J Plant Nutr 7:449.

20. Kroll H, Knell M, Powers J, Simonian J (1957). A phenolic analog of ethylenediaminetetra acetic acid. J Am Chem Soc 79:2024.

21. Miles AA, Khimji PL (1975). Enterobacterial chelators of iron: their occurrence, detection and relation to pathogenicity. J Med Microbiol 8:477.

22. Smith JM, Shoolery JN, Schwyn B, Holden I, Neilands JB (1985). Rhizobactin, a structurally novel siderophore from Rhizobium meliloti. J Am Chem Soc 107:1739.

23. Wagner I, Musso H (1983). New naturally occurring amino acids. Angew Chem Intl Ed 22:816.

24. Hider RC (1984). Siderophore mediated absorption of iron. Structure & Bonding 58:25.

25. Storey KB, Miller DC, Plaxton WC, Storey JM (1982). Gas liquid chromatography and enzymatic determination of alanopine and strombine in tissues of marine invertebrates. Anal Biochem 125:50.

26. Chang C-C, Chen C-M (1983). Evidence for the presence of N^2-(1,3-dicarboxypropyl)-L-amino acids in crown gall tumors induced by Agrobacterium tumefaciens strains 181 and Eu6. FEBS Lett 162:432.

27. Chilton WS, Tempé J, Matzke M, Chilton M-D (1984). Succinamopine: a new crown gall opine. J Bacteriol 157:357.

28. Petit A, David C, Dahl GA, Ellis JG, Guyon P, Casse-Delbart F and Tempé J (1983). Further extension of the opine concept: plasmids in Agrobacterium rhizogenes cooperate for opine degradation. Mol Gen Genet 190:204.

29. Kneifel H, Bayer E (1973). Determination of the structure of the vanadium compound, amavadine, from fly agaric. Angew Chem Intl Ed 12:508.

30. Martel AE, Calvin M (1952). "Chemistry of Metal Chelate Compounds." New York: Prentice-Hall, p 134.

31. Otten LABM, Vreugdenhil D, Schilperoot RA (1977). Properties of D(+) lysopine dehydrogenase from crown gall tumor tissue. Biochim Biophys Acta 485:268.

32. Hack E, Kemp JD (1980). Purification and characterization of the crown gall specific enzyme, octopine synthase. Plant Physiol 85:949.

33. Fields JHA, Hochachaka PW (1981). Purification and properties of alanopine dehydrogenase from the adductor muscle of the oyster, Crassostrea gigas (Mollusca, Bivalvia). Eur J Biochem 114:615.

34. Davies D (1980). "The Biochemistry of Plants." New York: Academic, p 463.

35. Tempé J, Goldmann A (1982). Occurrence and biosynthesis of opines. In Kahl G, Schell J (eds): "Molecular Biology of Plant Tumors," New York: Academic, p 427.

36. Guyon P, Chilton M-D, Petit A, Tempé J (1980). Agropine
     in "null-type" crown gall tumors: evidence for generality
     of the opine concept. Proc Natl Acad Sci USA 77:2693.

37. Buck M, Ames BN (1984). A modified nucleoside in tRNA as
     s possible regulator of aerobiosis: synthesis of cis-2-
     methyl-thioribosylzeatin in the tRNA of Salmonella.
     Cell 36:523.

38. McLennan BD, Buck M, Humphreys J, Griffiths E (1981).
     Iron related modification of bacterial transfer RNA.
     Nucl Acids Res 9:2629.

39. Buck M, Griffiths E (1981). Regulation of aromatic amino
     acid transport by tRNA; role of 2-methylthio-N^6-(Δ^2-iso-
     pentenyl)adenosine. Nucl Acids Res 9:401.

40. Petrullo LA, Gallagher PJ, Elseviers D (1983). The role
     of 2-methylthio-N^6-isopentenyladenosine in readthrough
     and suppression of nonsense codons in Escherichia coli.
     Mol Gen Genet 190:289.

41. Tempé J, Petit A (1982). Opine utilization by Agrobac-
     terium. In Kahl G, Schell J (eds): "Molecular Biology
     of Plant Tumors." London: Academic p 451.

42. Tempé J (1983). personal communication.

43. Degreve H, Dhaese P, Lemmers M, Van Montagu M, Schell J
     (1982). Nucleotide sequence and transcript map of the
     Agrobacterium tumefaciens Ti plasmid-encoded octopine
     synthase gene. J Mol Appl Genet 1:499.

44. Depicker A, Stachel S, Dhaese P, Zambryski P, Goodman HM
     (1982). Nopaline synthase: transcript mapping and DNA
     sequence. J Mol Appl Genet 1:561.

45. Shine J, Dalgarno L (1975). Determinant of cistron
     specificity in bacterial ribosomes. Nature 254:34.

46. Ong SA, Peterson T, Neilands JB (1979). Agrobactin,
     a siderophore from Agrobacterium tumefaciens. J Biol
     Chem 254:1860.

47. Leong SA, Neilands JB (1982). Siderophore production by
     phytopathogenic microbial species. Arch Biochem
     Biophys 218:351.

48. Leong SA, Neilands JB (1981). Relationship of sidero-
     phore mediated iron assimilation to virulence in
     crown gall disease. J Bacteriol 147:482.

Molecular Strategies for Crop Protection, pages 169–185
© 1987 Alan R. Liss, Inc.

STUDIES ON TWO SETS OF SYMBIOTIC GENES IN RHIZOBIUM, ONE
INVOLVED IN EARLY STAGES OF INFECTION AND THE OTHER IN
EXOPOLYSACCHARIDE SYNTHESIS

A.W.B. Johnston, L. Rossen, C.A. Shearman,
I.J. Evans[1], J.L. Firmin, J.A. Downie[2], J.W. Lamb[3]
and D. Borthakur

John Innes Institute, Colney Lane, NORWICH NR4 7UH, UK

ABSTRACT Two aspects of recent work on symbiotic genes
of two Rhizobium species are presented in this paper.
In R. leguminosarum, the species that nodulates peas,
eight genes, nodABCIJDF and E, required for early stages
of nodulation were identified, analysed and sequenced.
Mutations in nodD,A,B and C abolish nodulation whereas
mutations in nodF,E,I and J only delay the onset of
nodule formation. Analysis of the sequence of these
genes indicates that nodF specifies a protein similar to
acyl carrier protein and that nodI resembles a family of
inner membrane transport proteins in enteric bacteria.
nodD is regulatory; it represses its own expression and,
in the presence of activating flavonoids present in the
exudate of legumes, causes active transcription of the
nodABCIJ and the nodFE operons.
    The second set of symbiotic genes is from R.
phaseoli, the species that nodulates Phaseolus beans. A
gene, psi (polysaccharide inhibition), was located near
nod genes on the symbiotic plasmid pRP2JI. When psi was
cloned in multicopy vectors, exopolysaccharide (EPS)
synthesis was less than 1% of normal and the strains

Present addresses: [1]Apcel Limited, 545/546 Ipswich Road,
Slough SL1 4EQ, Berkshire, UK
[2]Division of Plant Industry, C.S.I.R.O., ACT 2601
Canberra, Australia.
[3]Mikrobiologisches Institute, ETHZentrum LFW, CH-8092
Zurich, Switzerland.

failed to nodulate. Further, psi mutant strains
nodulated but failed to fix nitrogen. It is proposed
that in wild-type strains, psi is expressed only in the
bacteroids and that it functions to repress expression
of genes that are transcribed in the free-living state
but not in the nodule (bacteroids do not normally make
EPS). Another pRP2JI-located gene, psr, (polysaccharide
restoration) was found to inhibit trasncription of psi.
Also, a third gene, pss (polysaccharide synthesis),
which is not on pRP2JI, was identified. Mutations in
pss abolished EPS synthesis; pss::Tn5 mutant strains of
R. phaseoli induced nitrogen-fixing nodules on Phaseolus
beans but the same mutations in R. leguminosarum
abolished nodulaion of peas. These mutations were
corrected by cloned DNA from the phytopathogen
Xanthomonas campestris and the R. leguminosarum pss::Tn5
mutant containing the cloned DNA from Xanthomonas could
nodulate peas.

INTRODUCTION

Bacteria of the genus Rhizobium have the ability of
recognizing and invading legumes and eliciting the host to
make a complex, defined organ, the root (or stem) nodule
in which the bacteria fix nitrogen. The ammonia formed
allows these plants to grow without exogenous fixed nitrogen.
At a molecular level, little is known of what makes
Rhizobium 'special' in its ability to participate in this
interaction but there has been increasing interest in the
genes that allow it to nodulate and to fix nitrogen. Two
different approaches can be taken for such analyses. In one,
genes or mutations involved in the symbiosis are identified
by screening directly for effects on symbiotic phenotype
(also, as it is clear that symbiotic genes of different
Rhizobium species are conserved (see below), a shortcut to
such screening involves the use of probes containing genes of
one species to identify homologues in others). The second
approach is to identify genes or mutations with a scorable
phenotype on Petri dishes and then, secondarily, test their
effects on symbiotic phenotype. The advantage of this method
is that analysis of the biochemical basis of the defect is
simpler than if its only known effect is on the symbiotic
relationship itself.
Here, both these approaches are illustrated. In one, a
description is given of genes from the R. leguminosarum

plasmid, pRL1JI, which specify early stages of infection and which were identified initially by their effects on nodulation. In the second, genes from R. phaseoli, identified first because they affected exopolysaccharide (EPS) synthesis and which affected symbiotic nitrogen fixation are described.

RESULTS

Structure, Function and Regulation of a Cluster of Nodulation Genes on the Symbiotic Plasmid pRL1JI.

The R. leguminosarum symbiotic plasmid pRL1JI, 220kb in size, is transmissible to other Rhizobium species. Transconjugants of (for example) R. phaseoli (which normally nodulates Phaseolus beans) containing pRL1JI can nodulate peas and Vicia, the normal hosts of R. leguminosarum (Johnston et al., 1978). Like symbiotic plasmids in other Rhizobium species pRL1JI also has genes required for nitrogen fixation, several of which correspond to defined nif genes in Klebsiella pneumoniae (see Long, 1984 for review)..
    In pRL1JI, symbiotic genes are located in a 60kb region, the nodulation genes (nod) being located between two sets of nif genes (Downie et al. 1983b). A region of DNA, less than 10kb in size conferred the ability to nodulate peas when transferred to other species of Rhizobium cured of their own symbiotic plasmids (Downie et al., 1983a). In what follows, a description of the structure, function and regulation of the nod genes in this region of pRL1JI will be presented and compared with corresponding genes in other Rhizobium species.
    Identification of nodEFDABCI and J. By sequencing the nodulation region of pRL1JI, characterizing the phenotypes caused by mutations within it and determining the sizes of polypeptides specified by it, eight genes, nodEFDABCI and J were identified (Fig. 1 and Table 1; Downie et al., 1985; Evans and Downie, 1986; Rossen et al., 1984; Shearman et al., 1986).
    As judged by genetic and physical comparisons, many of the nod genes are conserved in Rhizobium species that nodulate different legume hosts. Thus in R. trifolii the sequences of the nodEFDAB and C genes are very similar to those of R. leguminosarum and their positions, relative to each other are essentially identical in these two species (P. Schofield and J. Watson, personal communication); at present it is not known if genes corresponding to nodI and nodJ exist in R. trifolii, but mutations in the equivalent region (i.e. downstream of nodC) in R. trifolii have the same phenotype as

TABLE 1
PROPERTIES OF nodABCIJF AND E GENES

Gene	Phenotype of mutant	Comments on gene product
A	Nod$^-$ Rhc$^-$	–
B	Nod$^-$ Rhc$^-$	–
C	Nod$^-$ Rhc$^-$	Membrane bound
I	Nod-delay Rhc$^+$	Transport protein?
J	Nod-delay Rhc$^+$	Hydrophobic
D	Nod$^-$ Rhc$^-$	Regulatory
F	Nod-delay Rhc$^+$	Similar to acyl carrier protein
E	Nod-delay Rhc$^+$	–

Nod$^-$ = fails to nodulate
Rhc$^-$ = fails to curl root hairs

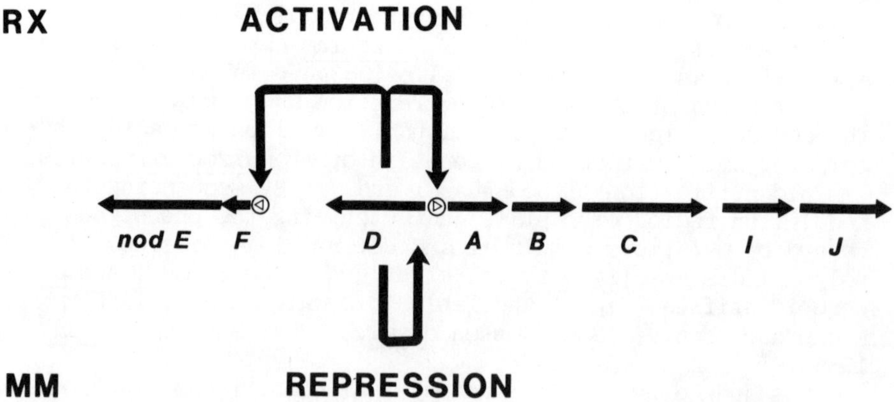

FIGURE 1. Location of eight nod genes in a 9kb region of
pRL1JI. The directions of trancription of the genes are
indicated. The circled open triangles show the location of
the two nod boxes – see also Figure 2 and text. Note that in
minimal medium (MM), nodD represses its own transcription but
that in the presence of root exudate (RX) nodD activates
nodABCIJ and nodFE.

those in nodI and nodJ (Innes et al., 1985; Evans & Downie, 1986). In R. meliloti the nodDABC genes are similar in their sequences and relative positions to the corresponding genes in R. leguminosarum and R. trifolii (Torok et al., 1984; Egelhoff et al., 1985) and nodI appears to start at a position similar to that found in R. leguminosarum (Jacobs et al., 1985). However, two R. meliloti genes that are similar in sequence to nodF and E are located approximately 10kb downstream of R. meliloti nodC (A. Kondorosi, personal communication) in a region that appears to be involved in the determination of host-range specificity (Kondorosi et al., 1984). This similarity in the sequences of nod genes in different Rhizobium species is reflected at a functional level. For example, mutations in nodDAB or C which block nodulation can be functionally corrected by cloned DNA of another species (see for example Fisher et al., 1985). The following provides a brief description of the properties of these nod genes (see also Table 1).

nodABC. These three genes, which are in the same transcript (see below) seem to be sufficient for root hair curling since strains that lack their native symbiotic plasmid but contain nodABC cloned in a wide host-range plasmid are able to induce root hair curling, one of the earliest steps in the infection process (Rossen et al., 1984). Interestingly, wild-type strains of R. leguminosarum containing the nodABC genes cloned in pKT230 are defective in nodulation suggesting that for normal nodulation, regulation of expression of these genes is important (Knight et al., 1986). No precise function has been allocated to these genes but there is good evidence that the nodC polypeptide of R. meliloti is associated with the outer membrane (John et al., 1985).

nodI and J. These genes, which are in the same transcriptional unit as nodABC (see below) also appear to be membrane-bound. Analysis of the deduced nodJ polypeptide showed it to be very hydrophobic and, in a search against polypeptide sequences in a data bank, it was found that the deduced nodI polypeptide was similar in sequence to those of malK, hisP, oppD and pstB (Evans & Downie, 1986). In enteric bacteria these four genes specify inner membrane, ATP-dependent, transport proteins which respectively are required for the transport of maltose, histidine, oligopeptides and phosphate; all four had been found to have sequences in common (Higgins et al., 1985) and it was at these conserved regions that the nodI gene product showed the greatest similarity to the other four.

It is interesting that the effects on nodulation (a two-day delay) of Tn5 insertions in nodI and nodJ (Downie et al., 1985), should be so minor.

nodD. Mutations in nodD abolish nodulation and severely inhibit root hair curling ability (Downie et al., 1985). As described below, nodD is regulatory, being capable of repressing its own transcription and, in the presence of activator molecules from legume exudates, of inducing transcription of the nodABCIJ and nodFE operons.

nodFE. In R. leguminosarum and R. trifolii, mutations in the nodFE region severely inhibit and delay nodulation of peas and clover respectively (Shearman et al., 1986; Innes et al., 1985). Interestingly, in R. trifolii, mutations in nodE actually allow nodulation of host legumes, such as pea, which are normally not nodulated by R. trifolii (Djordjevic et al., 1985). Thus a functional nodE gene of one species inhibits nodulation of heterologous host plants. Given that the nodE genes of R. trifolii and R. leguminosarum are structurally similar, this observation is somewhat unexpected and its molecular basis remains to be determined.

By comparing the deduced polypeptide sequence of R. leguminosarum nodF with sequences in a data bank, a similarity was found between the amino acid sequence of the nodF protein and that of the acyl carrier proteins of E. coli and barley (Shearman et al., 1986). However, the precise biochemical role of nodF remains to be established.

nod gene regulation. Studies on the regulation of R. leguminosarum nod genes have been facilitated by constructing translational fusions in which individual nod genes plus potential regulatory upstream sequences were fused to the E. coli lacZ gene that had been cloned in a wide host-range plasmid vector (Rossen et al., 1985; Shearman et al., 1986; unpublished observation).

In summary the following conclusions were drawn (and see Fig. 1).

(i)   nodD is transcribed as a single gene transcriptional unit and is expressed at relatively high level in free-living culture in minimal medium (Rossen et al., 1985). Such constitutive expression of nodD has also been found in R. meliloti (Mulligan & Long, 1985) and R. trifolii (Innes et al., 1985).

(ii)   In R. leguminosarum, nodD is autoregulatory; transcription of nodD is inhibited by the presence of an intact nodD gene (Rossen et al., 1985). For reasons that are not clear, such autoregulation of nodD was not observed in R. meliloti (Mulligan & Long, 1985).

(iii) nodFE and nodABCIJ are in two separate transcriptional units (Rossen et al., 1985; Shearman et al., 1986; unpublished observations) both of which are expressed only at low levels in free-living culture.

(iv) When R. leguminosarum cells are exposed to exudate obtained from pea roots, the nodFE and nodABCIJ operons were both expressed at high level (a 70-fold increase over background in the case of nodC). This induction was absolutely dependent on the presence of the regulatory nodD gene (Rossen et al., 1985; Shearman et al., 1986). Analogous results have been found in R. meliloti and R. trifolii; in both species, transcription of nodABC requires an activator from root exudate and it was also shown that, in R. meliloti, this induction depended on nodD.

(v) Upstream of the nodF and nodA genes of R. leguminosarum and R. meliloti is a repeated sequence; (Figures 1 and 2). Rossen et al. (1985) showed that a small region of DNA that spanned this sequence upstream of nodA was required for nodD-mediated activation of nodABC. Since transcription of both nodFE and nodABCIJ requires nodD plus a factor in pea exudate, it is possible that the nodD gene product modified by the inducer(s) in root exudate, can interact with these so-called 'nod-boxes' to promote transcription (Shearman et al., 1986). Significantly, similar sequences are also found a upstream of R. meliloti nodA (Mulligan & Long, 1985) and of nodA and nodF in R. trifolii (P. Schofield & J. Watson, personal communication).

(vi) The activators of R. leguminosarum nod genes, shown initially to be present in pea root exudate, were also found in other pea tissues; namely leaves, callus and seeds. nodD-dependent activation of nodFE and nodABCIJ was also found using extracts from several legumes (clover, alfalfa and cowpea) which are not nodulated by R. leguminosarum. It is apparent that there are at least three active components which are separable by HPLC. The chromatographic properties of these inducers indicated that they were phenolic in nature and likely to be flavanoids. Although the precise chemical structures of the authentic inducers from peas have not been completely established it was clear that several commercially available flavanoids could activate expression of nodABCIJ and nodFE. These included the flavones, 4',5,7-trihydroxyflavone "Apigenin" and 3',4',7-trihydroxyflavone; the flavanones,

4',5',7-trihydroxyflavanone "Naringenin", 3',5,7-
trihydroxy-4'methoxyflavanone (Hesperetin) and the
flavanone-glycoside Apigenin 7-0-glycoside.

Conclusions on the nod genes.

Perhaps the most striking feature concerning the
organization, structure and regulation of nod genes is the
degree to which they are conserved in Rhizobium strains with
different host-range specifities. There is now detailed
information on the structure of these genes in three
Rhizobium species and significant insights on how their
expression is regulated. However, the key question of the
way in which the information content in these genes
determines at a biochemical level the ability of Rhizobium to
infect legumes remains unanswered.

Identification of psi, psr and pss, three R. phaseoli genes
that affect exopolysaccharide (EPS) synthesis.

In R. phaseoli, genes for nodulation and nitrogen
fixation are also on large plasmids. Since strains of R.
phaseoli, lacking a symbiotic plasmid can make EPS, the genes
involved in its synthesis must lie elsewhere. However,
recent observations showed that in R. phaseoli strain 8002,
two sets of genes in the symbiotic plasmid pRP2JI are
involved in the regulation of EPS synthesis.
    Characterization of psi. Lamb et al. (1985) constructed
a clone bank of R. phaseoli strain 8002 DNA by ligating 30kb
genomic fragments into the wide host range vector pLAFR1
(Friedman et al., 1982). Two recombinant plasmids, pIJ1097
and pIJ1098 (see Fig. 3) both corrected non-nodulating
mutants of R. phaseoli with deletions in pRP2JI. The DNA
cloned in pIJ1097 and pIJ1098 was from plasmid pRP2JI and
overlapped by 14.9kb. Strains of R. phaseoli containing
pIJ1098 (but not pIJ1097) had a non-mucoid appearance on
complete medium (Borthakur et al., 1985). This suggested
that DNA in pIJ1098 but not pIJ1097 contained a gene or
genes, termed psi (polysaccharide inhibition) which, when
cloned, repressed EPS synthesis. This was confirmed by two
methods. pIJ1098 was mutagenized with the transposon Tn5 and
derivatives which failed to inhibit EPS synthesis were
isolated and mapped; these mutants contained the transposon
in a region of pIJ1098 which was not shared with pIJ1097
(Fig. 3).

FIGURE 2   Conservation of 'nod box' sequences upstream of nodD-activated nod genes

R.leguminosarum nodF A T C C A T A G T G T G G A T G C T T T T G A T C C A C A C A A T C A A T T T T A C C A A T G A T G C
R.leguminosarum nodA A T C C A T T C C A T A G A T G A T T G C C A T C C A A A C A A T C A A T T T T A C C A A T C T T T C
R.meliloti nodA   A T C C A T A T C G C A G A T G A T C G T T A T C C A A A C A A T C A A T T T T A C C A A T C T T G C

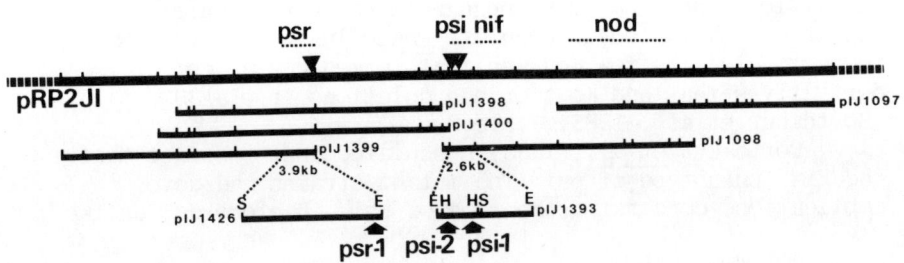

FIGURE 3   Location of psi and psr genes on the symbiotic plasmid pRP2JI and recombinant plasmids containing them. pIJ1399, pIJ1398, pIJ1400, pIJ1098 and pIJ1097 each contain approximately 30kb of pRP2JI cloned in the vector pLAFR1. pIJ1426 contains the psr gene cloned as a 3.9kb SstI (S) to EcoRI fragment cloned in pKT230 and pIJ1397 contains psi genes in a 2.6kb EcoRI fragment cloned in pKT230. H in pIJ1393 refers to HindIII sites EcoRI sites in pRP2JI are indicated by vertical lines. The triangles in pRP2JI indicate the locations of psr::Tn5 and psi::Tn5 mutations; the arrows under pIJ1426 and pIJ1393 show more precisely their positions. The previously identified pRP2JI regions containing nitrogen fixation (nif) or nodulation (nod) genes are shown.

Secondly, EcoRI fragments from pIJ1098 were subcloned into the wide host-range plasmid pKT230 (Bagdasarian et al, 1981); a 2.6kb EcoRI fragment, cloned in either orientation, caused inhibition of EPS synthesis. This fragment was the same one in which the psi::Tn5 mutations were located (Borthakur et al., 1985; Fig. 3).

Effects of psi on symbiotic nitrogen fixation. Strains of R. leguminosarum and R. phaseoli containing plasmid pIJ1393 (the EcoRI fragment containing psi gene(s) cloned in pKT230 - see Fig. 3) were constructed and used to inoculate peas and Phaseolus beans respectively. Also, the psi-1::Tn5 and psi-2::Tn5 mutations were transferred by marker exchange (Ruvkun & Ausubel, 1981) from their locations on pIJ1098 to the corresponding positions on the symbiotic plasmid pRP2JI. These psi::Tn5 mutant strains made normal amounts of EPS in free-living culture. They nodulated Phaseolus, but the nodules failed to fix nitrogen. In contrast, the R. phaseoli and R. leguminosarum transconjugants containing multiple copies of psi failed to nodulate Phaseolus beans or peas respectively; the few nodules that formed contained derivatives that had lost or had deletions in pIJ1393 (Borthakur et al., 1985; Table 2).

Borthakur et al. (1985) speculated that the fact that the Fix⁻ phenotype of psi::Tn5 mutant strains had novel implications concerning the role of psi. Bacteroids, unlike

TABLE 2
EFFECTS OF MULTIPLE COPIES OF psi AND psi MUTATIONS ON
SYMBIOTIC NITROGEN FIXATION

psi status	R. leguminosarum	R. phaseoli
psi::Tn5	Not tested	Fix⁻ on Phaseolus
multicopy psi (pIJ1393)	Nod⁻ on peas	Nod⁻ on Phaseolus

Fix⁻ = Fails to fix nitrogen
Nod⁻ = Fails to nodulate

the free-living form of Rhizobium, do not make EPS (Tully and
Terry, 1985). It was suggested that psi functions to inhibit
expression of genes that are transcribed in the free-living
state but not in the nodule. Thus, psi itself would be 'off'
in the free-living state and 'on' in the nodule. By cloning
psi, it would escape its normal control and its effects
are thus seen in free-living culture. In this model,
psi::Tn5 mutants would not repress EPS synthesis in the
bacteroid; this might account for their Fix⁻ phenotype.
Interestingly, a reduction in oxygen tension reduces
synthesis of EPS in the symbiont of soybeans (Tully and
Terry, 1985); since the nitrogen-fixing zone of the nodule is
microaerobic, expression and/or activity of psi may be
influenced by oxygen tension.

Identification of psr, a gene which represses transcription
of psi. Wild-type strains of R. phaseoli containing psi
cloned in pIJ1098 made some EPS but derivatives (e.g. strain
8401, Lamb et al., 1982) lacking the symbiotic plasmid pRP2JI
were completely non-mucoid (Borthakur et al., 1985). This
suggested that pRP2JI contained a gene or genes that
counteracted the inhibition of EPS synthesis in strains
carrying multicopy psi. This was confirmed.
    The wild-type R. phaseoli strain 8002 containing pIJ1393
(psi cloned in pKT230) was used as a recipient in a cross in
which the plasmids comprising the R. phaseoli clone bank were
mobilized en masse from E. coli. and transconjugants were
screened for any that were mucoid. Three such colonies were
found and the recombinant plasmids were isolated from them.
    Restriction mapping and hybridization experiments showed
that these three plasmids, pIJ1398, pIJ1399 and pIJ1400
contained DNA exclusively from pRP2JI, that they overlapped
with each other, and that pIJ1400 shared a 0.8kb fragment
with the previously identified pIJ1098 (see Fig. 3).
    The term, psr (polysaccharide restroration) was given
the gene, cloned in these three plasmids, which overcame the
inhibition of EPS synthesis in strains carrying multicopy
psi. The DNA containing psr was shown as follows to be
approximately 13kb from psi (see Fig. 3). EcoRI fragments of
pIJ1399 were subcloned into pKT230 and the resultant
plasmids were mobilized into Rhizobium containing pIJ1098
(psi cloned in pLAFR1). One such derivative, termed pIJ1415,
containing a 9kb EcoRI fragment restored EPS synthesis to
this strain. Also, pIJ1399 was mutagenized with Tn5 and a
psr::Tn5 mutation was isolated; the Tn5 was located in the
same EcoRI fragment cloned in pIJ1415 (see Fig. 3).

The psr::Tn5 mutation was transferred by marker exchange to the corresponding location in pRP2JI. The colony morphology of this strain was slightly less mucoid than the wild-type but the strains induced apparently normal nitrogen-fixing nodules on Phaseolus beans. Further, it was found that the presence of multicopy psr not only overcame the EPS defect in strains with multiple copies of psi but also restored their ability to nodulate Phaseolus beans (see Table 3).

psr reduces the transcription of psi. One explanation for the behaviour of psr is that it represses the transcription of psi. This explanation was confirmed using the lac transcriptional fusion transposon Tn3 lac (Stachel et al., 1985).

Plasmid pIJ1098 was mutagenized with Tn3lac and two psi::Tn3lac mutations were isolated. Both were in the 2.6kb EcoRI fragment in which psi had been located and their orientations were such that lacZ was transcribed from right to left as drawn in Fig. 3. These mutant derivatives of pIJ1098 were mobilized into the cured R. phaseoli strain 8401 and also into a derivative containing pIJ1415 (psr cloned in pKT230). In the cured strain, psi was actively expressed (500 units of β-galactosidase), but in strain with multicopy psr β-galactosidase activity was near background level (50 units). This confirmed that psr can somehow inhibit the transcription of psi.

Identification of pss, a gene required for EPS synthesis. In the experiment used to isolate the recombinant plasmids containing psr (i.e. in which the plasmids containing the R. phaseoli clone bank were introduced into a strain of Rhizobium containing multicopy psi), some colonies became slightly mucoid after 5–6 days. A recombinant plasmid, pIJ1427 was isolated from one such colony.

By DNA hybridization experiments it was shown that, unlike psi and psr, the DNA cloned in pIJ1427 was not from the symbiotic plasmid pRP2JI. The DNA in pIJ1427 responsible for overcoming the inhibitory effects of multicopy psi was localised; a 3.5kb EcoRI fragment from it was cloned into pKT230 to form the recombinant plasmid, pIJ1481 which, when transferred to a Rhizobium strain carrying pIJ1098 (psi cloned in pLAFR1), restored the mucoid phenotype. Also, Tn5-induced mutations in pIJ1427 which abolished its ability to overcome the inhibitory effect on EPS synthesis caused by multicopy psi were isolated; these mutations were located in this same 3.5kb EcoRI fragment. For reasons given below, the gene(s) in this region of pIJ1427 was termed pss (polysaccharide synthesis).

Two pss::Tn5 mutations in pIJ1427 were transferred by marker exchange into two near-isogenic strains which differed only in the identify of their symbiotic plasmids. One was the wild-type R. phaseoli strain 8002; the other was a strain of R. leguminosarum which was derived from strain 8002 in two steps. First, a derivative of strain 8002 cured of the symbiotic plasmid pRP2JI was isolated. Then into this strain, termed 8401, was introduced pRL1JI, a transmissible symbiotic plasmid that specifies the ability to nodulate peas. Thus, de facto, strain 8401 pRL1JI (Downie et al., 1983a) is a strain of R. leguminosarum.

The pss::Tn5 mutant derivatives of both the R. phaseoli and R. leguminosarum strains were non-mucoid showing that gene(s) cloned in pIJ1427 are required for EPS synthesis. One possible explanation for the finding that multicopy pss overcomes the inhibitory effects on EPS synthesis due to multicopy psi is that psi could inhibit the expression of pss or somehow interfere with the activity of the pss gene product. Thus, by cloning pss on a multicopy plasmid, the copy numbers of psi and pss would be restored to a ratio approaching that in the normal wild type strains; hence strains carrying both psi and pss in multiple copies could make EPS.

Symbiotic Phenotypes of pss::Tn5 Mutant Strains and Correction of Defects by Cloned Xanthomonas DNA.

Although the pss::Tn5 mutant derivatives of R. phaseoli and R. leguminosarum were both non-mucoid, there was a striking difference in their symbiotic phenotypes; whereas the R. phaseoli Pss⁻ strains induced normal nitrogen-fixing nodules on Phaseolus beans, the R. leguminosarum pss::Tn5 mutant strains failed to nodulate peas.

Other studies have shown that different mutant strains of several Rhizobium species which are defective in EPS synthesis are variously abolished in nodulation (Sanders et al., 1981), in nitrogen fixation (Chakravorty et al., 1982; Leigh et al., 1985) or apparently have no effect on the symbiosis (Sanders et al., 1981). The finding here that the same mutation that prevents EPS synthesis in both R. phaseoli and R. leguminosarum, yet only abolishes nodulation ability in the latter, adds another dimension to the complexity and subtlety of the role of EPS in the nodulation process (Borthakur et al., 1986).

As expected, the reintroduction of pIJ1427 into the pss::Tn5 mutant strains of R. phaseoli and R. leguminosarum restored their ability to make EPS and, for the latter species, to nodulate peas. Of more interest was the fact that a pLAFR1-based recombinant plasmid, pIJ3040, containing DNA from the phytopathogen Xanthomonas campestris pv. campestris (which causes black-rot on crucifers) can also correct both the EPS and the symbiotic defects of these mutant strains (Borthakur et al., 1986).

Daniels et al. (1984) constructed a clone bank of X.c. pv. campestris DNA in pLAFR1, and by mobilizing the pooled recombinant plasmids into a mutant strain of Xanthomonas defective in EPS synthesis, detected a plasmid, pIJ3040, which conferred a mucoid phenotype to the mutant. Plasmid pIJ3040 was mobilized from E. coli into the R. phaseoli pss::Tn5 and R. leguminosarum pss::Tn5 mutants. For both species, the transconjugants became mucoid and in the case of R. leguminosarum, induced normal numbers of nitrogen-fixing nodules on peas. Thus at least one of the steps involved in

TABLE 3

EFFECTS OF MUTATIONS AND MULTIPLE COPIES OF psi, psr AND pss ON EPS SYNTHESIS AND SYMBIOTIC NITROGEN FIXATION

psi, psr or pss status	R. phaseoli[+]	R. leguminosarum[+]
mc psi	EPS$^-$ Nod$^-$	EPS$^-$ Nod$^-$
mc psi + mc psr	EPS$^+$ Nod$^+$	NT
mc psi + mc pss	EPS$^{\pm}$ Nod$^+$	NT
psi::Tn5	EPS$^+$ Fix$^+$	NT
psr::Tn5	EPS$^-$ Fix$^+$	NT
pss::Tn5	EPS$^+$ Fix$^+$	EPS$^+$ Nod$^-$
pss::Tn5 + mc pss	EPS$^+$ Fix$^+$	EPS$^+$ Fix$^+$

[+]R. phaseoli and R. leguminosarum were used to inoculate Phaseolus beans and peas respectively
mc = multiple copy, genes cloned in vectors pLAFR1 or pKT230
NT = not tested
EPS$^{\pm}$ = semi-mucoid

the synthesis of EPS is functionally equivalent in
Xanthomonas and Rhizobium; indeed the chemical structure of
the EPS molecules are similar in the two genera, both
containing a backbone with the same 1→ 4-linked glucan.
    Conclusions on the genetic studies on EPS synthesis.
The data presented here (see Table 3) stemmed from a chance
observation that DNA, close to the nodulation region of R.
phaseoli, when cloned, inhibits EPS synthesis. We suggest
that psi plays a central role in determining the transition
of Rhizobium from the free-living state into the bacteroid
form and that normally is expressed only in the latter state.
Just as there are genes, which are not expressed in the
free-living state but which are actively transcribed in the
nodule (the most obvious being the nif genes), so there may
be others (e.g. for EPS synthesis) in which the reverse is
true. We would suggest that psi serves to switch off such
genes in the bacteroid. Moreover, it is apparent that psi is
itself subject to regulation, its transcription being
repressed by psr, another gene on the symbiotic plasmid of R.
phaseoli. Could the purpose of psr be to repress genes (e.g.
psi or more speculatively nif genes) that may be 'off' in the
free-living state but 'on' in the nodule?

REFERENCES

1.   Bagdasarian M, Lurz R, Ruckert B, Franklin FCH,
      Bagdasarian MM, Frey J, Timmis KN (1981). Specific-
      purpose plasmid cloning vectors. II. Broad host-range,
      high copy numbers, RSF1010-derived vectors, and a host-
      vector system for gene cloning in Pseudomonas. Gene
      16:237-347.
2.   Beynon JL, Beringer JE, Johnston AWB (1980). Plasmids
      and host range in Rhizobium leguminosarum and Rhizobium
      phaseoli. J Gen Microbiol 120:421-429.
3.   Borthakur D, Downie JA, Johnston AWB, Lamb JW (1985).
      psi, a plasmid-linked Rhizobium phaseoli gene that
      inhibits exopolysaccharide production and which is
      required for symbiotic nitrogen fixation. Mol Gen Genet
      200:278-282.
4.   Borthakur D, Barber CE, Lamb JW, Daniels MJ, Downie JA,
      Johnston AWB (1986). A mutation that blocks exopoly-
      saccharide synthesis prevents nodulation of peas by
      Rhizobium leguminosarum but not of beans by R. phaseoli
      and is corrected by cloned DNA from Rhizobium or the
      phytopathogen Xanthomonas. Mol Gen Genet (in press).

5.   Chakravorty AK, Zurkowski W, Shine J, Rolfe BG (1982).
     Symbiotic nitrogen fixation: molecular cloning of
     Rhizobium genes involved in enopolysaccharide synthesis
     and effective nodulation. J Mol Appl Genet 1:585-596.
6.   Daniels MJ, Barber CE, Turner PC, Sawzyc MK, Byrde RJW,
     Fielding AH (1984). Cloning of genes involved in
     pathogenicity of Xanthomonas campestris pv. campestris
     using the broad host range cosmid pLAFR1. EMBO J
     3:3323-3328.
7.   Djordjevic MA, Schofield PR, Rolfe BG (1985). Tn5
     mutagenesis of Rhizobium trifolii host-specific
     nodulation genes result in mutants with altered host-
     range specificity. Mol Gen Genet 200:463-471.
8.   Downie JA, Hombrecher G, Ma Q-S, Knight CD, Wells B,
     Johnston AWB (1983a). Cloned nodulation genes of
     Rhizobium leguminosarum determine host range
     specificity. Mol Gen Genet 190:359-365.
9.   Downie JA, Knight CD, Johnston AWB, Rossen L (1985).
     Identification of genes and gene products involved in
     the nodulation of peas by Rhizobium leguminosarum.
     Molec Gen Genet 198:255-262.
10. Downie JA, Ma Q-S, Knight CD, Hombrecher G, Johnston AWB
     (1983b). Cloning of the symbiotic region of Rhizobium
     leguminosarum; the nodulation genes are between the
     nitrogenase genes and a nifA-like gene. EMBO J
     2:947-952.
11. Egelhoff TT, Fisher RF, Jacobs TW, Mulligan JT, Long SR
     (1985). Nucleotide sequence of Rhizobium meliloti 1021
     nodulation genes: nodD is transcribed divergently from
     nodABC. DNA 4:241-248.
12. Evans IJ, Downie JA (1986). The nodI gene product of
     Rhizobium leguminosarum is closely related to
     ATP-binding bacterial transport proteins; nucleotide
     sequence analysis of the nodI and nodJ genes. Gene (in
     press).
13. Fisher RT, Tu JK, Long SR (1985). Conserved nodulation
     genes in Rhizobium meliloti and Rhizobium trifolii.
     Appl Env Microbiol 49:1432-1435.
14. Friedman AM, Long SR, Brown SE, Buikema WJ, Ausubel FM
     (1982). Construction of a broad host range cosmid
     cloning vector and its use in the genetic analysis of
     Rhizobium mutants. Gene 18:289-296.
15. Higgins CF, Hiles ID, Whalley K, Jamieson DJ (1985).
     Nucleotide binding by membrane components of bacterial
     periplasmic binding protein-dependent transport systems.
     EMBO J 4:1023-1040.

16. Innes RW, Kuempel PL, Plazinski J, Cander-Cremens H, Rolfe BG, Djordjevic MA (1985). Plant factors induce expression of nodulation and host-range genes in Rhizobium trifolii. Molec Gen Genet 201:426-432.

17. Jacobs TN, Egelhoff TT, Long SR (1985). Physical and genetic map of a Rhizobium meliloti nodulation gene region and nucleotide sequence of nodC. J. Bacteriol. 162:469-476.

18. John M, Schmidt J, Wieneke V, Kondorosi E, Kondorosi A, Schell J (1985). Expression of the nodulation gene nodC of Rhizobium meliloti in Escherichia coli: role of the nodC gene product in nodulation. EMBO J 4:2425-2430.

19. Johnston AWB, Beynon JL, Buchanan-Woolston AV, Setchell SM, Hirsch PR, Johnston AWB (1978). High frequency transfer of nodulation ability between strains and species of Rhizobium. Nature 276:634-636.

20. Knight CD, Rossen L, Robertson JG, Wells B, Downie JA (1986). Inhibition of nodulation by multicopy nodABC in Rhizobium leguminosarum and analysis of early stages of plant infection. J Bacteriol 166:552-558.

21. Kondorosi E, Banfalvi Z, Kondorosi A (1984). Physical and genetic analysis of a symbiotic region of Rhizobium meliloti: identification of nodulation genes. Mol Gen Genet 193:445-452.

22. Lamb JW, Hombrecher G, Johnston AWB (1982). Plasmid determined nodulation and nitrogen fixation abilities in Rhizobium phaseoli. Mol Gen Genet 186:449-452.

23. Lamb JW, Downie JA, Johnston AWB (1985). Cloning of the nodulation (nod) genes of Rhizobium phaseoli and their homology to R. leguminosarum nod genes. Gene 34:235-241.

24. Leigh JA, Signer ER, Walker GC (1985). Enopoly-saccharide-deficient mutants of Rhizobium meliloti that form ineffective nodules. Proc Natl Acad Sci USA 82:6231-6235.

25. Long SR (1984). Genetics of Rhizobium nodulation. In Kosage T, Nester EW (eds): "Plant Microbe Interactions, Vol 1, Molecular and Genetic Aspects", New York: MacMillan.

26. Mulligan JT, Long SR (1985). Induction of Rhizobium meliloti nodC expression by plant exudate requires nodD. Proc Natl Acad Sci USA 82:6609-6613.

27. Rossen L, Johnston AWB, Downie JA (1984b). DNA sequence of the Rhizobium leguminosarum nodulation genes nodAB and C required for root hair curling. Nucl Acids Res 12:9497-9508.

28. Rossen L, Shearman CA, Johnston AWB, Downie JA (1985). The nodD of Rhizobium leguminosarum is autoregulatory and in the presence of plant exudate induces the nodABC genes. EMBO J 13:3369-3373.
29. Ruvkun GB, Ausubel FM (1981). A general method for site-directed mutagenesis in procaryotes. Nature 289:85-88.
30. Sanders R, Raleigh E, Signer E (1981). Lack of correlation between entracellular polysaccharide and nodulation ability in Rhizobium. Nature 292:148-149.
31. Shearman CA, Rossen L, Johnston AWB, Downie JA (1986). The Rhizobium leguminosarum gene nodF encodes a polypeptide similar to acyl-carrier protein and is regulated by nodD plus a factor in pea root exudate. EMBO J (in press).
32. Stachel ES, Gynhcung A, Flores C, Nester EW (1985). A Tn3 lacZ transposon for the random generation of B-galactosidase gene fusions: application to the analysis of gene expression in Agrobacterium. EMBO J 4:891-898.
33. Torok I, Kondorosi E, Slepkowski T, Pasfai J, Kondorosi A (1984). Nucleotide sequence of Rhizobium meliloti nodulation genes. Nucl. Acids Res. 12:9509-9524.
34. Tully R, Terry ME (1985). Decreased enopolysaccharide synthesis by Anaerobic and symbiotic cells of Bradyrhizobium japonicum. Plant Physiol 79:445-450.

Molecular Strategies for Crop Protection, pages 187–191
© 1987 Alan R. Liss, Inc.

WORKSHOP SUMMARY: REGULATION OF FUNGAL PATHOGENICITY
AND HOST RESISTANCE

H. D. Sisler

Department of Botany,
University of Maryland,
College Park, Maryland 20742

This workshop focused mainly on the manipulation of host
defense systems or of fungal pathogenicity mechanisms in ways
that result in disease control or reduction of disease severity. When the use of chemicals or other agents were discussed, the focus was on novel actions that lead to pathogen
control through host resistance systems or suppression of
pathogenic mechanisms and possible exclusion of the pathogen
by microbial competition.

E. W. B. Ward discussed the manipulation of basic host
resistance mechanisms by compounds that are not toxic in the
normal sense and that are active only at the sites of host-
pathogen interaction. Compounds of this type might act by
preventing the fungal pathogen from avoiding or defeating
the triggering mechanism for host resistance or by preventing the pathogen from suppressing the defense mechanism.
With such compounds, either the host or pathogen could be
the primary target, but antifungal action would be based on
host resistance systems. Dr. Ward discussed several compounds reported to control fungal diseases by indirect
mechanisms. Among these were 2,2-dichloro-3,3-dimethyl-
cyclopropane carboxylic acid and probenazole both of which
are reported to control rice blast disease caused by
Pyricularia oryzae through stimulation of host defense responses. When these compounds are present in host tissue,
the response to challenge by the pathogen is similar to that
ordinarily observed in incompatible host/parasite interactions. The specific basis for the action of these compounds has not been resolved.

The disease control activity of another compound,
aluminum tris-O-ethylphosphonate, is believed by some
investigators to be based on stimulation of host defenses
because a hypersensitive reaction results with accumulation
of phytoalexin when tissue containing this compound is challenged by the pathogen. Other investigators contend that
direct fungitoxicity is the mode of action of aluminum tris-
O-ethylphosphonate. The compound controls several diseases

caused by Peronosporales. Phosphorous acid released from the compound is probably responsible for disease control activity.

Systemic fungitoxic compounds are often used to eradicate fungi within plant tissues several days after infection has occurred. The role, if any, of host participation in pathogen eradication is not known in most cases, but may be quite important in the practical success of some fungicides. As an example, Ward discussed the acylalanine fungicide, metalaxyl, which is highly active in vitro and in vivo against various Peronosporales. Disease control by metalaxyl in some cases appears to be better than would be expected from the fungitoxic activity of the compound toward the pathogen in vitro. At marginally inhibiting concentrations of metalaxyl for control of <u>Phytophthora megasperma</u> in soybean, the level of glyceollin and metalaxyl in lesions both exceeded in vitro ED_{90} levels. Metalaxyl was not as effective when plants were heat shocked or treated with glyphosate, which inhibits accumulation of glyceollin. Therefore, it appears that the host plant does contribute to disease control at marginally inhibitory concentrations of metalaxyl. An explanation of these results is that the action of metalaxyl on the pathogen may allow time for the host to develop its defenses or it may prevent the pathogen from producing products which suppress these defenses.

H. D. Sisler discussed control of pathogenicity in <u>Pyricularia oryzae</u> and <u>Colletotrichum</u> species by chemical or genetic blocks in the pentaketide pathway to melanin. This pathway plays a critical role in appressorial mediated penetration of epidermal barriers by <u>Pyricularia</u> and <u>Colleto-</u><u>trichum</u> species. After initial cyclization of the pentaketide to form 1,3,6,8-tetrahydroxynaphthalene (1,3,6,8-THN), present evidence indicates the following pathway to melanin in these fungi: 1,3,6,8-THN $\xrightarrow{\{H\}}$ scytalone $\xrightarrow{-\{H_2O\}}$ 1,3,8-trihydroxynaphthalene $\xrightarrow{\{H\}}$ vermelone $\xrightarrow{-\{H_2O\}}$ 1,8-dihydroxynaphthalene (1,8-DHN). The 1,8-DHN is then converted to melanin precursors and melanin. These products may cross-link appressorial wall polymers or infiltrate between the wall fibrils. The resulting modifications of the appressorial wall appear to confer the rigidity and architecture needed for epidermal penetration.

Genetic or chemical blocks in the pathway prior to the first cyclized pentaketide, 1,3,6,8-THN, results in albino strains with unmelanized appressoria that fail to penetrate epidermal or nitrocellulose barriers in the case of either

Pyricularia or Colletotrichum species. In the case of
Colletotrichum species, albino appressoria germinate to form
hyphae which grow laterally along the barrier surface.
Lateral germination due to lack of wall melanization may be
totally responsible for penetration failure in Colletotrichum species. Melanization and penetration in non-pig-
mented appressoria of C. lagenarium can be restored by
scytalone or vermelone.

Lateral germination is not typical of appressoria of
albino strains of Pyricularia; however, these fail to penetrate (or melanize) unless supplemented with scytalone,
vermelone or 1,8-DHN. In buff (tan) mutants of P. oryzae
and in wild type P. oryzae treated with specific melanin
biosynthesis inhibitors (MBI) such as tricyclazole
pyroquilon or fthalide, the NADPH dependent reduction of
1,3,8-trihydroxynaphthalene to vermelone is blocked. These
appressoria fail to melanize or penetrate epidermal barriers
unless supplemented with vermelone or 1,8-DHN. Scytalone
does not restore melanization or penetration in these or
buff appressoria or in albino appressoria treated with MBI
inhibitors. Both melanization and penetration can be restored by 1,8-DHN in albino appressoria treated with 0.1,
1.0 or 10 µg/ml of the MBI, tricyclazole or fthalide. In
wild type appressoria, restoration of melanization by 1,8-
DHN is successful at these concentrations of MBI, whereas
restoration of penetration is good at 0.1 µg/ml but poor or
unsuccessful at 1.0 and 10 µg/ml. Failure to restore penetration at higher MBI concentrations is believed to be due
to accumulation of pentaketide metabolites which inhibit
penetration. This phenomenon is believed to result from
secondary blocks by tricyclazole and fthalide between
1,3,6,8-THN and scytalone or in detoxication pathways which
branch off the main melanin biosynthetic pathway at 1,3,6,8-
THN and at 1,3,8-THN. Thus, in P. oryzae, blocking the
polyketide pathway to melanin may result in a mechanism of
antipenetrant action based on a deficiency of melanin or of
melanin precursors and another based on the accumulation of
inhibitory polyketide metabolites. Additional studies are
needed to clarify the precise role that appressorial wall
melanization or late metabolites of the pathway play in the
appressorial penetration process and to identify the inhibition sites and accumulated metabolites associated with the
action of higher concentrations of MBI.

C. P. Paul discussed prospects for reduction of virulence in plant pathogens by cytoplasmically transmissible
agents. The natural occurrence of hypovirulent strains of

Endothia parasitica has allowed the survival of chestnut
trees in Europe and in Michigan. The introduction of hypovirulent strains into areas infected with virulent strains
of Endothia has been successful in controlling chestnut
blight in Europe. For successful establishment of hypovirulent strains, the strains must not be so severely debili-
tated that they cannot spread in the field. They must also
be able to transmit the hypovirulent phenotype to the virulent strains already present in the infected area.

Using a genetically marked strain containing double
stranded RNA (dsRNA) it has been shown that dsRNA can spread
within a stand of chestnut trees and that the dsRNA can be
transferred to the naturally occurring virulent strains in
the stand.

The presence of dsRNA in Endothia correlated with the
hypovirulent phenotype, but the mechanism by which dsRNA may
cause hypovirulence is unknown. The dsRNA in hypovirulent
strains exhibits a variety of banding patterns on polyacrylamide gels. Using the dsRNA from two Michigan hypovirulent strains as probes to blots of dsRNA from other strains,
it was shown that dsRNA from one Michigan strain (GH2) shares
homology with dsRNA from all Michigan strains tested, with
the exception of dsRNA from strain RC1. When RC1 dsRNA is
used as the probe, it does not share homology with dsRNA from
any other Michigan strains. Neither GH2 nor RC1 dsRNA share
homology with dsRNA from other states or Europe. The homology found among dsRNAs from one region suggests that natural
spread of dsRNA has occurred. The lack of homology between
dsRNAs from different areas would seem to indicate that dsRNA
may cause hypovirulence by a nonspecific mechanism. However,
dsRNAs found in other fungi do not cause a decrease in virulence. The dsRNAs found in some fungi are known to produce
specific protein products. Virulence is a complex trait and
the possibility for more than one type of hypovirulence
exists. Until the mechanism of hypovirulence is understood,
its value in controlling other plant pathogenic fungi cannot
be evaluated. However, evidence indicates that hypovirulent
strains of Endothia can be used successfully to control
chestnut blight.

Elias Shahin discussed the use of fusaric acid, a non-
specific phytotoxin produced by Fusarium species, as a
selecting agent for the production of wilt resistant mutants
of tomato using protoplasts of tomato variety UC-82, which is
resistant to race 1 but not to race 2 of the Fusarium wilt
pathogen. Eight mutants resistant to Fusarium race 2 were
selected from fusaric acid resistant calli while five other

mutants were recovered from control calli (somaclonal variation). In both cases, the resistance was conferred by a
single dominant gene, but it is not known whether the mutants
arose from the same allele. Moreover, it is not known at
present whether these genes differ from the single dominant
I-2 gene which controls resistance to race 2 of the Fusarium
wilt pathogen. When answers to these questions are known,
the role, if any, that fusaric acid plays as a primary disease determinant and its value as a selecting agent for
Fusarium wilt resistance should be apparent.

J. D. Paxton discussed the use of preparations of an
Erwinia species normally associated with roots of soybean
plants as a seed treatment to increase soybean yields. By
marking the bacterial isolates with resistance to the antibiotics rifampicin and naldixic acid, it was established that
the strains used for seed treatment became established on the
root system of soybean plants under field conditions. Seed
treatments with these bacterial preparations have produced
increases in soybean yields averaging 10 percent.

These bacteria are good colonizers of soybean roots and
can also inhibit growth of Phytophthora megasperma f. sp.
glycinea on media in the laboratory. The bases of protective
action of soybean plants is not known, but it may be due to
interference with pathogen colonization of the soybean roots.

Molecular Strategies for Crop Protection, pages 193–197
© 1987 Alan R. Liss, Inc.

Workshop: **PHYSIOLOGY AND GENETICS OF FUNGAL PATHOGENS AND THEIR HOSTS**

Convener: R. W. Michelmore, University of California, Davis.

The workshop focused on specificity in pathogenic fungi: Why do some combinations of fungus and plant result in disease while others do not and what progress has been made in understanding such specificity?

Different levels of specificity in plant-pathogen interactions must be recognized. Determinants of pathogenicity control the ability of one species to cause disease upon another. In many cases these will require positive fungal attributes which will be determined by dominant alleles; variation for these attributes is not usually found naturally within a species but may be areated by mutation in the laboratory. Within a species, pathotypes vary in their ability to cause disease on specific host lines. These determinants of virulence are often monogenic with avirulence dominant to virulence. Within pathotypes isolates of the same virulence phenotype may vary in aggressiveness, i.e. in the amount of disease they cause. When considering the molecular basis of specificity, the level at which specificity is operating must be considered.

In several diseases intraspecific variation has been shown to be determined by a gene-for-gene interaction (Flor, 1956). Every dominant resistance gene in the host is specifically matched by a dominant gene for avirulence in the pathogen. While this is undoubtedly an oversimplification, the gene-for-gene hypothesis has withstood rigorous testing in several systems (Crute, 1986). The apparently simple genetic basis suggests that such interactions will be amenable to characterization and manipulation at the molecular level, at least for those interactions where incompatible and compatible phenotypes are unambiguous. The goal of several programs is to clone the genes determining resistance and avirulence from host and pathogen respectively.

Ideally there are several prerequisites for cloning genes determining specificity. Both host and pathogen should be easy to manipulate; compatible and incompatible interactions should have unambiguous phenotypes; both host and pathogen should have well developed genetics; transformation systems should be available for host and pathogen; and, methods for identification of genes with unknown products should be available. No plant pathogen interaction fulfills all these criteria. Five presentations were made which illustrate the diversity of interactions being studied and the various approaches being employed.

Towards a molecular understanding of lettuce downy mildew.
R. W. Michelmore; S. H. Hulbert, B. S. Landry and T. Ilott, University of California, Davis.
The interaction between lettuce (Lactuca sativa) and Bremia lactucae fulfills some but not all of the above criteria. A gene-for-gene interaction is determined by at least eighteen pairs of resistance and avirulence genes. Most gene-for-gene interactions determine unambiguous incompatible or compatible phenotypes. Extensive genetics is possible in both host and pathogen as detailed below. Lettuce can be transformed using Agrobacterium tumefaciens vectors (Michelmore, unpublished). A transformation system for B. lactucae and methods for gene identification in lettuce are in the initial stages of development.

The genetics of resistance and virulence has been characterized in detail for both lettuce and B. lactucae. Initially, there seemed to be several inconsistencies with a gene-for-gene interpretation. Most of these were, however, resolved by making simultaneous genetic studies of host and pathogen. All of the 14 resistance genes so far studied in detail map to one of three linkage groups. Cosegregation of resistance can indicate closely linked genes, alleles at a single locus or the identical allele. In B. lactucae, all avirulence genes segregate indepently except two which are loosely linked. Therefore, digenic segregation of avirulence to monogeneticly inherited resistance indicates the presence of two linked resistance genes; cosegregation of avirulence to different cultivars indicates that the cultivars have a resistance allele in common. When two linked genes are indicated, large progenies of lettuce have been screened for rare recombinants to test for the presence of separate loci. Simultaneous genetics has allowed a precise characterization of the genes determining specificity in lines of L. sativa and B. lactucae. This was a prerequisite for molecular characterization.

Several cloning strategies would be aided by a detailed genetic map. Restriction fragment length polymorphism (RFLP) analysis is being used to developed maps for L. sativa and B. lactucae. There are 3 stages in RFLP analysis: (i) developing a source of probes; (ii) identifying polymorphisms; and (iii) analyzing segregation of polymorphisms and developing a linkage map. Two sources of probes, low copy number random genomic fragments cloned in pUC13 (Landry and Michelmore, 1986) and cDNA clones, are being used for lettuce. cDNA clones seem to detect polymorphisms slightly more frequently than random genomic fragments. F_2 progeny are being analyzed and a map including RFLPs, isozymes, morphological markers and resistance genes is being developed. This will provide molecular markers flanking resistance genes for use in transposon mutagenesis studies. Low copy number, random genomic fragments are being used to generate a genetic map of B. lactucae. Large numbers of RFLPs have been detected which are inherited in a simple Mendelian manner. The genetic markers in B. lactucae will be used as starting points to chromosome walk to avirulence genes and will provide tools to study the mechanisms by which variation in virulence phenotype is generated.

During the preparation of random genomic probes for RFLP analysis of B. lactucae many clones were found to contain repeated sequences. The frequency of clones with repeated sequences increased with insert size indicating that the repeats were dispersed. This was confirmed by hybridizing Southern blots of random lambda clones of genomic B. lactucae DNA to total labelled DNA. Over 80 percent of the clones contained repeated sequences and the majority contained low copy as well as repeated sequences. Subsequent analysis of dot blots demonstrated the existence of several non-ribosomal, non-mitochondrial families of repeated sequences in the B. lactucae genome. This type of genomic organization is very different from that so far described for fungi. No plant pathogens, however, have been studied in detail. Repeated sequences may have a role in variability either as transposable elements or as sites of homologous recombination. Mechanisms of variation in plant pathogens is largely unexplored. Whether strategies for obtaining resistance rely on classical or molecular techniques, an understanding of the ability of pathogens to respond to the resistance strategy is needed.

Immunochemical approaches to the identification of avirulence gene products. A.R. Ayers, Harvard University, Cambridge.

While Phytophthora spp. do not have well defined genetics of virulence, they do have the advantage compared to B. lactucae that they can be cultured independently of their host. This has allowed an immunochemical investigation for components of P. megasperna f. sp. glycines (Pmg) that initiate defence responses in its host, soybean. Monoclonal antibodies (mabs) which are specific for glycoproteins and wall components of Pmg have been identified. Glycomoieties of extracellular glycoproteins were the immunodominant antigens. The majority of mabs which had been raised to glycoprotein antigens to one race, had the same affinity for equivalent antigens from other races. Some mabs, however, had different affinities for equivalent antigens indicating structural differences in the glycomoities of these antigens.

Immunosuppression with cyclophosphomide (modified from Matthew and Patterson, 1983) is being employed to develop mabs to antigens which are produced only in the presence of specific avirulence genes. In preliminary experiments with a regimen of injections involving two different races of Pmg, race specific differences were readily detectable with polyclonal sera and mabs. These experiments illustrate the potential of immunosuppression for identifying specific antigens of other pathogens such as expressed in tissue infected by an obligate biotroph where the genetics of virulence is better characterized.

Mabs may be an important tool for studying other aspects of plant-pathogen interactions. Mabs have been generated to soybean microsomal membranes and to glucan elicitor preparations. Also biotinylated Pmg glycoproteins have been shown to bind to membrane-associated receptors.

Mutants of Erysiphe graminis with increased virulence of barley
J. E. Sherwood and S. C. Somerville, Michigan State University, East Lansing.

Erysiphe graminis exhibits a gene-for-gene interaction with barley. Mutational analysis of the fungus has been initiated to develop strains with different virulence phenotypes in a similar genetic background.

After mutagenesis of E. graminis strain CR3 with EMS, virulence on lines of barley with Reg1a1 resistance allele could only be increased from 0 to 1 infection type rather than a 4 infection type (0 = no infection, 4 = fully compatible interactions). This small increase in virulence was generally specific to that allele, with few increases in virulence on other isogenic lines of barley carrying other resistance alleles.

When one of the mutants with increased virulence was remutagenized, virulence on Reg1a1 increased to a 2 infection type. This was concommitant with increases in virulence on 3 other barley lines carrying different barley alleles at the Reg1 locus.

Avirulence can therefore be mutated to increased virulence. However, the interaction seems more complicated than a simple gene-for-gene interaction. The stepwise increases in virulence suggest a specific sequence of genes act together to determine avirulence. Further mutagenesis and segregation analysis will test this hypothesis.

Vectors for cloning virulence genes from fungi. R. C. Garber, B. G. Turgeon and O. C. Yoder, Cornell University, Ithaca.

Unlike biotrophic pathogens such as B. lactucae and E. graminis, some necrotrophic pathogens produce toxins which specifically cause extensive host-necrosis. Such fungi can generally be cultured in vitro and so the

determinants of specificity and virulence may be amenable to analysis at the
molecular level. One example of a "specificity factor" is the Tox1 locus
in the Ascomycete Cochliobolus heterostrophus, a pathogen of maize. Strains
of C. heterostrophus with the Tox1 allele (race T) product T toxin and
exhibit high virulence on Texas male sterile cytoplasm maize. Isolates
with the tox1 allele cause indistinguishable levels of disease on
male-sterile and male-fertile cytoplasms, and are less virulent on
male-sterile cytoplasm maize than are T isolates. The Tox1 locus is thus
both a virulence factor and a specificity determinant. Our goal is to clone
the Tox1 locus from C. heterostrophus by transferring the Tox1 allele into an
isolate with the alternate allele. Such a transfer requires the ability to
transform C. heterostrophus. A transformation system for C. heterostrophus
has been developed based on the prokaryotic hygromycin B phosphotransferase
gene (Hm^R) fused to a C. heterostrophus sequence that enables Hm^R to
function in fungal cells. The approach is particularly well suited to fungi
which are relatively undeveloped genetically, as cloned genes and their
corresponding mutations are not required. It is only necessary for the
fungus to be sensitive to the aminoglycoside antibiotic hygromycin B and to
be able to obtain suitable regulatory sequences.

Sequences capable of promoting the hygromycin gene were selected
directly by transformation of C. heterostrophus. A library of random 0.5-1.5
kb C. heterostrophus DNA fragments, the source of "promoters", was inserted
at the 5' end of an Hm^R gene lacking both a translation-initiating ATG
and any upstream regulatory sequences. Wild type C. heterostrophus
protoplasts were transformed (Turgeon et al., 1985) with the "promoter"
library. Colonies resistant to hygromycin arose at low frequency. Probing
DNA from resistant colonies with radiolabelled vector DNA demonstrated that
the transforming plasmid had integrated into the C. heterostrophus genome,
and that the integrated DNA had 0.8-1.0 kb inserts--presumptive
promoters--immediately 5' to Hm^R. The integrated copy of Hm^R with
its fused C. heterostrophus "promoter" sequence was recovered from a lambda
library of transformant DNA, and used to transform wild type C.
heterostrophus protoplasts to hygromycin resistance. As expected,
hygromycin-resistant colonies arose with greater frequency than in the
original library transformation. Plasmid constructions containing 5 kb of C.
heterostrophus chromosomal DNA that flanked the locus of integrations
transformed at a frequency approximately one order of magnitude higher (3/ug)
than plasmids without flanking DNA.

Molecular and genetic analyses of C. heterostrophus transformants showed
that: (1) transformation in C. heterostrophus is always integrative and
occurs at a single locus in each transformant; (2) the frequency of
homologous integration is correlated with the amount of homology between the
vector and the C. heterostrophus genome. With 5 kb of homology, for example,
the frequency of homologous integration is 100%.

The Hm^R gene and fused C. heterostrophus "promoter" were used as the
basis for construction of a cosmid with the capacity to accept 40-45 kb
fragments of genomic DNA. The cosmid will serve as a vector for cloning of
genes from C. heterostrophus including those for specificity and virulence
determinants.

Transformation vectors developed for C. heterostrophus also transform
the Ascomycetes Colletotrichum lindemuthianum, a pathogen of beans
(R. Rodriguez, G. Turgeon, and O. C. Yoder, unpublished data), and
Leptosphaeria maculans, a pathogen of crucifers (R. Garber, G. Turgeon and
O. Yoder, unpublished data). Thus, this system for gene cloning may be
applicable to a number of fungi.

Plasmid-like DNAs in Fusarium oxysporum. H.C. Kistler[1] and
S.A. Leong[2], University of Florida, Gainesville[1]; University of
Wisconsin, Madison[2].

Several races of Fusarium oxysporum f. sp. conglutinans, a wilt pathogen
of crucifers, exist which exhibit different host ranges. Two races (1 and 5)
infect cabbage but not radish while one race (race 2) infects radish but not
cabbage. A linear plasmid-like DNA (pl DNA) is correlated with this level of
pathogenic specialization. Plasmid-like DNA elements, 1.9 kb in size, were
found in all 30 strains of a world wide collection of F. oxysporum f. sp.
conglutinans. Southern hybridization analysis however showed these pl DNAs
to be of two types: one type was homologous to pl DNA found in one cabbage
strain (pFOXC1) and was found only in cabbage strains; the second type was
homologous to pl DNA found in one radish strain (pFOXC2) and was found only
in radish strains. Whether these pl DNA encode product(s) which determine
this host range specificity is being determined.

Conclusion

There is no ideal system for studying specificity in plant-fungus
interactions. Since diverse organisms and modes of nutrition occur, diverse
determinants of the various levels of specificity can be anticipated. The
genetics of virulence is best understood for biotrophic fungi but the
inability to culture them obstructs a rapid molecular analysis of
specificity. Transformation systems for several necrotrophic fungi is
becoming routine and the determinants of specificity in such pathogens should
soon be amenable to analysis.

Cloning of the genes involved may or may not directly reveal the
molecular basis of specificity. It will, however, definitely be a critical
step in that direction. More work may also be needed to determine how the
initial events are transduced to elicit compatible and incompatible
responses. This would finally link the genetic studies of specificity
(recognition?) with the biochemical studies of response, which are discussed
in other workshops.

Crute, I.R. (1986). The genetic bases of relationships between
microbial parasites and their hosts. In: Mechanisms of Plant Disease, Ed.
R.S.S. Fraser. Martinus Nijhoff/W. Junk Publ., Boston. pp. 80-142.

Flor, H.H. (1956). The complementary genic systems in flax and flax
rust. Adv. Genet. **8**, 29-54.

Landry, B.S. and Michelmore, R.W. (1986). Selection of probes for
restriction fragment length analysis from kplant genomic clones. Pl. Mol.
Biol. Reptr. **3**, 174-179.

Matthew, W.D. and Patterson, P.H. (1983). The production of a
monoclonal antibody that blocks the action of a neurite outgrowth promoting
factor. Cold Spring Harbor Symp. Quant. Biol. **48**, 625-631.

Turgeon, B.G., Garber, R.C. and Yoder, O.C. (1985). Transformation of
the fungal maize pathogen, Cochliobolus heterostrohpus using the Aspergillus
nidulans amdS gene. Mol. Gen. Genet. **201**, 450-453.

Molecular Strategies for Crop Protection, pages 199–202
© 1987 Alan R. Liss, Inc.

WORKSHOP SUMMARY: BACTERIAL DETERMINANTS

Dallice Mills

Department of Botany and Plant Pathology
Oregon State University
Corvallis, Oregon 97331

This workshop addressed four general questions. They
were: 1) What types of genes are determinants of
pathogenesis? 2) How are the genes organized and
expressed? 3) What, if any, functions are ascribed to
these genes? and 4) How may these genes be exploited to
develop molecular strategies for crop protection? A brief
summary of strategies and work in progress involving
several systems is presented below. As some of the work is
preliminary, it can not be found in the abstracts of this
meeting (Journal of Cellular Chemistry, Supplement 10C,
1986).

S. Leong reported on experiments undertaken in
collaboration with V. Morales, P. Xu and L. Sequeira to
initiate a molecular/genetic analysis of the determinants
of host-pathogen recognition in the Pseudomonas
solanacearum/tobacco system. To investigate the mechanism
of pathogen recognition and HR induction, a series of Tn5
insertions was generated in the spontaneous, avirulent HR-
inducing strains BINC and K2R. Ten Tn5-containing HR
negative mutants were isolated in BINC and six in K2R. The
sizes of the EcoRI fragments bearing the Tn5 insertions
were found to vary from mutant to mutant, suggesting that
many genes may be involved in pathogen recognition. In a
parallel study, ten avirulent Tn5 mutants of the wild-type
virulent strain K60 were isolated. In addition, one
virulent EPS-deficient mutant was identified suggesting
that EPS production may not be required for virulence as
was previously thought. It should be emphasized that the
association of the Tn5 insertions with the observed
phenotypes has not yet been rigorously proven by
complementation and marker exchange analysis. The Tn5-
containing fragments are currently being cloned for this
purpose.

L. Glass discussed experiments conducted in T.
Kosuge's laboratory that indicate a role for IAA-lysine in
tryptophan secondary metabolism and virulence of P.

<u>syringae</u> pv. <u>savastanoi</u>. Gall formation is dependent upon
the bacterial production of the phytohormone indoleacetic
acid (IAA). She has observed that bacteria isolated from
oleander plants can further metabolize IAA to 3-indole-
acetyl-ε-lysine, a compound that does not accumulate
appreciably in olive and privet isolates. A 6.25 Kb <u>Eco</u>RI
fragment bearing the IAA-lysine synthetase gene was cloned
from the oleander isolate and subjected to Tn<u>5</u> mutagenesis.
Constructs with Tn<u>5</u> inserted outside the <u>IAA1</u> locus were
fully capable of restoring IAA-lysine synthesizing capacity
when transferred into <u>P</u>. <u>savastanoi</u> mutants lacking IAA-
synthesizing activity. A Tn<u>5</u> insert within the IAA-lysine
synthetase gene failed to restore activity. Concomitant
with the introduction of the IAA-lysine synthetase gene
into an olive isolate was its ability to convert IAA to
IAA-lysine, which resulted in a reduction of the IAA pool
size by one-third. An oleander mutant constructed by
insertion of Tn<u>5</u> into the IAA locus was incapable of
converting IAA to IAA-lysine, and its IAA pool size was
increased 5-fold. This mutant produces less severe
symptoms than the wild type parent strain, suggesting that
the conversion of IAA to IAA-lysine regulates IAA pool
size, which may modulate virulence as assayed by gall size.

S. Patil reported on his experiments directed towards
understanding the pathogenesis of <u>Pseudomonas</u> <u>syringae</u> pv.
<u>phaseolicola</u>, the causal agent of halo blight of bean. One
aim of his research is to define genes that are involved in
phaseolotoxin production. To determine the number and
organization of genes involved in phaseolotoxin production,
genomic libraries were made of toxin-producing strains.
Toxin-deficient mutants have been identified using
ultraviolet light, Tn<u>5</u>, and ethyl methanesulfonate
mutagenesis. One UV-induced and four Tn<u>5</u> mutants have been
complemented with library clones. One fragment that
complements has been characterized using restriction
analysis, λ::Tn<u>5</u> mutagenesis and marker exchange. To study
race-specific incompatibility in pathovar <u>phaseolicola</u>, a
genomic library of race 1 was mated into a strain
designated race 2. One of 82 transconjugants analyzed
showed a change in phenotype from a compatible water-
soaking reaction to a hypersensitive, incompatible reaction
when assayed in the cultivar Red Mexican. The change in
phenotype is associated with the presence of a 19 Kb
fragment which is currently being analyzed by physical and
functional mapping.

S. Farrand reported on recent experiments designed to

make agrobacteria strains that are used in the biological
control of Agrobacterium tumefaciens, the etiologic agent
of crown gall disease of plants, more stable in nature. A.
radiobacter strain K84 produces a low molecular weight
adenine nucleotide analogue, designated agrocin 84, which
kills certain strains of A. tumefaciens. A locus (ca. 7
Kb) on the Ti plasmid has been mapped which confers
sensitivity to agrocin 84. This locus also is associated
with the transport of agrocinopines A and B, opines that
are catabolized by agrocin-sensitive strains and strain
K84. Agrocin 84 is encoded from a 20 Kb region of a 48 Kb
self-conjugal plasmid that is present in K84 strains. The
tra region (ca. 3 Kb) from this plasmid has been mapped
near its origin of replication. As K84-mediated biological
control of crown gall can fail due to transfer of the
plasmid from strain K84 into tumorigenic strains of A.
tumefaciens, transfer-deficient mutant plasmids were
constructed and tested for control of A. tumefaciens.
These strains control crown gall as effectively as strain
K84 when the pathogen has the biovar 2 chromosomal
background. Biovar 1 biological control is significantly
less effective. Time course studies show that the biovar 2
biological control strains colonize host plant roots
significantly better than their biovar 1 counterparts.

    B. Kearney working in the laboratory of B. Staskawicz
reported on a strategy they are developing to clone a plant
resistance gene for bacterial spot disease of pepper,
Capsicum annuum, caused by Xanthomonas campestris pv.
vesicatoria. Using a well defined gene-for-gene system,
near isogenic plants have been developed after eight
generations of backcrossing that are homozygous for either
the dominant resistance gene Bs1 or its recessive allele
bs1. Upon infiltration with a race of X.c. pv. vesicatoria
that carries the corresponding dominant avr Bs1 gene,
pepper plants homozygous for Bs1 condition a hypersensitive
response, whereas those homozygous for the recessive allele
are susceptible to this race. Both plant genotypes will be
infiltrated with two races, one of which carries the avr
Bs1 gene, and one that does not. Among the resulting four
interactions only one will be incompatible (Bs1/avr Bs1).
A complementary DNA (cDNA) library will be made to mRNA
isolated from plants that exhibit the incompatible
interaction. The mRNAs derived from each of the other
interactions will be used to make cDNA probes to probe the
cDNA library. The detection of clones that show
differential hybridization to the probes may identify the

Bs1 locus as well as other genes that are expressed only during the incompatible interaction. The cDNA clone homologous to Bs1 will be identified by restriction fragment length polymorphism analysis. Resistant C. annuum plants will be crossed with C. chinense, a sexually compatible species that does not carry Bs1. An F2 population segregating for Bs1 will be analyzed on Southern blots, using the selected cDNAs as probes. Since the hybridizing restriction fragments of the two species are likely to be of different sizes, potential Bs1 clones will be those which show cosegregation of a C. annuum band with the hypersensitive plant phenotype in F2 individuals.

D. Mills reported on pathogenicity mutants of Pseudomonas syringae pv. syringae, the causal agent of brown spot disease of Phaseolus vulgaris. Two of four Path⁻ mutants obtained by Tn5 mutagenesis have been partially characterized by molecular analysis, and in terms of their growth properties in the susceptible bean cultivar, 'Eagle'. One mutant, designated PS9021, has been complemented with an 8.5 Kb sequence from the cosmid library of the wild type parent and a protein of approximately 85 Kd has been expressed from this region in Escherichia coli maxi cells. This mutant fails to grow in planta but like the other Tn5 Path⁻ mutants, its growth on minimal agar medium is indistinguishable from its parent. A second mutant, PS9024, carries the Tn5 insertion in an EcoRI fragment that is not present in all other pathogenic strains that cause brown spot disease. The growth rate in planta and the number of cells attained at stationary phase of this mutant is indistinguishable from its Path⁺ parent. Also, unlike its parent, this mutant fails to produce symptoms after reaching stationary phase. That pathogenicity was adversely affected by Tn5 mutagenesis in PS9024 awaits confirmation by complementation. These mutants are providing an opportunity to identify gene functions that appear essential for disease expression in bean.

II. Protection Against Viruses

Molecular Strategies for Crop Protection, pages 205–213

TRANSGENIC PLANTS THAT EXPRESS THE COAT PROTEIN GENE OF TMV ARE RESISTANT TO INFECTION BY TMV[1]

Roger N. Beachy,*[2] Patricia P. Abel,*[3]
Richard S. Nelson,* Steven G. Rogers,[+]
and Robert T. Fraley[+]

*Plant Biology Program
Department of Biology, Box 1137
Washington University, St. Louis, MO 63130

[+]Monsanto Company, 700 Chesterfield Village
Parkway, St. Louis, MO 63198

ABSTRACT   A cloned cDNA encoding the coat protein (CP)
gene of tobacco mosaic virus (TMV) was ligated to the
CaMV 35S promoter and the NOS 3' polyadenylation signal.
The chimeric gene was introduced into tobacco cells on a
disarmed Ti-plasmid in <u>Agrobacterium</u> <u>tumefaciens</u>, and
plants were regenerated. R_1 seedlings were inoculated
with TMV and observed for virus infection and disease
symptoms. Transgenic plants that expressed the chimeric
CP gene either developed disease symptoms significantly
later than the control plants, or did not develop
symptoms. Plants that did not produce symptoms
accumulated a low or undetectable amount of virus. This
type of resistance bears striking resemblence to the
resistance in plants that are cross-protected from virus
infection by previous infection with a mild
(attenuated) strain of the virus.

[1]This work was supported by a research grant from The
Monsanto Company.
[2]To whom inquiries should be addressed.
[3]Supported by pre-doctoral training grants from the NIH
(GM07067 and GM08036), the Division of Biology and
Biomedical Sciences at Washington University, and a
Plant Biology Graduate Fellowship from the Monsanto
Company.

INTRODUCTION AND PROPOSED MECHANISMS
FOR CROSS-PROTECTION

Cross-protection is a phenomenon in which infection of a host with a virus prevents, or reduces, the probability of superinfection by a related virus or virus strain. It was initially proposed that cross-protection is effective only against closely related viruses or virus strains, and has been used by plant virologists as a measure of the degree of relatedness between virus strains (1). However, cross-protection between virus strains is often not reciprocal, as in the case of tobacco streak virus (2). In other cases, cross-protection occurs between virus strains that are minimally related to each other serologically (3,4).

In practical terms cross-protection has been used to protect tomato plants from severe strains of tobacco mosaic virus (TMV; 5), citrus plants from disease symptoms caused by citrus tristeza virus (6), and potato plants from severe infection by potato spindle tuber viroid (7). In some cases, super-infection may not occur, i.e., infection or replication of the second virus may be blocked. More often, however, replication of the super-infecting virus is delayed in plants that are cross-protected compared with those that are not. In such cases plants that are cross-protected produce virus at a delayed rate with symptoms or cytopathic effects of the infection developing later in time than in plants that are not cross-protected (8).

A number of different mechanisms have been proposed, in several reviews, to explain cross-protection (8-11). In these reviews three basic mechanisms, with variations, are proposed. In the first, replication of the superinfecting virus is prevented because of competition for host factors that may be utilized in the viral replicase or for "sites" of virus replication. A related proposal suggests that one viral RNA template binds all of the replicase molecules, thereby preventing replication of a superinfecting viral RNA. A second mechanism, proposed by Zaitlin (12) and Palukaitis and Zaitlin (10), suggests that RNA:RNA interactions may be responsible for cross-protection. These authors suggest that annealing of sense and anti-sense RNA molecules of protecting and super-infecting viruses prevents replication or translation of the latter. These authors cite the results of experiments with viroids (7) and strains of TMV that produce insoluble (defective) capsid protein (12) as support for their hypothesis. The third mechanism includes the direct or indirect involvement of capsid protein in cross-protection. This model suggests a possible role of CP in a) the plasma membrane or apoplastic space to reduce virus infection (13), or b) the encapsidation of viral RNA from the super-infecting virus by CP in the protected cell (14), or c) preventing the formation and functioning of stripasomes in early stages of virus infection (15). Although these models propose different mechanisms for cross-protection, one mechanism need not preclude another, i.e., cross-protection could be the result of a number of sequential or simultaneous reactions.

In a review article published in 1980, prior to development of the necessary techniques, R. I. Hamilton (8) suggested that cloned cDNAs representing single viral genes might be expressed in transformed cells, and that such expression might result in a protective response by the plants. "If protection genes are present in viral genomes, then the corresponding gene products may produce protection against infection by the challenger, either directly or by inducing the synthesis of host factors" . . . "Moreover, cDNA-mediated protection should be transmitted from generation to generation via the seed, an advantage in those crops that are not propagated vegetatively or for which the isolation of viable protoplasts is very difficult." With the recent development of gene cloning and gene transfer technologies described below, it is now possible to engender a type of cross-protection in transgenic plants, an advance that should be of both scientific value in determining the molecular mechanisms involved, and of agricultural value in crop protection.

CROSS PROTECTION IN TRANSGENIC PLANTS

Recently we reported the results of our experiments to express the CP gene of TMV in transgenic tobacco plants, and the effect of this expression on the subsequent infection of these plants and their progeny with TMV (16). TMV was chosen for these experiments because it is physically and biochemically well characterized (17), because cross-protection against TMV has been used commercially (5), and because hosts of TMV include solanaceous plants that can be readily transformed by Agrobacterium tumefaciens and regenerated to fertile plants (18). The research approach and the results of our experiments are briefly summarized below:

1.   A cDNA clone containing the cistron encoding the
TMV-coat (capsid) protein (CP) from the common (U_1) strain
of TMV was generated. The clone contained nucleotides 5707
to 6395, and represents the 3' end of TMV-RNA (19).
2.   The cDNA was ligated into a polylinker region on an
intermediate plasmid to create a chimeric gene consisting of
the cauliflower mosaic virus (CaMV) 35S transcript promoter,
the TMV-CP cDNA, and the polyadenylation signal from the
nopaline synthase gene (NOS 3' end).
3.   The chimeric gene was transferred by bacterial
triple mating to the T-DNA (transferred DNA) region of a
disarmed Ti-plasmid in A. tumefaciens. A. tumefaciens
colonies containing the chimeric gene were selected on
appropriate antibiotics.
4.   Tobacco cells were transformed with the modified A.
tumefaciens and transformed cells were regenerated into
intact plants. Six of the eight transgenic plants
contained 1 to 3 copies of the chimeric gene, with the
remaining two plants each containing greater than 5 copies.
5.   Transgenic plants produced both virus related mRNA,
and TMV-CP. The chimeric gene was expressed in progeny as a
Mendelian trait.
    Progeny raised from seeds of self-fertilized transgenic
plants appeared normal in morphology, growth, and fertility.
Seedlings in the three leaf stage were assayed for presence
or absence of TMV-CP and nopaline and were subsequently
inoculated with TMV. The control plants in these
experiments included seedling progeny not expressing the CP
gene (approximately ¼ of the progeny). In these experiments
plants were either held in a growth chamber under conditions
that were optimal for the development of symptoms, or in a
greenhouse, in which symptoms developed somewhat more
slowly. Plants were then scored for the appearance of
disease symptoms on systemically infected leaves, i.e.,
scored for the combined results of infection, virus spread,
and symptom development. In the seedlings that expressed
the chimeric gene the appearance of symptoms was delayed by
2 to 10 days when compared with the control seedlings. The
delay in symptoms was highly correlated with expression of
the chimeric viral gene. To date more than 400 plants from
five different transgenic plant lines have shown a delay in
symptom development compared with control plants. A subset
of these plants failed to develop symptoms during the course
of the experiment. In one such experiment plants in the
greenhouse failed to develop symptoms within 29 days post-
infection. In leaves of plants that had not developed
symptoms no virus was detected.

The results of these experiments led us to conclude that transgenic plants expressing the CP-gene of TMV, and plants that are cross-protected by a mild-strain of TMV, are similar in their response to TMV. Therefore we have initiated experiments to further compare the resistance in transgenic plants with those of plants that are cross-protected.
    1. The effect of increasing the concentration of TMV in the inoculum. Plants that are cross-protected are less resistant to high concentrations of virus in the inoculum than to low concentrations (20). Similar results were observed with transgenic tobacco plants inoculated with TMV (Table 1) (16) and with transgenic tomato plants inoculated with TMV (Beachy et al., unpublished).

TABLE 1

Effect of Increasing Virus ($TMV-U_1$) Inoculum on Resistance in Transgenic Plants. Seeds of transgenic lines 3773 and 3404 were germinated in a greenhouse and seedlings were assayed for the presence of coat protein (CP) by western blot analysis. Approximtely 4 days later, the two youngest leaves of each plant were inoculated with TMV (U_1 strain). Plants were scored as having symptoms when leaves above the inoculated leaves showed vein clearing which subsequently led to severe chlorosis N.E. = segregated progeny from transgenic plants which did not express CP. E = segregated progeny from transgenic plants which expressed CP. Numbers in parentheses indicate sample size for each plant type.

TMV Inoculum Concentration	Plant Type	\multicolumn{4}{c}{Days After Inoculation}			
		5	7	15	19
–μg/ml–		\multicolumn{4}{c}{% of Plants Showing Symptoms}			
0.4	N.E. (4)	50	75	100	100
	E. (11)	0	0	18	36
0.8	N.E. (4)	100	100	100	100
	E. (12)	0	8	58	67
2.0	N.E. (4)	75	100	100	100
	E. (12)	8	33	67	92

2. Inoculation with viral RNA vs. virions. Tomato plants, which had been cross-protected by a mild strain of cucumber mosaic virus, were more susceptible to infection by CMV-RNA than to CMV virions (22). We have found that the transgenic tobacco plants used in our experiments are more susceptible to infection after inoculation with TMV-RNA than with TMV (Table 2) (21).

TABLE 2

Symptom Development in Transgenic and Non-Transgenic Plants Inoculated with TMV (U_1 Strain) or TMV-RNA. Seeds of transgenic line 3646 and parent line Nicotiana tabacum cv. Xanthi were germinated in a greenhouse. Leaves of 3646 plants were assayed for presence of coat protein (CP) as described in Table 1. Only plants of line 3646 that express the CP gene were utilized. Approximately 4 days later, the two youngest expanding leaves of each plant were inoculated with either viral RNA (4µg/ml) or virions (0.25 µg/ml). The virion inoculum produced four times as many infectious units on local lesion plants as did the RNA inoculum. Plants were placed in a growth chamber under conditions of 14h/24°C light periods and 10h/19°C dark periods. Plants were scored for positive symptoms as described in Table 1. Numbers in parentheses indicate sample size for each plant type-infectious agent combinations.

Plant Type	Infectious Agent	Days After Inoculation				
		4	5	6	7	12
		% of Plants Showing Symptoms				
Xanthi	Virion (15)	7	87	93	93	93
Xanthi	RNA (15)	13	93	100	100	100
3646	RNA (13)	8	54	92	100	100
3646	Virion (12)	0	8	62	85	85

Thus, by these two criteria, the transgenic plants used in our experiments have characteristics similar to those of plants that are cross-protected by infection with a mild strain of TMV.

Discussion of the mechanism of the cross-protection against TMV in transgenic tobacco plants. To date we have not determined the mechanism of resistance of the transgenic plants against TMV. Preliminary evidence indicates that one effect of the expression of the CP-gene is to reduce the number of sites of infection on leaves inoculated with TMV (Nelson et al, unpublished). In this experiment the number of chlorotic lesions produced by a severe strain of TMV were significantly decreased in inoculated leaves of transgenic plants compared with control plants. This decrease in lesion number could result from fewer sites of virus entry into the leaf, or from fewer number of sites of successful replication.

The observation that infection by viral RNA overcomes the resistance in transgenic plants suggests that RNA:RNA interactions (such as those proposed earlier, 10) are not the only mechanism involved in the resistance observed in transgenic plants. Further experiments are necessary to determine whether RNA:RNA interactions are involved at all in the protection.

More extensive and diverse experimental approaches are being undertaken to address the involvement of the CP itself in the resistance reaction. For example, to address the possibility that RNA of the superinfecting virus is encapsidated by CP in the cell, we searched for the purported building block of TMV assembly, the 20S disc, which is comprised of 34 capsid molecules (17). We were only able to detect CP molecules of less than 6S, with the majority of the protein being approximately 4S (16). Assuming that CP aggregates were not disrupted during the extraction procedure, this result indicates that aggregates containing 4 to 6 CP molecules predominate in these cells. This result does not necessarily rule out encapsidation as a mode of interference with infection, since there is a controversy about whether the 20S or 4S unit is actually utilized in TMV assembly (23). Likewise it does not address the possibility that CP in transgenic plants blocks co-translational stripping of TMV during infection as suggested by Wilson and Watkins (15).

It is clear from the discussion above that more experiments are necessary to identify the molecular mechanism by which the transgenic tobacco plants are resistant to infection by TMV. Likewise it will be important to determine the breadth of application of this approach to other viruses and groups of viruses, and to other plants. Only then will the role of this type of virus disease resistance in agriculture be established.

REFERENCES

1. Kado, CI, Agrawal, HO (1972). Principles and Techniques in Plant Virology, New York: Van Nostrand Reinhold Co.
2. Fulton, RW (1978). Superinfection by strains of tobacco streak virus. Virology 85:1.
3. Bawden, FC, Kassanis, B (1945). The suppression of one virus by another. Ann Appl Biol 32:52.
4. Fulton, RW (1975). Unilateral cross-protection among some NEPO viruses. Acta Hortic 44:29.
5. Rast, AThB (1972). MII-16, an artificial symptomless mutant of tobacco mosaic virus for seedling inoculation of tomato crops. Neth J Plant Pathol 78:110.
6. Costa, AS, Müller, GW (1980). Tristeza control by cross-protection: a US-Brazil cooperative success, Plant Dis. 64:538.
7. Niblett, CL, Dickson, E, Fernow, KH, Horst, RK, Zaitlin, M (1978). Cross-protection among four viroids. Virology 91:198
8. Hamilton, RI (1980). Defenses triggered by previous invaders: Viruses. In Horsfall, JG, Cowling, EB (eds): "Plant Disease: An Advanced Treatise," Vol V, New York: Academic Press, Inc, p 279.
9. Fulton, RW (1980). The protective effects of systemic virus infection. In Wood, RKS (ed): "Active Defense Mechanisms in Plants," New York: Plenum Press, p 231.
10. Palukaitis, P, Zaitlin, M (1984). A model to explain the "cross-protection" phenomenon shown by plant viruses and viroids. In Kosuge, T, Nester, EW (eds): "Plant-Microbe Interactions: Molecular and Genetic Perspectives," New York: MacMillan Publishing Co, p 420.
11. Sequeira, L (1984). Cross-protection and induced resistance: their potential for plant disease control. Trends in Biotechnology 2:25.
12. Zaitlin, M (1976). Viral cross-protection: more understanding is needed. Phytopathology 66:382.
13. DeZoeten, GA, Gaard, G (1984). The presence of viral antigen in the apoplast of systemically virus-infected plants. Virus Research 1:716.
14. DeZoeten, GA, Fulton, RW (1975). Understanding generate possibilities. Phytopathology 65:221.

15. Wilson, TMA, Watkins, PAC (1986). Influence of exogenous viral coat protein on the cotranslational disassembly of tobacco mosaic virus (TMV) particles in vitro. Virology 149:132.
16. Abel, PP, Nelson, RS, De, B, Hoffmann, N, Rogers, SG, Fraley, RT, Beachy, RN (1986). Cross-protection in transgenic plants that express the coat protein gene of TMV. Science (in press).
17. Hirth, L, Richards, KE (1981). Tobacco mosaic virus. Model for structure and function of a simple virus. In Lauffer, MA, Bang, FA, Maramorosch, K, Smith KE (eds): "Advances in Virus Research," New York: Academic Press, Inc, Vol 26, p 145.
18. Caplan, A, Herrera-Estrella, L, Inzé, D, Van Haute, E, Van Montagu, M, Schell, J, Zambryski, P (1983). Introduction of genetic material into plant cells. Science 222:815.
19. Goelet, P, Lomonosoff, GP, Butler, PJG, Ekam, ME, Gait, MJ, Karn, J (1982). Nucleotide sequence of tobacco mosaic virus RNA. Proc Natl Acad Sci USA 79:5818.
20. Cassells, AC, Herrick, CC (1977). Cross-protection between mild and severe strains of tobacco mosaic virus in doubly inoculated tomato plants. Virology 78:253.
21. Nelson, RS, Abel, PP, Beachy, RN. Characterization of the virus resistance in transgenic tobacco plants expressing TMV coat protein (in preparation).
22. Dodds, JA, Lee, SQ, Tiffany, M (1985). Cross-protection between strains of cucumber mosaic virus: Effect of host and type of inoculum of accumulation of virions and double-stranded RNA of the challenge strain. Virology 144:301.
23. Shire, SJ, Stechert, JJ, Schuster, TM (1981). Mechanism of tobacco mosaic virus assembly: Incorporatin of 4S and 20S protein at pH 7.0 and 20°C. Proc Natl Acad Sci USA 78:256.

Molecular Strategies for Crop Protection, pages 215–219
© 1987 Alan R. Liss, Inc.

GENETIC ENGINEERING OF PLANTS FOR TOBACCO MOSAIC VIRUS
RESISTANCE USING THE MECHANISMS OF CROSS PROTECTION

M.W. Bevan and B.D. Harrison[1]

Plant Breeding Institute, Maris Lane, Trumpington
Cambridge CB2 2LQ, England

ABSTRACT   Tobacco plants have been engineered using an
Agrobacterium binary vector expression cassette to
produce tobacco mosaic virus coat protein. We have
tested for in vivo packaging of a mutant TMV strain
defective in coat protein in these plants, and we have
assayed for symptom formation (local lesions) in the
engineered host plants. We observed no attenuation of
symptom production and no packaging, and we concluded
that higher levels of coat protein expression must be
obtained before packaging can be observed.

INTRODUCTION

Cross protection is the protection of a virus-infected
plant against subsequent infection by a related virus. It
has been described for a wide variety of plant RNA viruses
and several different theories have been advanced to account
for this general phenomenon (1,2). A favoured virus for
cross protection studies has been tobacco mosaic virus (TMV),
and contrasting experimental results have been obtained when
investigators have studied the role of TMV coat protein in
the cross-protection phenomenon. Initially, it was observed
(3) that superinfection of TMV-infected N. sylvestris with
TMV virions lead to attenuation of symptoms typical of the
superinfecting virus but that superinfection with viral RNA
did not cause symptom attenuation.
It was postulated that the free coat protein of the

[1]Scottish Crop Research Institute, Invergowrie, Dundee
DD2 5DA, Scotland

resident virus slowed down the uncoating of the super-
infecting virus, and that differing degrees of cross-
protection were conferred by coat protein subunits with
different affinities for other subunits and RNA molecules.
However, two lines of evidence suggested that coat protein
may not be specifically responsible for abating the symptoms
of a super-infecting TMV virus.

These studies (4,5) utilised mutants of TMV with
defective coat proteins that did not assemble correctly or
were not produced at all. Both strains of resident viruses
protected against symptom development caused by super-
infecting wild-type TMV strains. It was concluded that cross
protection was probably due to a number of different
phenomenon, such as the titration of essential replication
and transport machinery by the resident virus. This model
did not exclude coat protein from a role in cross-protection.

We wish to exploit the phenomenon of cross-protection in
the genetic engineering of plants for TMV resistance, and we
have initially concentrated our efforts on evaluating the
role of TMV coat protein in cross protection. We reasoned
that the expression of TMV coat protein as a nuclear gene
would critically test the role of coat protein in cross
protection as most other factors would be absent from the
challenged plant.

EXPERIMENTAL DESIGN

The coat protein cistron of TMV OM strain was obtained
as a cDNA clone (6). It was digested with Dra1 and Apa1
which cleaved 4 bp 5' of the NH_2 terminus and 2 bp 5' of the
3' end of TMV respectively. This fragment was ligated into
the BamH1 site of the expression vector pRok1 based on a
binary Agrobacterium transformation vector (7,8). This
placed the ORF of the TMV coat protein under the
transcriptional control of the powerful constitutive
cauliflower mosaic virus 35S promoter. Leaf discs of N.
tabacum Samsun NN and N. tabacum White Burley (a systemic
host for TMV) were transformed with pRok1-TMVcp in
Agrobacterium tumefaciens LBA4404 and selected on kanamycin.
Transformed plants were selected and coat protein levels were
measured using polyclonal TMV coat protein antibodies on
Western blots of crude plant extracts.

The levels of coat protein in transformants varied
widely between 0.5 μg/gm fresh weight in young newly expanded
leaves. Several elite lines that produced high levels of

coat protein were selected from among 50 individual
regenerants for further study.

N. tabacum Samsun NN (a local lesion host) expression
coat protein was selfed and the F_1 progeny were tested for
kanamycin resistance. This trait segregated 2 resistant:1
sensitive which was consistent with Southern blot data which
showed multiple T-DNA insertions at one locus. Young progeny
plants were germinated in soil and challenged with TMV MRC
wild-type virus at various concentrations. The number of
local lesions were measured and a sample of leaf tissue was
taken to test for the kanamycin resistance trait. Both
Samsun NN and White Burley (a systemic host) expressing coat
protein were infected with TMV strain PM_2 (9) a strain of TMV
that makes coat protein defective in viral assembly and the
levels of packaged virus were assayed by immunosorbent
electron microscopy (IEM) and a micrococcal nuclease protein
assay.

RESULTS

One hundred F_1 progeny of Samsun NN were inoculated with
TMV vulgare and the number of lesions per inoculated half
leaf was recorded. Portions of uninoculated leaves were
plated on kanamycin to test for the transformed phenotype.
There was a large variation in the number of local lesions
produced on the engineered plants but there was no
correlation with the transformed phenotype. As a third of
the plants were expected to contain two copies of the T-DNA
locus, and a third to contain no copies, we interpreted our
results as showing that resident coat protein had no effect
on the infectivity of TMV vulgare.

In a parallel series of experiments involving both
Samsun NN and White Burley expressing nuclear-encoded TMV
coat protein, we inoculated plants with the defective coat
protein mutant PM_2 (8). At various times after inoculation
samples of tissue were taken for analysis by IEM, for
infectivity, and for micrococcal nuclease protection.
Despite an intensive search no encapsidated TMV could be
observed (although PM_2 is competent for packaging) and no
increased infectivity due to encapsidation could be
detected. These findings were supported by our observations
that upon digestion of plant extracts with micrococcal
nuclease and Northern blot analysis of remnant RNA no TMV RNA
could be detected. In the absence of nuclease treatment
abundant PM_2 RNA could be detected by this method.

CONCLUSIONS

While naturally disappointed by the collection of negative data, we are able to draw several conclusions from this work that will help in the design of subsequent experiments. First, we believe that one must obtain substantially higher levels of coat protein expression than those reported here. This presents several difficulties as the CaMV 35S promoter is the most active constitutive promoter known. Other light-regulated promoters have similar activities but are only transcribed in photosynthetic tissue. In order to make a plant resistant to systemic TMV infection it is thought that all cell types should express an anti-viral factor. We predict that the construction of hybrid promoters based on CaMV 35S and heterologous enhancer sequences will help in increasing coat protein levels. While nuclear encoded genes cannot be expected to synthesize coat protein to the same level as a replicating mRNA, it may be possible to obtain biologically meaningful levels of coat protein if the effective pool of cross-protecting coat protein is the cytoplasmic population of intermediate discs.

Second, if coat protein cross-protects by displacing the equilibrium between assembly and disassembly, particularly during the early stages of infection in which "striposomes" are present (10), it may be possible to engineer the nuclear-encoded coat protein to have a higher affinity for the 5' end of the RNA and lower affinity for other coat proteins and thus block "ordinary" early unpackaging. This may lead to a lower multiplication of the virus. Third, the coat-protein defective strain PM_2 used in these studies synthesizes a coat protein that assembles into stringy structures in infected plant cells. It is feasible that our negative results on packaging this virus may have been due to preferential agglutination of the nuclear coat protein and PM_2 coat protein. The use of different TMV strains would avoid this problem in future experiments.

ACKNOWLEDGEMENTS

The authors gratefully acknowledge the help and advice given by Dave Zimmern and Ron Fraser.

REFERENCES

1.   Matthews REF (1970). Plant Virology, Academic Press, New York, p 414.
2.   De Zoeten GA, Fulton RW (1975). Understanding generates possibilities. Phytopathology 65:221-222.
3.   Sherwood JL, Fulton RW (1982). The specific involvement of coat protein in TMV cross protection. Virology 119:150-158.
4.   Zaitlin M (1976). Virus cross protection: More understanding is needed. Phytopathology 66:382-383.
5.   Sarkar S, Smitamana P (1981). A proteinless mutant of TMV: evidence against the role of a viral coat protein for interference. Mol Gen Genet 184:158-159.
6.   Meshi T, Takamatsu N, Ohno T, Okada Y (1982). Nucleotide sequence of the 3' end of TMV. Virology 118:64-75.
7.   Bevan MW (1984). A binary Agrobacterium vector for plant transformation. Nucl Acids Res 12:8711-8721.
8.   Bevan MW, Mason SE, Goelet P (1985). Expression of TMV coat protein by a CaMV promoter in transformed plants. EMBO J 4:1921-1926.
9.   Seigel A, Zaitlin M, Seghal OP (1962). The isolation of defective TMV strains. Proc Natl Acad Sci USA 48:1845-1851.
10. Shaw JG, Plaskitt KA, Wilson TMA (1986). Evidence that TMV particles disassemble co-translationally in vivo. Virology 148:326-336.

Molecular Strategies for Crop Protection, pages 221–234
© 1987 Alan R. Liss, Inc.

EXPRESSION OF ALFALFA MOSAIC VIRUS COAT PROTEIN GENE
AND ANTI-SENSE cDNA IN TRANSFORMED TOBACCO TISSUE

L. S. Loesch-Fries, E. Halk, D. Merlo,
N. Jarvis, S. Nelson, K. Krahn, L. Burhop

Agrigenetics Advanced Science Company
5649 E. Buckeye Road
Madison, Wisconsin 53716

ABSTRACT   cDNA copies of the AMV coat protein gene
cloned either in the sense or anti-sense orientation
were transferred to tobacco cells using a binary
vector containing the neomycin phosphotransferase II
gene. Sixty-three percent of the kanamycin-resistant
regenerants from transformation with AMV cDNA in the
sense orientation expressed coat protein at levels up
to 340 ng per mg soluble protein. The protein expression level varied between plants and between leaves
of individual plants; however, amounts of AMV-
specific RNA per   µg total RNA were similar. The
coat protein expressed in transformants was biologically active in the initial steps of virus replication when transformed protoplasts were inoculated
with AMV genomic RNAs. In contrast, the presence of
coat protein in transformed protoplasts protected the
protoplasts from infection with AMV virions.
Transformation with the binary vector containing
anti-sense AMV cDNA resulted in regenerants that
expressed AMV-specific RNAs. The RNAs, however, were
larger than expected and much less abundant than the
AMV RNAs in regenerants expressing coat protein.

INTRODUCTION

    Expression of virus genes and their complements in
plants using recently developed transformation systems
allows analysis of the role of individual genes in the
development of disease symptoms. Additionally, the role

of virus genes in a type of disease resistance known as cross protection may be examined. Cross protection, the reduced susceptibility of a virus-infected plant to infection by related strains of the same virus, has been studied with many different viruses but the mechanisms of protection are not yet understood. The protection is strain-specific; therefore, it is likely that one or more virus gene products are involved. Virus coat protein has been implicated by some studies (1,2,3,4) but not by others (5,6).

Strains of alfalfa mosaic virus (AMV), a bacilliform plant virus, have been shown to cross protect (7). We are studying the replication of the three genomic RNAs of this virus. Replication requires coat protein or its encapsidated messenger, RNA 4. Thus, coat protein of AMV has a role both in replication and virion structure. To determine if coat protein is involved in cross protection, RNA 4 cDNA was transferred to tobacco cells. Here we report on the expression of AMV coat protein in transformed plants and its effect on virus infection.

METHODS

An *Eco*RI–*Sma*I fragment containing RNA 4 cDNA sequences was excised from pSP65A4 (8), the *Eco*RI ends were filled in by reaction with the Klenow fragment of *E. coli* polymerase I, and the fragment was cloned into pDOB513. pDOB513, a derivative of pDOB512 that was kindly provided to us by Ken Richards (9), contains the cauliflower mosaic virus (CaMV) 19S promoter and polyadenylation sites separated by a *Sma*I restriction site and flanked by *Bgl*II sites. The RNA 4 cDNA fragment was ligated into the *Sma*I site in the sense and anti-sense orientations with respect to the CaMV promoter. A fragment containing promoter/RNA 4 cDNA/polyadenylation sequences was isolated by *Bgl*II digestion and inserted into pH400, a micro Ti plasmid binary vector (10). The plasmid with AMV cDNA in the sense orientation was designated pH400A4 and that with AMV cDNA in the inverted anti-sense orientation was designated pH400A4I (Fig. 1). Both plasmids contained the left and right borders of T-DNA, an octopine synthase gene under its own promoter and a neomycin phosphotransferase II gene (NPTII) under the CaMV 19S promoter. pH400A4 and

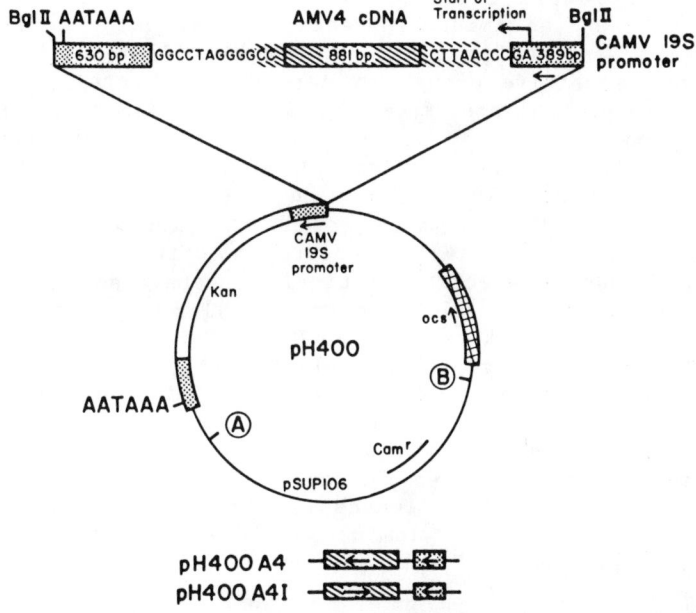

FIGURE 1.   AMV Micro Ti.   Only the orientation shown above was found; the construct containing the two 19S promoters as inverted repeats was not found.

pH400A4I were mobilized into *Agrobacterium tumefaciens* strain LBA4404 as described (11).

Leaf pieces of axenic *Nicotiana tabacum* var. Xanthi-nc were transformed by exposing them to a suspension of *A. tumefaciens* LBA4404 containing pH400A4 or pH400A4I essentially as described (12).   The tissue was placed on a medium containing a tobacco cell feeder layer, then transferred to shooting medium for selection of kanamycin-resistant transformed shoots (Firoozabady *et al.*, manuscript in preparation).   Morphologically normal shoots were excised, grown on a rooting medium containing kanamycin, then potted into a soil mix and placed in the greenhouse.   A cutting of each plant was also kept axenically growing in a medium containing kanamycin.

Protoplasts were isolated from axenic nontransformed Xanthi-nc and from regenerants and inoculated with AMV (strain 425-Madison) RNA or virions as described (8).

Protoplasts were cultured for 24 hours, then scored for infection by an immunofluorescent assay using anti-coat protein antibodies (13). The immunofluorescent assay was not sensitive enough to detect coat protein expression in protoplasts from transformed plants prior to infection. Only infected protoplasts, which usually contain 50-100 times the amount of coat protein present in transformed cells, could be detected.

DNA was isolated from greenhouse plants essentially as described (14), treated with restriction endonucleases, electrophoretically separated in 0.8% agarose, blotted to Genetran nylon membrane and hybridized to nick-translated DNA fragments (15). RNA was isolated from greenhouse plants (16), separated electrophoretically in 1.0% agarose, blotted to Genetran in distilled water and hybridized to nick-translated DNA fragments. Coat protein was quantitated in transformed plants by enzyme-linked immunosorbant assay (ELISA) of plant sap from crushed leaves (17) and analyzed by electrophoresis in polyacrylamide gels, followed by blotting to nitrocellulose and immunodetection (18).

RESULTS

Transformation

From 761 leaf pieces originally inoculated with either pH400A4 or pH400A4I, 226 morphologically normal shoots were excised. Of these, 78 shoots rooted when placed on a medium containing 200 mg/l kanamycin. Fifty-one kanamycin-resistant shoots were from transformations with pH400A4, which contains the coat protein gene in the sense orientation, and 27 were from transformations with pH400A4I, which contains the coat protein gene in the anti-sense orientation. Morphology of most transformants was normal except for a few plants that showed variations, such as short internodes, sterility or poor seed set.

Approximately 35% of the plantlets contained detectable levels of octopine (>3 ng octopine/mg tissue, Bookland and Paaren, manuscript in preparation). ELISA showed that 63% of the plants transformed with pH400A4 expressed coat protein. All pH400A4 transformants that expressed octopine, expressed coat protein. The range of expression was from 5 to 2500 ng coat protein/ml plant sap. From analysis of the regenerants, eight transformants expressing high levels of coat protein and six transformants containing anti-sense AMV cDNA were selected for presentation here.

DNA Analysis

DNA from transformed plants was analyzed to determine which genes had been transferred from the *Agrobacterium* to the tobacco genome. All 14 selected transformants contained a full-length NPTII gene (data not shown). Tissue transformed with the pH400 vector without a viral insert also contained DNA which hybridized, as expected with the NPTII probe, whereas DNA extracted from untransformed tissue contained no detectable homology to the probe.

Tobacco DNA from transformed plants hybridized to RNA 4 cDNA (Fig. 2A). Bands at 3.3 kbp and 0.6 kbp or at 3.5 kbp and 0.3 kbp indicate the presence of a full-length coat protein gene or anti-sense gene plus a full-length octopine synthase gene. DNA of either regenerants transformed with the pH400 vector or of untransformed plants revealed no hybridization to the AMV probe. Using laser densitometry of similar autoradiographs containing genome reconstructions, we estimate that 1 to 3 copies of alfalfa mosaic virus coat protein gene were integrated per haploid genome.

In addition, other bands hybridized to the AMV probe. Of special interest is a band of approximately 2 kbp in plants 14-1, 37-6, 27-10 and 38-2 (Fig. 2A), which also hybridized to an octopine synthase cDNA probe. Dhaese *et al.* (19) have reported that an internal T-DNA sequence upstream from the major octopine synthase polyadenylation site can serve as an internal border. Our data suggest that the internal border was used in these transformations because, in such a case, the fragments containing the AMV cDNA plus a portion of octopine synthase would be about 2 kbp for both pH400A4 and pH400A4I transformants. The presence of the 2kbp band in plants 14-1 and 37-6, along with the expected full length band at 3.2 kbp, indicates that multiple insertion events have taken place. We conclude from these results that all the plants represented in Figure 2A contain all or part of the octopine synthase gene.

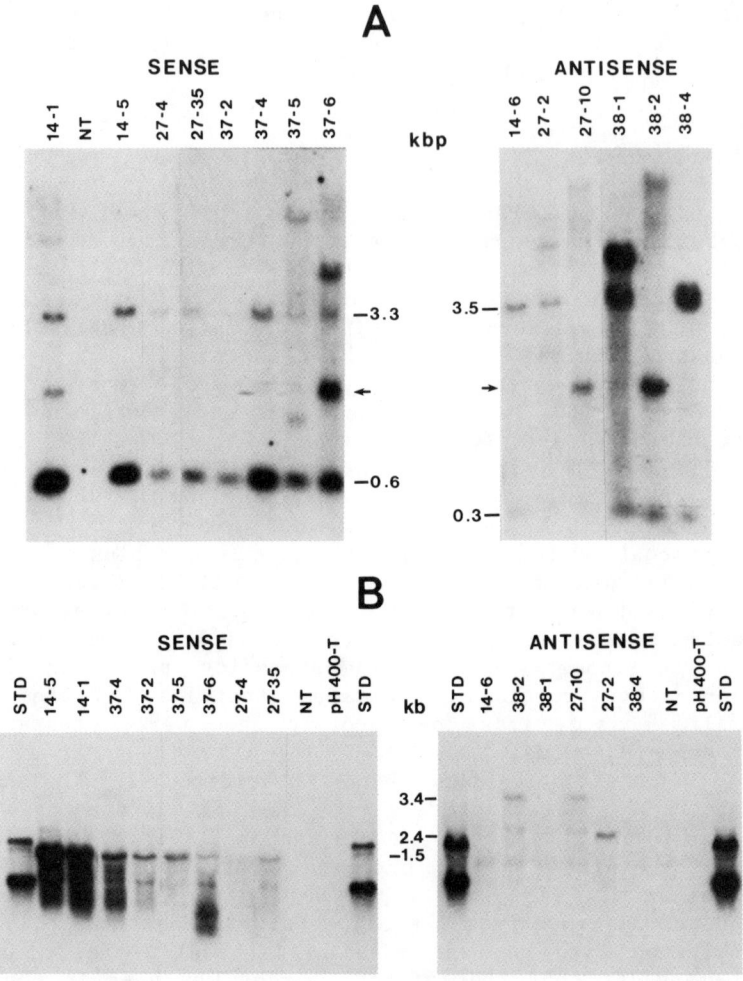

FIGURE 2. Analysis of DNA and RNA from transformants. (A) Autoradiograph of a DNA blot hybridized to a RNA 4 cDNA fragment. The lanes contain DNA, cut with *Bam*HI, from transformants or from untransformed tobacco. The arrows indicate the position of a band of approximately 2 kbp. (B) Autoradiograph of an RNA blot hybridized to a RNA 4 cDNA fragment. The lanes contain equal amounts of RNA from transformants. pH400-T lanes contain RNA from plants transformed with pH400 vector. The STD lane shows the position of AMV RNA 3 and RNA 4.

RNA Analysis

RNA from 25 of 44 pH400A4 transformants (57%) contained distinct and prominent species which hybridized to AMV cDNA. These RNAs could not be detected in pH400-transformed or untransformed plants. Patterns of RNA from the eight selected transformants (Fig. 2B) showed a strong band at approximately 1.5 kb. This is the expected size of the AMV RNA 4-CaMV transcript which includes 9 bases between the transcription start signal and the AMV RNA 4 region, 0.88 kb of AMV RNA 4 and 0.6 kb of CaMV-derived sequence at the 3' end. The nature of the minor bands which hybridized to the AMV probe in Figure 2B is, as yet, unknown. Comparison of the hybridization intensity of the 1.5 kb transcript to the hybridization intensity of known quantities of AMV RNA 4 indicated that the transformed plants contain up to 10 pg of AMV sequences per μg of total RNA.

Analysis of pH400A4I transformants showed that RNA from 10 of 25 plants (40%) hybridized to nick-translated RNA 4 cDNA. RNA from the six selected transformants is shown in Figure 2B. The bands, approximately 2.4 kb or 3.4 kb, were larger than expected. The hybridization indicates that these bands contain AMV sequences. However, we have not yet defined the initiation and termination points of the transcripts.

Expression of coat protein

Coat protein expression for the eight selected pH400A4 transformants which consistently had high levels of expression are shown in Table 1. The amount of coat protein varied from 21 to 340 ng/mg soluble protein. Electrophoretic analysis of proteins from young leaves showed that the size of coat protein in transformants corresponded to that of virion coat protein (data not shown). Expression level depended on leaf age (Fig. 3). The greatest concentration of coat protein was in young, expanding leaves; coat protein was often not detected in older leaves more than 10 nodes from the apical meristem. In contrast, the amount of AMV-specific RNA from leaves of different ages was similar (Fig. 3 inset).

TABLE 1

EXPRESSION OF AMV COAT PROTEIN IN TRANSFORMED TOBACCO

| | | Amount of Coat Protein[b] | |
Transformant	Octopine[a]	(ng/ml sap)	(ng/mg protein)
14-1	−	3000	100
14-5	+	2300	200
27-4	+	130	21
27-35	+	590	48
37-2	+	1300	100
37-4	+	1400	180
37-5	+	500	51
37-6	−	2300	340

[a] A (+) indicates that the sample contained greater than 3 ng of octopine per mg of tissue.

[b] Coat protein concentration in leaf sap was determined by ELISA and soluble protein concentration was determined by the Bradford dye-binding assay (20).

Biological Activity of Coat Protein Expressed in Transformed Tobacco Protoplasts

Alfalfa mosaic virus can infect tobacco plants or protoplasts when the inoculum contains the three genomic RNAs plus either coat protein or RNA 4, the subgenomic messenger for coat protein. For virus replication, the coat protein in the inoculum, or that translated in the cell from RNA 4, must bind to the 3' ends of the AMV RNAs. To test whether the coat protein expressed in transformed plants could initiate replication of genomic RNAs, protoplasts from transformants were inoculated with AMV RNAs 1,2 and 3. Table 2 shows that RNAs 1,2 and 3 did not infect untransformed protoplasts. However, all four AMV RNAs did infect these protoplasts. In contrast, AMV RNAs 1,2 and 3 were able to replicate in protoplasts from plants transformed with pH400A4. These protoplasts contained, before inoculation, an average of 3×10^5 molecules of coat protein per protoplast. This is approximately 10-fold less than the number of RNA molecules per protoplast in the inoculum. Nevertheless, it is sufficient for initiation of replication. Note that, as expected, transformed protoplasts are suscep- tible to infection by unfractionated AMV RNA.

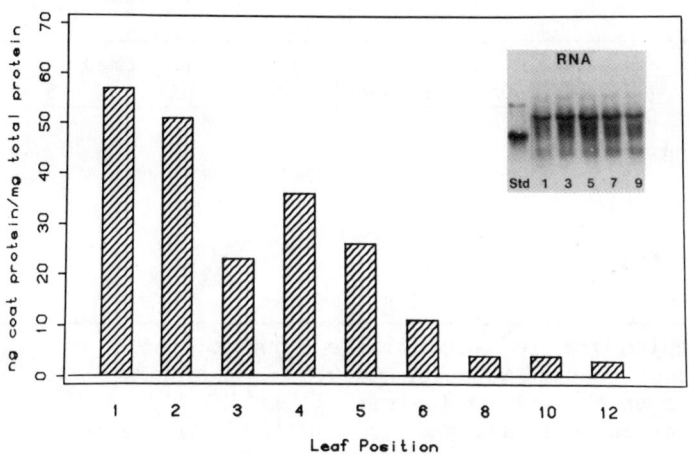

FIGURE 3. The effect of leaf position on the expression of viral coat protein. The amount of coat protein was determined for individual leaves from plant 37-4 transformed with pH400A4. Leaves were numbered down the 45 cm plant beginning with a young leaf about 4 cm long. Coat protein concentration was determined by ELISA and soluble leaf protein was determined by the Bradford dye-binding procedure (20). The autoradiograph (insert) shows RNAs that were isolated from individual leaves of another cloned 37-4 plant, separated by electrophoresis in agarose, blotted to Genetran, and hybridized to a nick-translated RNA 4 cDNA. Leaf position is indicated for each RNA sample. The positions of AMV RNA 3 and RNA 4 are shown in the Std lane.

Effect of Coat Protein in Transformed Tobacco Protoplasts Challenged with Alfalfa Mosaic Virus

To determine if coat protein decreased virus susceptibility, protoplasts from transformed plants expressing coat protein were inoculated with alfalfa

TABLE 2
BIOLOGICAL ACTIVITY OF COAT PROTEIN EXPRESSED IN
TRANSFORMED TOBACCO

Inoculum[a]	Infected Protoplasts (%)[b]		
		Transformed	
	Untransformed	14-1	14-5
AMV RNA 1+2+3	0	63	30
AMV RNA 1+2+3+4	33	61	66
Unfractionated AMV RNA	69	85	77

[a] Inoculum for 10^5 protoplasts consisted of 0.6 μg of unfractionated AMV RNA or 0.6 μg of AMV RNAs 1+2+3 alone or in combination with 1 μg AMV RNA 4.
[b] Infection was assayed by an immunofluorescent assay that detected only infected protoplasts.

mosaic virus virions. Table 3 shows that fewer transformed protoplasts were infected than were untransformed
protoplasts. In experiment 1, transformant 101 had the
highest level of coat protein and the lowest percentage
of infection. There was no difference in susceptibility
when untransformed or transformed protoplasts were inoculated with AMV RNA (Table 2 and other data not shown).
Thus, infection was suppressed only when AMV virions were
the inoculum.

DISCUSSION

Following transformation of tobacco leaf pieces with
pH400A4 and pH400A4I, 78 shoots were kanamycin-resistant,
while only 35% of these contained detectable levels of
octopine. Thus, kanamycin selection was a much better
method to identify transformants than was our octopine
assay. DNA analysis showed that AMV sequences, neomycin
phosphotransferase II sequences and octopine synthase
sequences usually were present. A few plants contained
no functional octopine synthase because a sequence 3' to

TABLE 3
PROTECTIVE EFFECTS OF COAT PROTEIN IN
TRANSFORMED TOBACCO

Protoplast Source	Coat Protein[a] (ng per 10^7 protoplasts)		Infected Protoplasts[b] (%)	Protection[c] (%)
Experiment 1				
Untransformed				
Tobacco		0	62	–
Transformant 14-1		38	32	48
101		130	5	92
102		56	9	86
107		45	17	73
108		10	26	58
Experiment 2				
Untransformed				
Tobacco		–	98	–
Transformant 14-1		–	21	79
14-5		–	44	55

[a] Coat protein was determined by ELISA.
[b] Inoculum for 10^5 protoplasts consisted of 1-3 µg AMV virions. Infection was assayed by an immunofluorescent assay that detected only infected protoplasts.
[c] Protection was calculated as 1-(% infection of transformed protoplasts ÷ % infection of untransformed protoplasts) x 100.

the octopine synthase coding region was used as a right border.

RNA analysis showed that over half of the plants transformed with pH400A4 expressed an expected 1.5 kb RNA containing AMV sequences. This RNA was easily detected in preparations of total leaf RNA that contained primarily ribosomal RNA. Detection of coat protein in transformants was more reliable in younger leaves. Older leaves contained lower amounts of coat protein per mg soluble protein, but contained nearly the same amount of AMV-specific RNA per mg total RNA as did younger leaves. This suggests that, in older leaves, there is an accelerated degradation of coat protein or a lower translational efficiency than in younger leaves. Plants expressing AMV coat protein showed no obvious virus symptoms. However, the concentration of coat protein in the transformants was about 50-100 times lower than that in infected plants. Thus, we cannot rule out a role for coat protein in symptom production.

Analysis of the regenerants from transformations with pH400A4I showed that nearly all contained expected integration patterns for the AMV cDNA. The levels of AMV-specific RNA expression in pH400A4I transformants were considerably lower than in the pH400A4 transformants. This may be due to degradation of the AMV complementary RNA. As far as is known, the complementary RNA is not a messenger RNA in wheat germ cell-free translation systems or in *Xenopus* oocytes (Loesch-Fries and Nelson, manuscript in preparation); thus, this RNA may not be stable in tobacco cells.

Infection of protoplasts with AMV genomic RNAs showed that the coat protein expressed in tobacco initiates infection of RNA. This result is significant because it shows that the coat protein is stable in the absence of infection and is neither inactivated by the cell nor partitioned in a nonfunctional fashion. As far as we know, this is the first report of a functionally active virus protein synthesized from a gene inserted into a plant genome.

Although coat protein in protoplasts from transformants enabled genomic AMV RNAs to replicate, coat protein may protect transformed protoplasts from infection with AMV virions. The finding that AMV RNA replicated in transformed protoplasts indicates that transformed cells are able to support virus infection. Resistance to infection by virions suggests that attachment or entry of virions through the plasma membrane is altered in transformed protoplasts or that virions are unable to release the RNA into the protoplast. Comparison of results from protoplasts to that from whole plants will be needed to determine if the mechanism of resistance is similar.

ACKNOWLEDGEMENTS

We are especially grateful to Ebrahim Firoozabady for helpful suggestions regarding transformation and selection protocols. Also we thank Jane Cramer and Thomas Zinnen for critically reading the manuscript.

This is Agrigenetics Advanced Science Company Publication No. 61.

REFERENCES

1. Sherwood JL, Fulton RW (1982). The specific involvement of coat protein in tobacco mosaic virus cross protection. Virology 119:150.
2. de Zoeten GA, Gaard G (1984). The presence of viral antigen in the apoplast of systemically virus-infected plants. Virus Research 1:713.
3. Wilson TMA, Watkins PAC (1986). Influence of exogenous viral coat protein on the cotranslational disassembly of tobacco mosaic virus (TMV) particles *in vitro*. Virology 149:132.
4. Zinnen TM, Fulton, RW (1986). Cross protection between sunn-hemp mosaic and tobacco mosaic viruses. J. gen Virol (in press).
5. Zaitlin M (1976). Viral cross protection: more understanding is needed. Phytopathology 66:382.
6. Sarkar S, Smitamana P (1981). A proteinless mutant of tobacco mosaic virus: evidence against the role of a viral coat protein for interference. Mol. Gen. Genet. 184:158.
7. Schmelzer K (1963). Untersuchungen an Viren der Zier - und Wildgeholze. Phytopath. Z. 46:2
8. Loesch-Fries LS, Jarvis NP, Krahn KJ, Nelson SE, Hall TC (1985). Expression of alfalfa mosaic virus RNA 4 cDNA transcripts *in vitro* and *in vivo*. Virology 146:177.
9. Balázs E, Bouzoubaa S, Guilley H, Jonard G, Paszkowski J, Richards K (1985). Chimeric vector construction for higher-plant transformation. Gene 40:343.
10. Merlo D, Talbot D, Stratton H, Sutton D, Stock C, Staffeld G (1985). Construction and utilization of kanamycin resistance-eukaryotic promoter cassettes in disarmed Ti and binary vector systems. Abstracts of First International Congress of Plant Molecular Biology, Savannah, GA.
11. Kemp JD, Sutton DW, Fink C, Barker RF, Hall TC (1983). Agrobacterium-mediated transfer of foreign genes into plants. In Owens LD (ed): Beltsville Symposia in Agriculture 7: "Genetic Engineering

Applications to Agriculture" Ottawa: Rowman and Allenheld, p 215.

12. Horsch RB, Fry JE, Hoffmann NL, Eichholtz D, Rogers SG, Fraley RT (1985).   A simple and general method for transferring genes into plants.   Science 227:1229.

13. Loesch-Fries LS, Hall TC (1980).   Synthesis, accumulation and encapsidation of individual brome mosaic virus RNA components in barley protoplasts.   J. gen. Virol. 47:323.

14. Murray MG, Thompson WF (1980).   Rapid isolation of high molecular weight plant DNA.   Nuc. Acids Res. 8:4321.

15. AMF, Inc. (1984).   Capillary transfer of DNA, DNA hybridization, and DNA rehybridization using *Zetabind*.   Technical data 20.7, Microfiltration Products Division, 400 Research Parkway, Meriden, CT.

16. Murray MG, Hoffman LM, Jarvis NP (1983).   Improved yield of full-length phaseolin cDNA clones by controlling premature anticomplementary DNA synthesis.   Plant Mol. Biol. 2:075.

17. Loesch-Fries LS, Halk EL, Nelson SE, Krahn KJ (1985).   Human leukocyte interferon does not inhibit alfalfa mosaic virus in protoplasts or tobacco tissue.   Virology 143:626.

18. Halk EL (1986).   Serotyping plant viruses with monoclonal antibodies.   In Weissbach A, Weissbach H (eds): "Methods in Enzymology 118,"   Orlando: Academic Press, p 766.

19. Dhaese P, De Greve H, Gielen J, Seurinck J, Van Montagu M, Schell J (1983).   Identification of sequences involved in the polyadenylation of higher nuclear transcripts using *Agrobacterium* T-DNA genes as models.   The EMBO J. 2:419.

20. Bradford MM (1976).   A rapid and sensitive method for the quantitation of microgram quantities of protein utilizing the principle of protein dye binding.   Anal. Biochem. 72:248.

Molecular Strategies for Crop Protection, pages 235–242
© 1987 Alan R. Liss, Inc.

GENETIC ENGINEERING OF VIRAL PESTICIDES:
EXPRESSION OF FOREIGN GENES IN NONPERMISSIVE CELLS[1]

Luis F. Carbonell,[2] and Lois K. Miller[2]

Department of Bacteriology and Biochemistry
University of Idaho, Moscow, Idaho 83843

ABSTRACT In a model study of baculovirus-mediated
foreign gene expression in various cell types,
substantial   levels   of   chloramphenicol acetyltransferase   (CAT)   activity   were   observed in
lepidopteran and nonpermissive dipteran cell lines
infected with a recombinant Autographa californica
nuclear polyhedrosis virus, AcNPV L1LC-galcat, which
carries the CAT gene under the control of a
mammalian-active promoter.   Baculovirus-mediated
expression of foreign genes in insect cells suggests
a possible route for improving the efficacy and
acceptability of viruses as pest control agents.
Extremely low levels of CAT activity were found in
AcNPV L1LC-galcat-infected mouse cells both in the
presence and absence of cycloheximide demonstrating
that this CAT activity is not due to active
expression of the baculovirus-borne gene in the
mouse cell line. The inability of the CAT gene to
be expressed in this cell line provides additional
assurance of the safety of viral pesticides with
respect to mammalian species.

[1]This material is based upon work supported by the
U.S. Dept. of Agriculture, Office of Grants and Program
Systems under competitive research grant number
85-CRCR-1-1796.   Research paper 8655 of the Idaho
Agricultural Experiment Station.
[2]Present address: Departments of Entomology and
Genetics, The University of Georgia, Athens, Georgia
30602.

INTRODUCTION

In response to the need for decreasing our dependence on chemicals for insect pest control, insect-specific microbial pathogens are being developed as pesticides (1,2). Insect viruses of the baculovirus group have been intensively studied in this regard and several different baculoviruses have been registered as pesticides by the U.S. Environmental Protection Agency (EPA) for controlling specific pest populations (2,3). Despite these efforts to bring viruses to the marketplace and the field, adoption of this form of pest control by either industry or the farmer is hampered by: (a) the cost of producing and marketing individual pesticides for each different pest species and (b) the delayed action of these viruses in killing the pest compared to the rapid "knock-down" effect of chemical insecticides.

The application of recombinant DNA technology to enhance the efficacy and acceptability of viral pesticides holds considerable promise (2,4,5). Specifically, it may be possible to engineer a baculovirus to express a gene encoding an insect-specific neurotoxin or other protein that elicits a dramatic biological response and consequently enhance the rate at which the virus affects insect feeding behavior. At the same time, expression of foreign genes of this nature might expand the effective host range of the recombinant virus (4). Although expansion of the host range to other pest insects could dramatically decrease production and marketing costs and thereby make the recombinant virus more acceptable to industry and users, dramatic expansion of the virus host range could make the recombinant virus unacceptable to the EPA or to regulatory agencies controlling the release of genetically engineered organisms into the environment.

The bacterial chloramphenicol acetyltransferase (CAT) and β-galactosidase genes were inserted into a bac-ulovirus genome and the resulting recombinant virus, AcNPV L1LC-galcat, was employed in model studies to assess the possible extension of baculovirus host range (4). Regulated expression of both genes was observed in permissive lepidopteran cells. Expression of baculovi-rus-borne genes in nonpermissive dipteran cells was dependent on the type of promoter controlling transcrip-tion; the β-galactosidase gene, controlled by the very late polyhedrin promoter was not expressed; whereas, the

CAT gene, controlled by the promoter of the Rous sarcoma virus long terminal repeat RSV-LTR, was expressed. Low levels of CAT activity were also found in mouse cells upon continuous exposure to high titers of AcNPV L1LC-galcat which contains the CAT gene under the control of the mammalian-active RSV-LTR promoter. In this report, we now trace the low levels of CAT activity found in AcNPV L1LC-galcat-infected mouse cells to enzymes associated with the virus particles used to infect cells; the results corroborate the relative safety of recombinant viruses to mammalian species.

METHODS

The cells and virus used were described previously as were procedures for infection of cell monolayers (4). The stock of recombinant virus, AcNPV L1LC-galcat, used for infecting cells in this current study was purified from the growth medium of infected permissive cells by pelleting the virus by centrifugation, resuspending the virus in medium, banding the virus by isopycnic sucrose density gradient centrifugation (6), repelleting, and suspending the final pellet in appropriate media. Infections in the presence of cycloheximide included pretreatment of the cells with 100 µg/ml cycloheximide for 1 hr. Cycloheximide was also present during the 1 hr virus adsorption and throughout the remaining time of incubation. CAT synthesis and total protein were assayed as described (4). Specific activities are reported as nmol chloramphenicol acetylated per min and mg of protein (nmol CM/min mg).

RESULTS

Expression of the CAT gene in Drosophila melanogaster cells. Previous studies demonstrated that a recombinant baculovirus, AcNPV L1LC-galcat, which contains the CAT gene under RSV-LTR control, expresses approximately the same level of CAT activity in permissive lepidopteran cells as in the nonpermissive dipteran cells of Drosophila melanogaster (4). The presence of the CAT activity in D. melanogaster cells

infected with the recombinant virus AcNPV L1LC-galcat is
dependent on active protein synthesis as demonstrated in
Figure 1.   Cycloheximide,   an   inhibitor   of eukaryotic
cytoplasmic protein synthesis, effectively   inhibits   the
expression of the CAT gene borne by the virus AcNPV
L1LC-galcat in D. melanogaster cells. This confirms our
previous conclusion that AcNPV efficiently enters and
expresses   at   least   a   portion   of its genome in a wide
range of insect cells.

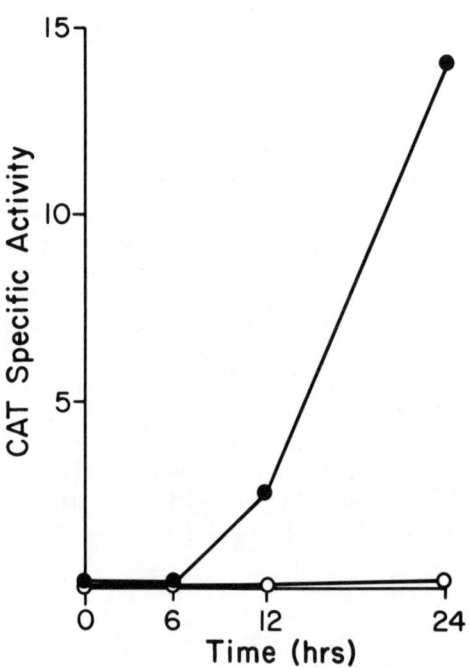

FIGURE 1.   Effect of cycloheximide on CAT gene
expression in Drosophila melanogaster (DL-1, WR) cells
infected with AcNPV L1LC-galcat [20 plaque forming units
(pfu) per cell].   The specific activity of CAT (nmol
CM/min mg) was determined at various times post-infection
from CAT assays and protein determinations on crude cell
lysates of infected cells of D. melanogaster treated (•)
or untreated (o) with cycloheximide.

Expression of the CAT gene in mouse cells. Previous studies demonstrated that CAT activity was inefficiently expressed in AcNPV L1LC-galcat-infected mouse cells and was detected at 48 hrs only if the virus inoculum was allowed to remain in contact with the cells during the infection process (4). Continual exposure of the L929 cell line with AcNPV L1LC-galcat in the absence of cycloheximide resulted in a gradual increase in CAT activity during the infection (Figure 2).

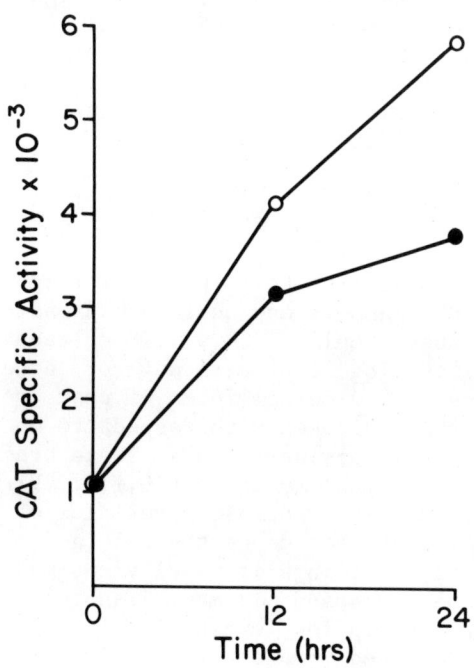

FIGURE 2. Appearance of CAT in Mus musculus cells continuously exposed to AcNPV L1LC-galcat: Effect of cycloheximide. Monolayers of M. musculus (L929) cells were continually exposed to AcNPV L1LC-galcat (100 pfu per cell) in the presence (●) or absence (o) of cycloheximide. At various times after the 1 hr initial exposure period, cells were washed with phosphate buffered saline (6), pelleted, disrupted by freeze-thawing, and CAT specific activity determined.

However, even after 24 hrs of infection, only very low
levels of CAT activity were associated with the cells
(Figure 2, note difference in scale of CAT specific
activity from Figure 1). Similar levels of CAT activity
were observed in cycloheximide-treated infected mouse
cells indicating that the CAT activity observed in these
cells is brought in with the virus particle. At 0 hrs
after infection (equivalent to the end of the 1 hr virus
adsorption period), a low level of CAT activity is
already observed associated with the cells. Specific
activity of CAT increased in both the presence and
absence of cycloheximide but appeared slightly higher in
the inhibitor-treated cells (Figure 2). This may be due
to decreasing levels of total protein in extracts of
cycloheximide-treated infected cells compared to
increasing levels of total protein in extracts of
untreated cells (data not shown).

DISCUSSION

It is possible to insert foreign genes into
baculovirus DNA genomes and achieve expression of those
genes in insect cells (7,8,9). Such technology can be
useful in improving the efficacy or acceptability of
these viruses as biological forms of pest control. The
safety of these viruses with respect to nontarget hosts
remains a central consideration. There are reports that
AcNPV can enter mammalian cells (10,11). Virus particles
are usually found in cytoplasmic vesicles of the cell.

In initial studies using the CAT gene as a model
system for determining if baculovirus-mediated foreign
genes could be expressed in mammalian cells, we reported
the presence of very low levels of CAT activity in AcNPV
L1LC-galcat-infected mouse cells (4). We show in this
report that this low level of CAT activity is not due to
the expression of the recombinant viral DNA in mammalian
cells but is brought into the cell with the virus. The
virus used for inoculation was extensively purified so
the CAT activity is closely associated with the virus
particles.

The results have important implications regarding
the safety of these viruses to nontarget mammals.
Although mammalian cells may slowly take up the viruses
by endocytosis or phagocytosis, the viral DNA fails to be
expressed in the nucleus. In contrast, AcNPV effectively

enters nontarget dipteran cells and efficiently expresses the RSV-LTR-CAT gene. This observation may have implications for the expansion of the effective host range of a recombinant virus. A recombinant virus carrying an insect neurotoxin gene under the control of an insect-active promoter might be able to paralyze an insect even if it does not replicate in the insect. This would require efficient expression of the toxin gene in midgut cells which are the primary site of virus infection in insects. We are continuing this line of inquiry by inserting an insect-specific neurotoxin gene into AcNPV, monitoring gene expression and observing the effects of the recombinant virus on insect behavior.

REFERENCES

1. Burges HD (ed) (1981). "Microbial Control of Pests and Plant Diseases 1970-1980." New York: Academic Press.
2. Miller LK, Lingg AJ, Bulla Jr LA (1983). Bacterial, viral, and fungal insecticides. Science 219:715.
3. Miltenburger HG, Krieg A (1984). Bioinsecticides: II. Baculoviridae. In Mizrahi A, van Wezel AL (eds): "Advances in Biotechnological Processes 3," New York: Alan R. Liss, p 291.
4. Carbonell LF, Klowden MJ, Miller LK (1985). Baculo-virus-mediated expression of bacterial genes in dipteran and mammalian cells. J Virol 56:153.
5. Kirschbaum JB (1985). Potential implication of genetic engineering and other biotechnologies to insect control. Ann Rev Entomol 30:51.
6. Lee HH, Miller LK (1978). Isolation of genotypic variants of Autographa californica nuclear polyhedrosis virus. J Virol 27:754.
7. Maeda S, Kawai T, Obinata M, Fujiwara H, Horiuchi T, Saeki Y, Sato Y, Furusawa M (1985). Production of human interferon α in silkworm with a baculovirus vector. Nature 315:592.
8. Pennock GD, Shoemaker C, Miller LK (1984). Strong and regulated expression of Escherichia coli β-galactosidase in insect cells with a baculovirus vector. Mol Cell Biol 4:399.

9. Smith GE, Summers MD, Fraser MJ (1983). Production of human beta interferon in insect cells infected with a baculovirus expression vector. Mol Cell Biol 3:2156.

10. Gröner A, Granados RR, Burand JP (1984). Interaction of <u>Autographa</u> <u>californica</u> nuclear polyhedrosis virus with two nonpermissive cell lines. Intervirology 21:203.

11. Volkman LE, Goldsmith PA (1983). In vitro survey of <u>Autographa</u> <u>californica</u> nuclear polyhedrosis virus interaction with nontarget vertebrate cells. Appl Environ Microbiol 45:1085.

Molecular Strategies for Crop Protection, pages 243–252
© 1987 Alan R. Liss, Inc.

CHARACTERIZATION OF MULTIMERIC FORMS OF CUCUMBER MOSAIC VIRUS SATELLITE RNA[1]

Nevin D. Young, Peter Palukaitis, and Milton Zaitlin

Department of Plant Pathology, Cornell University
Ithaca, New York 14853 U.S.A.

ABSTRACT Northern blot analysis of RNA prepared from plants infected with cucumber mosaic virus (CMV) plus its satellite-RNA (CMV-Sat) demonstrated that multimers of CMV-Sat were present in significant amounts in infected leaf tissue. Multimers of minus polarity were more prevalent than those of plus polarity. In order to characterize the generation of these multimeric forms of CMV-Sat, we developed a cell-free system which was capable of synthesizing CMV-Sat-specific RNA products in vitro. Upon non-denaturing agarose gel electrophoresis, these products consisted of RNAs with mobilities between double-stranded (ds) monomer and dimer CMV-Sat, while on denaturing gels, the CMV-Sat-specific products consisted predominantly of CMV-Sat monomer-length RNAs.

INTRODUCTION

Cucumber mosaic virus (CMV) is a plus stranded RNA virus comprised of three genomic RNA molecules (RNAs 1-3) and one subgenomic RNA molecule (RNA 4), the sequence of which is totally contained within RNA 3. A fifth RNA molecule, known as satellite RNA (CMV-Sat), is often associated with the RNAs of CMV. This small (1.1 x 10^5 MW) RNA contains little sequence homology with CMV, yet depends upon the CMV "helper virus" for both replication

[1]This research was supported by National Science Foundation Grant 84-09851.

and encapsidation. The presence of CMV-Sat has a dramatic
effect on the severity of symptoms cause by CMV infection
(17) and high levels of CMV-Sat in infected tissue are
associated with depressed levels of helper virus RNA in
virions (11).

Satellite RNAs have been found associated with many
other plant viruses (5). In several cases, multimeric
forms of these satellite RNAs have been described (8, 9).
In this report, we describe experiments using Northern
blot analysis to demonstrate that, as is the case with
these other satellite RNAs, multimeric forms of CMV-Sat
exist in infected plants. In addition, we have developed
a cell-free system capable of _in_ _vitro_ synthesis of
CMV-Sat-specific RNA products. The major products of this
cell-free system are labeled RNA species which migrate
upon non-denaturing gel electrophoresis between ds-monomer
and dimer length CMV-Sat RNAs.

MATERIALS AND METHODS

Virus and satellite RNA. The Fny-stain of CMV alone
was maintained in Cucumis pepo and Fny-CMV plus the
B-satellite (B-Sat) RNA of CMV (14) was maintained in
Nicotiana tabacum (cv. Xanthi).

Identification of CMV-Sat multimers in infected plant
tissue. Nucleic acids from tomatoes infected with Fny-CMV
and B-sat RNA were extracted according to the procedure
described by Zelcer et al. (19). The nucleic acids were
partitioned into 2 M LiCl soluble and insoluble fractions
(19), denatured in formaldehyde/formamide, and subjected
to electrophoresis in a 1.5% agarose gel according to
Palukaitis et al. (13). The nucleic acids in the gel were
blotted bidirectionally to nitrocellulose. The blots were
incubated with cDNA probes specific for either plus or
minus CMV-Sat RNA sequences (see below), washed, and
autoradiographed as previously described (13). The plus
strand of CMV-Sat refers to the RNA molecule which is
encapsidated into virions. The cDNA probes were prepared
using decanucleotide primers complementary to the 3'-ends
of the plus and minus B-Sat RNAs (13) and the template
RNAs were B-Sat RNA and heat denatured ds B-Sat RNA to
generate plus-sequence-specific and minus-sequence-
specific cDNAs, respectively (14).

Replication of CMV-Sat in vitro. Extracts prepared from infected C. pepo and N. tabacum were used to inoculate leaves of N. tabacum. Five or 6 days after inoculation, a crude 30,000 g membrane fraction was prepared from the primary inoculated leaves (18) and in vitro synthesis of CMV and CMV-Sat replication products was carried out under conditions identical to those used for the in vitro synthesis of tobacco mosaic virus replicative structures (18). In these experiments, replicase reactions were carried out for either 5 or 20 minutes with α-^{32}P-UTP as the radiolabeled nucleotide triphosphate. In vitro synthesized RNA products were analyzed by 2% agarose gel electrophoresis under non-denaturing (19) or glyoxal-denaturing conditions (10).

In order to determine the polarity of in vitro synthesized CMV-Sat products, labeled RNA molecules were electroeluted from non-denaturing agarose gels (18) and analyzed by two-dimensional RNA fingerprinting (12). Ds B-Sat RNA from tissue extracts and B-Sat RNA isolated from virions (14) were also analyzed by RNA fingerprinting for comparison with the results obtained with electroeluted CMV-Sat replication products.

RESULTS

Identification of CMV-Sat multimers in infected leaf tissue. Northern blotting using cDNAs specific to the plus and minus strands of B-Sat RNA demonstrated that tomato plants infected with CMV and its satellite RNA contain, in addition to CMV-Sat unit length RNA (Fig. 1a, lane 3), a series of multimeric forms of CMV-Sat of both plus and minus polarity (Fig. 1a, lane 1 and Fig. 1b, lane 1, respectively). The smallest and most prevalent form of both plus and minus strand RNAs comigrated with isolated and denatured single-stranded (ss)-monomer CMV-Sat, while the next larger species of both polarities comigrated with isolated and denatured ss-dimer CMV-Sat. The sizes of other multimers were not determined. Minus strand multimers were present as longer concatemers than plus strand multimers. Virtually all of the multimeric forms were present in the 2 M LiCl soluble fraction, (Fig. 1a, lane 1, 1b, lane 1) suggesting that the multimers were mostly present in ds, and not ss, forms (Fig. 1a, lane 2, 1b, lane 2). Some plus-stranded multimeric CMV-Sat RNAs

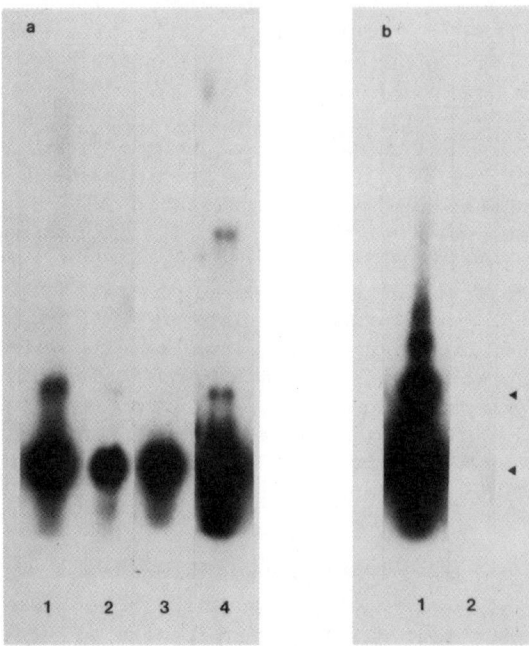

FIGURE 1. Northern blot analysis of multimeric forms of CMV-Sat RNA. RNAs from CMV plus CMV-Sat infected tomato leaves were fractionated by 2 M LiCl precipitation, denatured by formaldehyde-treatment, separated by 1.5% agarose gel electrophoresis, and transferred bidirectionally to nitrocellulose filters (see Methods). a. Negative-strand cDNA probe detecting plus-strand CMV-Sat RNAs: lane 1, LiCl supernatant; lane 2, LiCl pellet; lanes 3 and 4, purified virion RNA. Lanes 1, 2, and 3 were autoradiographed (with intensifying screen at -80°C) for 18 hours, while lane 4 was autoradiographed for 80 hours. b. Plus-strand cDNA probe detecting minus-strand CMV-Sat RNAs: lane 1, LiCl supernatant; lane 2, LiCl pellet. Lanes 1 and 2 were autoradiographed for 18 hours. Arrows mark migration of purified and denatured CMV-Sat monomer and dimer RNAs.

exist in a ss-form, however, as shown by their presence in virions (Fig. 1a, lane 4). Virions do not encapsidate minus strand CMV-Sat RNA molecules (results not shown).

RNA products of CMV-Sat replication in vitro. The products of CMV and CMV-Sat replication in vitro are shown in Fig. 2. Several labeled RNA products were specific to samples from CMV-Sat infected plants. In short labeling periods (5 minutes; Fig. 2a, lane 2) the major CMV-Sat specific RNAs were two heterogeneous species which migrated between the 399 and 517 bp molecular weight markers. In longer labeling periods (20 minutes; Fig. 2a, lane 4) a prominent RNA species which migrated slightly more slowly than authentic ds-CMV-Sat RNA was the major product of in vitro replication (labeled "A" in Fig. 2a). In addition to this RNA species, higher molecular weight RNAs which appear to be identical to those observed in short labeling periods were also present. In "pulse-chase" experiments (results not shown) radioactivity incorporated into the higher molecular weight species during a short pulse could be chased into an RNA species identical in mobility to that of RNA "A" in Fig. 2. Another, much more rapidly migrating CMV-Sat specific RNA comigrated with authentic ss-CMV-Sat RNA (Fig. 2a, lane 2). Synthesis of this RNA in vitro was often obscured by the large amount of labeled low molecular weight RNAs produced by the host RNA-dependent RNA polymerase (16).

The major RNA products of short labeling periods, when analyzed after denaturation by glyoxal, consisted of three equally labeled species, the largest of which comigrated with ss-CMV-Sat monomer (Fig. 2b, lane 2). In longer labeling periods, the only prominent CMV-Sat-specific RNA product (after denaturation) was an RNA comigrating with ss-CMV-Sat monomer (Fig. 2b, lane 4). Another CMV-Sat-specific RNA comigrating with the dimer form of CMV-Sat was also occasionally observed. Synthesis of this RNA species, however, was variable between separate experiments (results not shown).

The major RNA products specific to the CMV helper virus included two prominent high molecular weight species on non-denaturing gels (Fig. 2a, lanes 1 and 3). These two large RNA species probably correspond to replicative forms (RFs) of the two largest genomic RNAs of CMV, RNA-1 and 2. Less prominent bands corresponding to the RF forms of CMV RNA-3 and the subgenomic RNA-4 were also often

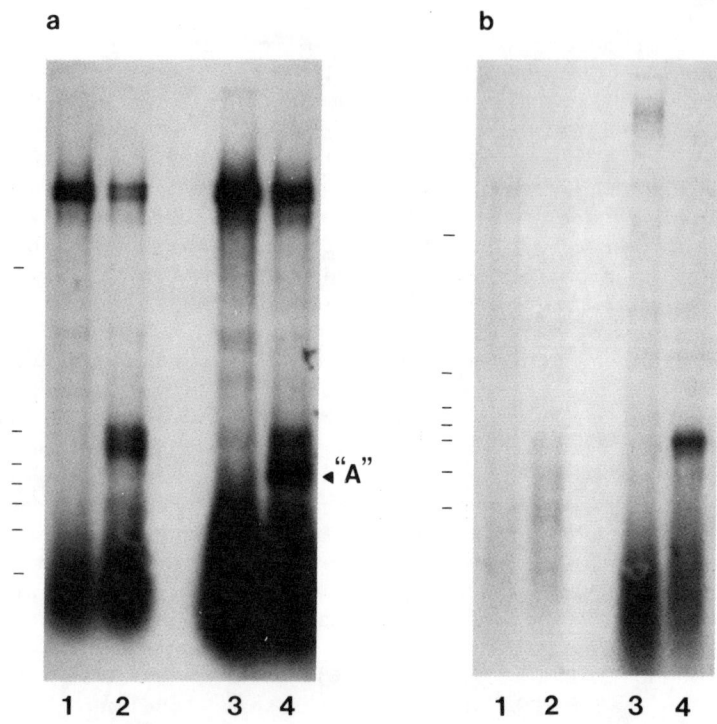

FIGURE 2. Products of CMV and CMV plus CMV-Sat replication in vitro. The 30,000 g pellet fractions were prepared from infected N. tabacum leaves and used in in vitro replication reactions. Labeled RNA products were isolated and separated (with or without denaturation) on 2% agarose gels. Radiolabeled RNAs were visualized by autoradiography for 18 hours (see Methods). a. Non-denatured RNA samples: lanes 1 and 3, CMV alone; lanes 2 and 4, CMV plus CMV-Sat. b. Glyoxal denatured RNA samples: lanes 1 and 3 CMV alone; lanes 2 and 4, CMV plus CMV-Sat. Labeling period was 5 minutes for lanes 1 and 2 and 20 minutes for lanes 3 and 4 in both a and b. "A" is described in text. Dashes mark the migration of Hinf I-digested pBR322 DNA molecular weight markers (1631, 517/506, 396, 344, 298, 221/220, 154 bp).

observed. Only faint bands corresponding to CMV RNAs 1-4 were observed in glyoxal denatured samples (Fig. 2b, lanes 1 and 3). In vitro synthesis of RNAs 1-4 of CMV has been observed and described in detail elsewhere (7).

RNA fingerprinting of the major CMV-Sat specific RNA product. In order to determine the polarity of the major labeled CMV-Sat specific RNA species, the RNA corresponding to species "A" in Fig. 2 was electroeluted from non-denaturing agarose gels and analyzed by RNA-fingerprinting (Fig. 3b). CMV-Sat RNA, isolated from virions, and ds-CMV-Sat RNA, isolated from infected leaf tissue, were also analyzed by RNA fingerprinting (Figs. 3a and 3c, respectively). The fingerprint of virion RNA is derived from the plus strand of CMV-Sat, while the fingerprint of ds-CMV-Sat is derived from both plus and minus strands. Fingerprinting of the major CMV-Sat specific RNA synthesized in vitro demonstrated that it was overwhelmingly composed of sequences of the plus strand. Several diagnostic spots (arrows on Fig. 3a and b) correspond almost exactly between the fingerprints of in vitro synthesized and virion CMV-Sat RNA. Differences do exist between these fingerprints, however. In some cases, spots are weaker or even absent from the fingerprint of in vitro synthesized RNAs. These differences may be due to the fact that the in vitro products were labeled internally into uridine residues, while radioactivity incorporated into fragments derived from virion or ds-CMV-Sat RNA was added by end-labeling (3). It is also possible that the in vitro replication system was capable of synthesizing only a subset of all possible CMV-Sat sequences since in vitro replication did not require (nor did it respond to) added CMV-Sat RNA template.

DISCUSSION

Northern blot analysis of CMV-Sat plus and minus strand RNA in infected tissue indicates that multimers of both polarities exist in significant levels. Multimers of minus polarity are more common, while most RNA of plus polarity is found in the monomer form along with a small fraction in the dimer form. Many other plant pathogenic RNAs, including viroids (1) and circular satellites, such as the satellite RNA of tobacco ringspot virus (TRSV) (8),

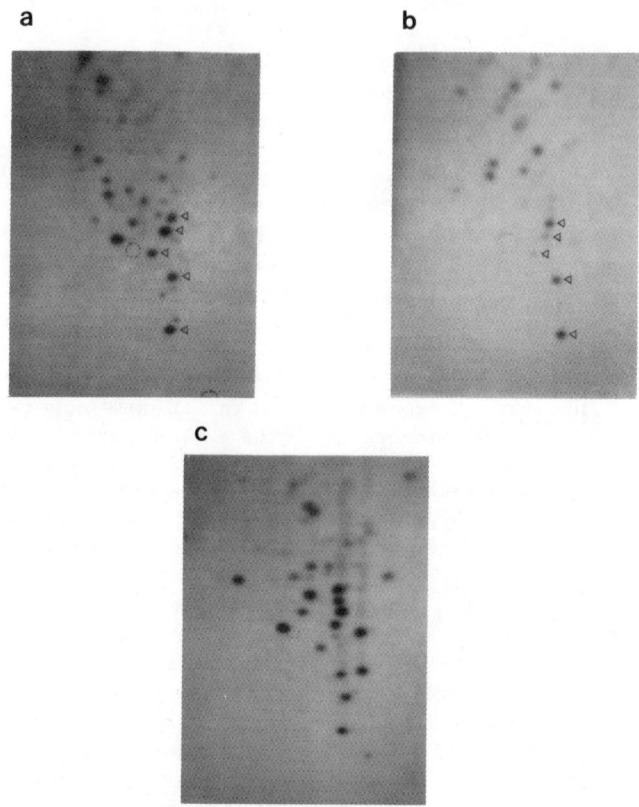

FIGURE 3. RNA fingerprinting of CMV-Sat RNAs.
a. Plus-strand B-Sat RNA was digested with RNase T1, end-
labeled with γ-^{32}P-ATP, analyzed by two dimensional
electrophoresis and autoradiographed (as described in
Methods). b. Radiolabeled RNA products of in vitro
replication (corresponding to Band "A" in Fig. 2) were
isolated by electroelution and analyzed as in 3a, with the
exception that there was no end-labeling treatment.
c. Double-stranded B-Sat RNA was analyzed as in 3a.
Fingerprints in 3a and 3c were autoradiographed for 30
mins. and the fingerprint in 3b was autoradiographed for 5
days.

produce multimers of both polarity during infection, as does the satellite of peanut stunt virus (PARNA-5) (9), a satellite which resembles CMV-Sat in having a cucumovirus helper virus.

In the case of viroids and circular satellites, multimers are thought to arise during replication by the "rolling circle" model (2). However, neither CMV-Sat nor PARNA-5 appears to exist in a circular form. Two dimensional gel electrophoresis failed to uncover circular forms of either satellite (9) and the termini of the ss and ds forms of CMV-Sat were found to be very similar to those of RNA-3 of CMV (4), suggesting that CMV-Sat might be replicated by the same mechanism as the genomic RNAs of CMV. If so, the generation of multimers of CMV-Sat during replication is difficult to understand. No multimers of the genomic RNAs of CMV are produced during their replication in vitro (7) and recent models for the replication of plus stranded RNA viruses do not include any multimeric forms of genomic RNAs (6).

In order to determine whether multimers are involved in the replication of CMV-Sat, we have developed a cell free system capable of CMV-Sat replication in vitro. This system neither requires nor responds to exogenous CMV-Sat RNA. However, this run-off form of replication synthesizes many CMV-Sat specific replicative structures. The major undenatured product of in vitro replication is a highly labeled, stable RNA species migrating slightly slower than ds-CMV-Sat RNA. The labeled portion of this RNA species consists of monomer CMV-Sat RNA of plus polarity. Rapidly labeled, higher molecular weight RNA complexes appear to act as precursors in the synthesis of this major product and the labeled portions of these precursor RNAs consist of monomer CMV-Sat RNA and other, smaller RNA molecules. It is possible that these complexes are composed of elongating plus-sense CMV-Sat RNA on a template of multimeric minus-sense CMV-Sat RNA. This model is supported by the fact that a large portion of the minus-sense CMV-Sat RNA is present as multimeric ds-molecules. Unfortunately, the in vitro replication system described here synthesizes CMV-Sat RNAs of plus polarity only and therefore provides no information on how such multimeric minus-strand CMV-Sat RNAs might have been generated or how these multimers are converted to monomers during replication. Multimers and monomers of TRSV-Sat RNA interconvert in a totally protein independent reaction

(15) and CMV-Sat RNAs appear to have the same properties (Palukaitis and Young, unpublished observations). RNA processing may, therefore, be involved in the generation of CMV-Sat multimers and we are currently performing experiments to determine the role of RNA-processing in CMV-Sat replication.

REFERENCES

1. Branch AD, Robertson HD, Dickson E (1981). Proc Natl Acad Sci USA 78:6381-6385.
2. Branch AD, Robertson HD (1984). Science 223:450-455.
3. Chaconas G, van de Sande JA (1980). Methods in Enzymology Vol. 65:75-85.
4. Collmer CW, Kaper JM (1985). Virology 145:249-259.
5. Francki RIB (1985). Ann Rev Microbiol 39:151-174.
6. Hull R, Maule AJ (1985). In RIB Francki (ed) "The Viruses: Plant Viruses Vol. I," New York: Plenum Press, pp 83-115.
7. Jaspers EMJ, Gill DS, Symons RH (1985). Virology 144:410-425.
8. Kiefer MC, Daubert SD, Schneider IR, Bruening G (1982). Virology 121:262-273.
9. Linthorst HJM, Kaper JM (1984). Virology 139:317-329.
10. McMaster GK, Carmichael GG (1977). Proc Natl Acad Sci USA 74:4835-4838.
11. Mossop DW, Francki RIB (1979). Virology 95:395-404.
12. Palukaitis P (1984). K Maramorosch, H Koprowski (eds) "Methods in Virology, Vol. VII," New York: Academic Press, pp 259-311.
13. Palukaitis P, Garcia-Arenal F, Sulzinski MA, Zaitlin M (1984). Virology 131:533-545.
14. Palukaitis P and Zaitlin M (1984). Virology 132: 426-435.
15. Prody GA, Bakos JT, Buzayan JM, Schneider IR, Bruening G (1986). Science 231:1577-1580.
16. Romaine CP and Zaitlin M (1978). Virology 86:241-253.
17. Waterworth HE, Kaper JM, Tousignant ME (1979). Science 204:845-847.
18. Young ND, Zaitlin M (1986). Plant Molecular Biology (In Press).
19. Zelcer A, Weaber KF, Balasz E, Zaitlin M (1981). Virology 113:417-427.

Molecular Strategies for Crop Protection, pages 253–265
© 1987 Alan R. Liss, Inc.

HOST RESPONSE TO CAULIFLOWER MOSAIC VIRUS (CaMV)
IN SOLANACEOUS PLANTS IS DETERMINED BY A 496 bp
DNA SEQUENCE WITHIN GENE VI[1]

J. E. Schoelz, R. J. Shepherd, and S. D. Daubert

Department of Plant Pathology
University of Kentucky, Lexington, Kentucky 40546

ABSTRACT Most strains of CaMV are limited to the
cruciferae in host range. However, some CaMV strains
replicate and cause hypersensitive (local) reactions
on a few solanaceous species. Recently an unusual
strain, D4, has been isolated which causes systemic
infections in various solanaceous plants, including
Datura stramonium and Nicotiana bigelovii. In the
present study we have constructed recombinant virus
genomes between D4 and ordinary CaMV strains to
determine which portion of the genome is responsible
for the expanded host range. A 496 bp fragment
consisting of the first half of the coding region of
gene VI was found to determine the type of host
response (localized hypersensitivity versus systemic
chlorotic mottle). Sequence comparisons between D4
and the incompatible strain CM1841 revealed that
within the 496 bp DNA segment there were 20
nucleotide differences which resulted in amino acid
exchanges.

INTRODUCTION

Cauliflower mosaic virus (CaMV) is a plant virus
whose genome is composed of circular, double-stranded DNA
(1). Many isolates of CaMV have been cloned in infectious

[1]This work was supported by NSF grant PCM-8342878, USDA
grants 84-CRCR-1-1505 and 85-CRCR-1-1743.

form and three of those isolates have been completely
sequenced, revealing a coding capacity organized into six
major open reading frames (ORF's) (2-4). The functions of
four of the six ORF's have been identified. The aphid
aquisition factor, coat protein, reverse transcriptase and
putative inclusion body protein have been identified with
regions II, IV, V, and VI, respectively (5-9).

In an effort to identify CaMV genes which determine
host specificity, we have been studying the ability of
three CaMV strains to infect the solanaceous hosts Datura
stramonium and Nicotiana bigelovii. The three viruses,
CM1841, Cabb-B, and D4, are very closely related to each
other. Their capsid proteins are virtually
indistinguishable serologically, their genomes have many
restriction enzyme sites in common, and they are able to
infect many of the same hosts. In spite of their great
similarities, CM1841 and Cabb-B induce a resistant
response in D. stramonium and N. bigelovii while D4 is
able to systemically infect the same hosts (10,11). In
this study, we constructed nine recombinant strains
between the three viruses in order to map the determinants
for host range and host response. We found that a 496 bp
DNA fragment within ORF VI controlled the ability of the
virus to systemically infect D. stramonium and N.
bigelovii. If this DNA segment was derived from D4, the
virus could systemically infect both hosts. If the DNA
segment was derived from CM1841 or Cabb-B, the recombinant
virus induced a resistant reaction.

RESULTS

Host response to infection by D4, CM1841, and Cabb-B.

Both D. stramonium and N. bigelovii were susceptible
to infection by D4. D4 induced chlorotic local lesions on
either host 10-14 days after inoculation (Fig. 1A).
Systemic symptoms, which consisted of a chlorotic mottle
and leaf distortion, appeared 3-4 weeks post-inoculation
(Fig. 1B).

CM1841 and Cabb-B induced a resistant response in
both hosts. D. stramonium responded to inoculation of
CM1841 or Cabb-B in a hypersensitive manner. Necrotic
local lesions appeared 12-16 days after inoculation

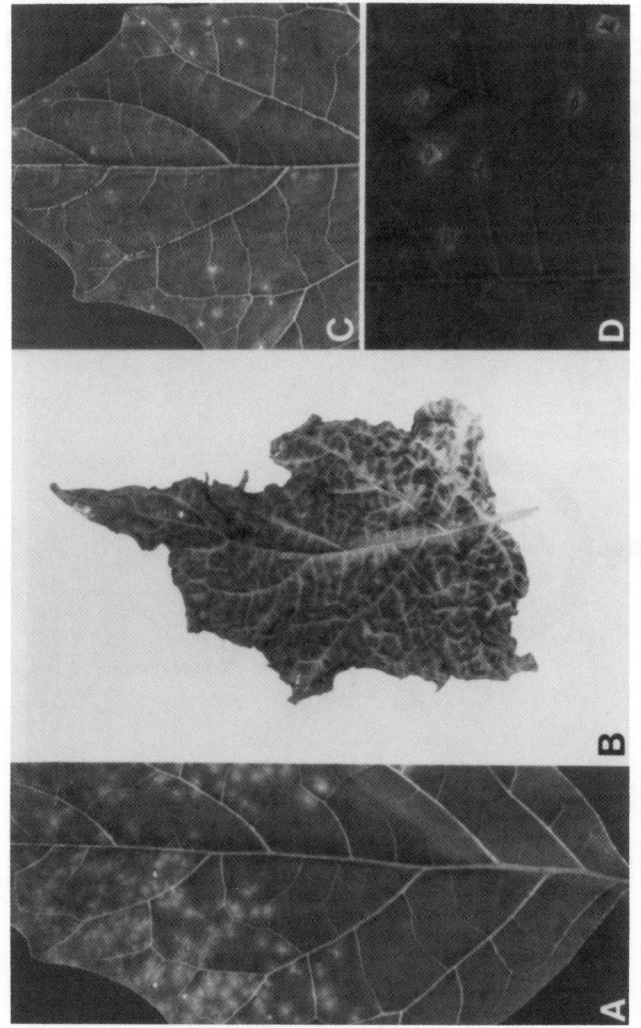

FIGURE 1. Response of Datura stamonium to infection by cauliflower mosaic virus strains D4, CM1841, and Cabb-B. 1A) D4 chlorotic local lesions; 1B) D4 systemic symptoms; 1C) Cabb-B necrotic local lesions; CM1841 local lesions are identical to Cabb-B; 1D) Cabb-B systemic symptoms.

(Fig. 1C). CM1841 did not cause any visible response on non-inoculated leaves while Cabb-B occasionally induced necrotic lesions on some of the older leaves (Fig. 1D). When N. bigelovii was inoculated with Cabb-B, chlorotic lesions appeared 12-14 days after inoculation but no systemic symptoms appeared thereafter and the virus could not be detected in non-inoculated leaves using an enzyme-linked immunosorbant assay (ELISA) (12). CM1841 did not cause any visible response in N. bigelovii.

Construction of recombinant viruses.

CM1841, Cabb-B, and D4 have each been cloned at a unique site and the full length clones are infectious (4,7,11). Fig. 2 illustrates restriction enzyme sites that were important for this study. The restriction

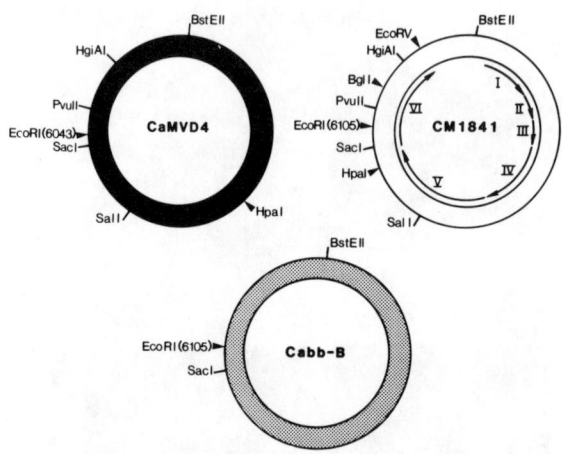

Fig. 2. Restriction enzyme maps of cauliflower mosaic virus strains D4, CM1841, and Cabb-B. The positions of the six open coding regions are indicated on the restriction enzyme map of CM1841, which has been completely sequenced (4). The circular maps are oriented with the zero position at the top. Restriction enzyme sites which were used for exchanges of DNA segments are indicated by a bar. Sites that were used for identifying a DNA segment are indicated by arrowheads.

enzyme maps of the three viruses are similar enough so
that DNA segments can be exchanged between them but
different enough so that any particular DNA segment can be
identified by restriction site polymorphisms.

The compositions of the recombinant viruses are
presented in Fig. 3. Recombinant viral genomes were
constructed in one of two ways using standard techniques
(13). In the first method, DNA segments delimited by
shared restriction enzyme sites were simply exchanged
between cloned strains. Recombinant viral clones were
identified by their unique restriction enzyme map. The
recombinant viral DNA would then be excised from the
vector and inoculated to turnip (Brassica campestris "Just
Right"). Most of the recombinant viruses were made in
this manner.

In the second method, viral DNA segments were
subcloned into a bacterial vector. Pairs of subclones
were then chosen which would constitute the recombinant
virus. The vector was cleaved from the viral DNA, the
complementing DNA fragments were mixed, and the DNA
inoculated onto young turnip plants. Recombinant viruses
H13, H14 and H17 were made in this manner. In all cases
the infectivity of the recombinant virus was confirmed by
purifying viral DNA from infected turnips and checking for
the presence or absence of key restriction enzyme sites.

Host response analysis.

The three CaMV strains and nine recombinant viruses
were propagated in turnips prior to inoculation to D.
stramonium and N. bigelovii. Virus that was partially
purified and concentrated from infected turnips was used
as the inoculum for solanaceous hosts. Inoculated plants
were observed for symptom development for seven weeks and
then tested by ELISA for the presence of the virus.

Table 1 summarizes the response of D. stramonium to
the twelve viruses. The local lesion response was
correlated with a DNA segment within ORF VI from Sac I
(position 5822) to Pvu II (position 6318). When this DNA
segment, which corresponds to the first half of ORF VI,
was derived from D4, the local lesion type was chlorotic.
When the first half of ORF VI was derived from CM1841 or
Cabb-B, the local lesion was necrotic.

258   Schoelz, Shepherd, and Daubert

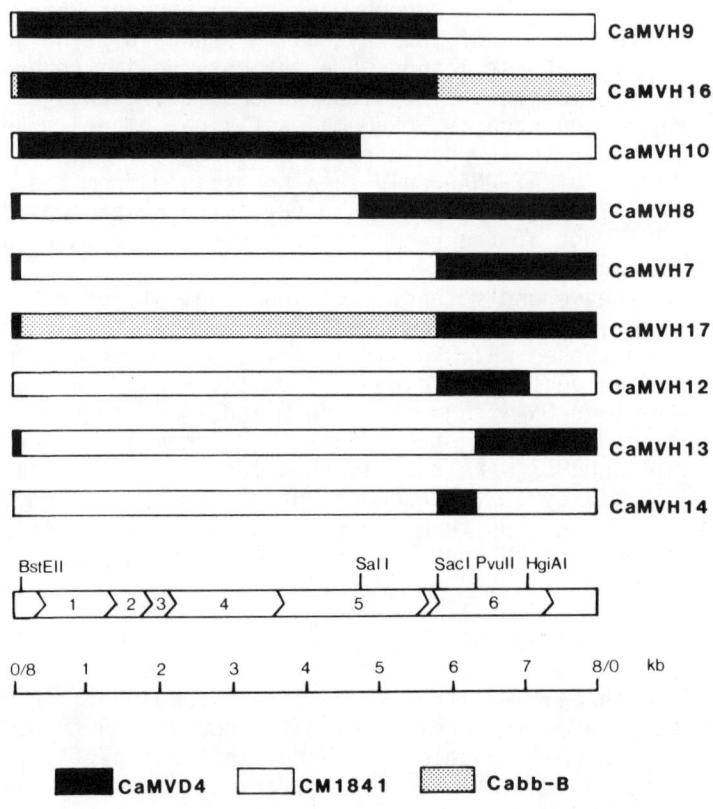

FIGURE 3. Composition of the nine recombinant viral genomes. The genomes are presented in a linear fashion to facilitate a comparison between them. The zero position of the circular map (Fig. 2) corresponds to each end of these linear diagrams.

TABLE 1

REACTION OF D. STRAMONIUM TO RECOMBINANT CaMV STRAINS

virus	Test 1			Test 2			Test 3		
	local lesions	systemic symptoms	A450	local lesions	systemic symptoms	A450	local lesions	systemic symptoms	A450
CaMVD4	10C[a]	10[b]	.492[c]	10C[a]	10[b]	1.024[c]	10C[a]	10[b]	.310[c]
Cabb-B	10N	4[d]	.032	10N	0	.032	10N	10[d]	.028
CM1841	10N	0	.017	6N	0	.016	-	-	-
CaMVH9	10N	0	.007	10N	0	.028	10N	1[d]	.036
CaMVH16	10N	0	.009	10N	0	.013	10N	1[d]	.021
CaMVH10	10N	0	.012	10N	0	.003	10N	0	.012
CaMVH8	10C	10	.805	10C	10	2.326	-	-	-
CaMVH7	10C	10	.787	10C	10	2.262	-	-	-
CaMVH17	10C	10	.269	10C	10	.167	-	-	-
CaMVH12	10C	10	.321	10C	10	2.281	-	-	-
CaMVH13	10N	9[d]	.006	10N	0	.013	10N	0	.025
CaMVH14	10C	10	.421	10C	10	.859	-	-	-
Healthy	0	0	.007	0	0	.057	0	0	.015

[a] Number of plants of 10 inoculated that gave the indicated reaction 2-3 weeks after inoculation with virus.
C = chlorotic local lesions, N = necrotic local lesions.
[b] Number of plants of 10 inoculated that reacted with systemic symptoms 4-5 weeks after inoculation.
[c] Virus concentration was determined by ELISA from tissue samples taken 6-7 weeks after inoculation. Antiserum was prepared against Cabb-B virus.
[d] Symptoms consisted of necrotic or chlorotic spots scattered on the older leaves (see text).

The ability of a virus to systemically infect D. stramonium was also correlated with the first half of ORF VI. When the first half of ORF VI was derived from D4, the test plants developed systemic infections. A chlorotic mottle developed on non-inoculated leaves and the virus was detected immunologically in those plants. Additionally D4, H7, H8, H12, H14, and H17 were purified from infected D. stramonium and their identities confirmed by restriction enzyme mapping.

When ORF VI was derived from CM1841 or Cabb-B, test plants were resistant to systemic infection. In most cases, no systemic symptoms developed on plants inoculated with CM1841, Cabb-B, H9, H10, H13, and H16. When symptoms did appear, they consisted of chlorotic and necrotic spots scattered on the older leaves. Even though non-inoculated

TABLE 2

REACTION OF N. BIGELOVII TO RECOMBINANT CaMV STRAINS

virus	local lesions	systemic symptoms	A450
CaMVD4	10[a]	10[b]	.156[c]
Cabb-B	6	0	.029
CM1841	0	0	.023
CaMVH9	5	0	.001
CaMVH16	9	0	.005
CaMVH7	2	7	.297
CaMVH17	9	9	.108
CaMVH12	9	9	.298
CaMVH13	4	0	.004
CaMVH14	8	8	.179
Healthy	0	0	.028

[a]Number of plants of 10 inoculated that reacted with local lesions 2-3 weeks after inoculation with virus.
[b]Number of plants of 10 inoculated that reacted with systemic symptoms 4-5 weeks after inoculation.
[c]Virus concentration was determined from tissue samples taken 6-7 weeks after inoculation.

leaves of some plants had symptoms, the concentration of virus in those plants was very low. When plants were sampled randomly and tested by ELISA for the presence of the virus, the absorbance was not different from that of the healthy plant extract (Table 1).

The results obtained with N. bigelovii were similar to those from D. stramonium (Table 2). The first half of ORF VI again determined whether the virus could systemically infect test plants. The only difference between the two hosts was that all local lesions on N. bigelovii were chlorotic.

Nucleotide and amino acid sequence comparison between CM1841 and D4.

Nucleotide sequencing of D4 was carried out on subclones from Sac I (5822) or Eco RI (6043) to Pvu II (6318) by the dideoxy terminator method (14). The DNA sequence of the first part of ORF VI from D4 is compared to that of CM1841 in Figure 4. Nucleotide differences which resulted in amino acid exchanges have been boxed. Because there were 20 amino acid exchanges, it was not possible to identify specific differences which resulted in the altered host range. It is interesting to note, however, that many of the substitutions are non-conservative. In 9 of 20 substitutions, a hydrophobic amino acid of CM1841 is replaced with a hydrophilic amino acid.

DISCUSSION

In this paper we constructed recombinant viruses between three CaMV strains to map host specificity and found that a 496 bp DNA segment within ORF VI was associated with local lesion type and ability of CaMV to systemically infect two solanaceous plants, D. stramonium and N. bigelovii. If this DNA segment was derived from D4, the virus induced chlorotic local lesions and systemically infected the two solanaceous hosts. Conversely, when this DNA segment was derived from CM1841 or Cabb-B, the virus induced necrotic local lesions and systemic spread was greatly impaired or could not be detected.

```
      5822
      GLU LEU ASP LEU VAL ARG  ALA LYS ILE SER LEU ALA ARG ALA ASN
1841  GAG CTC GAT CTA GTA AGA  GCA AAA ATA AGC TTA GCA AGA GCT AAC
  D4  GAG CTC GAT CTA GTA AAA  GCA AAA ATA AGC TTA GCA AGA GCT AAC
      GLU LEU ASP LEU VAL LYS  ALA LYS ILE SER LEU ALA ARG ALA ASN

      GLY SER SER GLN GLN GLY ASP  LEU PRO  LEU HIS ARG GLU THR PRO
1841  GGC TCT TCG CAA CAA GGA GAC  CTC CCT  CTC CAC CGT GAA ACA CCG
  D4  GGC TCT TCG CAA CAA GGA GAA  CTC TCT  CTC CAC CGT GAA ACA CCG
      GLY SER SER GLN GLN GLY GLU  LEU SER  LEU HIS ARG GLU THR PRO

      VAL  LYS GLU GLU ALA VAL HIS SER ALA LEU ALA THR PHE THR PRO
1841  GTA  AAA GAA GAA GCA GTT CAT TCT GCA CTG GCC ACT TTT ACG CCA
  D4  GAA  AAA GAA GAA GCA GTT CAT TCT GCA CTG GCC ACT TTT ACG CCA
      GLU  LYS GLU GLU ALA VAL HIS SER ALA LEU ALA THR PHE THR PRO

      THR GLN VAL LYS ALA ILE PRO GLU GLN THR ALA PRO GLY LYS GLU
1841  ACT CAA GTA AAG GCT ATT CCA GAG CAA ACG GCT CCT GGT AAA GAA
  D4  ACC CAA GTA AAA GCT ATT CCA GAG CAA ACG GCT CCT GGT AAA GAA
      THR GLN VAL LYS ALA ILE PRO GLU GLN THR ALA PRO GLY LYS GLU

      SER THR ASN PRO LEU MET ALA SER ILE LEU PRO LYS ASP MET ASN
1841  TCA ACA AAT CCG TTG ATG GCT AGT ATC TTG CCA AAA GAT ATG AAC
  D4  TCA ACA AAT CCG TTG ATG GCT AGT ATC TTG CCA AAA GAT ATG AAT
      SER THR ASN PRO LEU MET ALA SER ILE LEU PRO LYS ASP MET ASN

      PRO  VAL GLN THR GLY  ILE ARG LEU ALA VAL  PRO GLY  ASP PHE LEU
1841  CCA  GTT CAA ACT GGG  ATA AGG CTT GCA GTG  CCA GGG  GAC TTT TTA
  D4  TCA  GTT CAG ACT GAA  ATT AGG CTC AAA AGG  CCA TCG  GAC TTC TTA
      SER  VAL GLN THR GLU  ILE ARG LEU LYS ARG  PRO SER  ASP PHE LEU

      ARG PRO HIS  GLN GLY ILE PRO  ILE PRO GLN LYS SER GLU LEU SER
1841  CGT CCT CAT  CAG GGA ATT CCA  ATC CCA CAA AAA TCT GAG CTT AGC
  D4  CGT CCT TAT  CAG GGA ATT TCA  ATC CCA CAA AAA TCT GAG CTT AAC
      ARG PRO TYR  GLN GLY ILE SER  ILE PRO GLN LYS SER GLU LEU ASN

      SER ILE  VAL ALA PRO LEU ARG ALA  GLU SER GLY ILE HIS  HIS PRO
1841  AGC ATA  GTT GCT CCT CTC AGA GCA  GAA TCG GGT ATT CAC  CAC CCT
  D4  AGC ACA  GTT ACT CTT CAC GGA GTA  GAA TCG GGT ATT CAA  CAC CCT
      SER THR  VAL THR LEU HIS GLY VAL  GLU SER GLY ILE GLN  HIS PRO

      HIS ILE ASN TYR TYR VAL VAL TYR ASN GLY PRO HIS ALA GLY ILE
1841  CAT ATC AAC TAC TAC GTT GTG TAT AAC GGT CCA CAC GCC GGT ATA
  D4  CAT ATC AAC TAC TAC GTT GTG TAT AAC GGT CCA CAC GCC GGT ATA
      HIS ILE ASN TYR TYR VAL VAL TYR ASN GLY PRO HIS ALA GLY ILE

      TYR ASP ASP TRP GLY CYS THR LYS ALA ALA THR ASN GLY VAL PRO
1841  TAC GAT GAC TGG GGT TGT ACA AAG GGA GCA ACA AAC GGC GTT CCC
  D4  TAC GAT GAC TGG GGT TGT ACA AAG GCG GCA ACA AAC GGC GTT CCC
      TYR ASP ASP TRP GLY CYS THR LYS ALA ALA THR ASN GLY VAL PRO

      GLY VAL ALA TYR  LYS LYS PHE ALA THR ILE THR GLU ALA ARG ALA
1841  GGA GTT GCA TAC  AAG AAG TTT GCC ACT ATT ACA GAG GCA AGA GCA
  D4  GGA GTT GCA CAA  AAG AAG TTT GCC ACT ATT ACA GAG GCA AGA GCA
      GLY VAL ALA GLN  LYS LYS PHE ALA THR ILE THR GLU ALA ARG ALA
                                                                 6317
```

FIGURE 4.   Sequence comparisons between D4 and CM1841.

Our work demonstrates that ORF VI of CaMV mediates a defensive response in solanaceous hosts. It is probable that resistance to CM1841 and Cabb-B in solanaceous hosts is effective only in whole plants. Previous studies have shown that CM1841 can replicate and form inclusion bodies in inoculated D. stramonium leaves (10). The fact that individual cells are susceptible to CM1841 while the whole plant is resistant is in agreement with other host-virus systems (15,16).

A central question that could not be answered by our study is how ORF VI could determine the host response. Two models, based on fungal and bacterial systems, have been proposed to explain recognition of pathogen races. In the elicitor-receptor model, an elicitor present in the incompatible pathogen interacts directly with plant disease resistance genes (17,18). In the second model, the elicitor-suppressor model, all races of a pathogen contain a non-specific elicitor of host resistance genes. Compatible races are hypothesized to suppress the resistance response by producing an additional factor (19,20).

Our work could support either model. One interpretation of our results is that ORF VI of CM1841 and Cabb-B elicits a plant defensive response. Alternatively, it may be that ORF VI of D4 suppresses a defensive response. If this was the case, then it would be desirable to identify the non-specific elicitor.

It may be possible to determine which model is correct by inserting ORF VI into solanaceous hosts via an Agrobacterium vector. Transformed plants could then be challenged with different CaMV isolates to see if the defensive reaction of the host is suppressed.

ACKNOWLEDGEMENTS

The authors are grateful to Jo Ann Eason for skillful assistance in the greenhouse and to the Kentucky THRI for providing research facilities.

REFERENCES

1. Shepherd R (1979). DNA plant viruses. Annu. Rev. Plant Physiol. 30:405.

2. Balazs E, Guilley H, Jonard G (1982). Nucleotide sequence of DNA from an altered-virulence isolate D/H of cauliflower mosaic virus. Gene 19:239.

3. Frank A, Guilley H, Jonard G, Richards K, Hirth L (1980). Nucleotide sequences of cauliflower mosaic virus DNA. Cell 21:285.

4. Gardner R, Howarth A, Hahn P, Brown-Leudi M, Shepherd R, Messing J (1981). The complete nucleotide sequence of an infectious clone of cauliflower mosaic virus by M13mp7 shotgun sequencing. Nucleic Acids Res. 9:2871.

5. Armour SL, Melcher U, Pirone TP, Lyttle DJ, Essenberg RC (1983). Helper component for aphid transmission encoded by region II of cauliflower mosaic virus DNA. Virology 129:25.

6. Daubert S, Shepherd R, Gardner R (1983). Insertional mutagenesis of the cauliflower mosaic virus genome. Gene 25:201.

7. Daubert S, Richins R, Shepherd R, Gardner R (1982). Mapping of the coat protein gene of cauliflower mosaic virus by its expression in a prokaryotic system. Virology 122:444.

8. Toh H, Hayashida H, Miyata T (1983). Sequence homology between retroviral reverse transcriptase and putative polymerases of hepatitis B virus and cauliflower mosaic virus. Nature 305:82.

9. Odell JT, Howell SH (1980). The identification, mapping and characterization of mRNA for P66, a cauliflower mosaic virus-coded protein. Virology 102:349.

10. Lung MCY, Pirone TP (1972). Datura stramonium, a local lesion host for certain isolates of cauliflower mosaic virus. Phytopathology 62:1473.

11. Schoelz JE, Shepherd R, Richins R (1986). Properties of an unusual strain of cauliflower mosaic virus. Phytopathology (in press).

12. Clark MF, Adams AN (1977). Characteristics of the microplate method of enzyme-linked immunosorbant assay for the detection of plant viruses. J. Gen. Virol. 34:475.

13. Maniatis T, Fritsch EF, Sambrook J (1982). "Molecular Cloning: A Laboratory Manual," Cold Spring Harbor Laboratory: Cold Spring Harbor, New York.

14. Messing J (1983). New M13 vectors for cloning. In Wu R, Grossman L, Moldave K (eds): "Methods in Enzymology," New York: Academic Press, p 20.

15. Otsuki Y, Shimomura T, Takebe I (1972). Tobacco mosaic virus multiplication and expression of the N gene in necrotic responding tobacco varieties. Virology 50:45.

16. Beier H, Bruening G, Russell ML, Tucker CL (1979). Replication of cowpea mosaic virus in protoplasts isolated from immune lines of cowpeas. Virology 95:165.

17. Ellingboe AH (1976). Genetics of host-parasite interactions. In Heitefuss R, Williams PH (eds): "Encyclopedia of Plant Physiology, New Series, Vol 4: Physiological Plant Pathology," Berlin, Heidelberg, and New York: Springer-Verlag, p 761.

18. Keen N (1982). Specific recognition in gene-for-gene host-parasite systems. In Ingram DS, Williams PH (eds): "Advances in Plant Pathology," New York: Academic Press, 1:35.

19. Doke N, Tomiyama K (1980). Suppression of the hypersensitive response of potato tuber protoplasts to hyphal wall components by water soluble glucans isolated from Phytophthora infestans. Physiol. Plant Pathol. 16:177.

19. Kuc J, Tjamos E, Bostock R (1984). Metabolic regulation of terpenoid accumulation and disease resistance in potato. In Nes WD, Fuller G, Tsai L-S (eds). "Isopentenoids in Plants, Biochemistry and Function," New York: Mercel Dekker, p 103.

Molecular Strategies for Crop Protection, pages 267–293
© 1987 Alan R. Liss, Inc.

THE REPLICATION CYCLE OF CAULIFLOWER MOSAIC
VIRUS IN RELATION TO OTHER RETROID
ELEMENTS, CURRENT PERSPECTIVES

J.M. Bonneville(*), J.Fuetterer(*), K. Gordon(#),
T. Hohn(*), J. Martinez-Izquierdo(*),
P. Pfeiffer(#) and M. Pietrzak(*).

Friedrich Miescher-Institut, P.O. Box 2543, CH-4002,
Basel, Switzerland (*) and Institut de Biologie
Moleculaire et Cellulaire du C.N.R.S., Strasbourg,
France (#).

ABSTRACT Cauliflower mosaic virus,
albeit a DNA virus, uses reverse
transcriptase for genome replication
and its replication cycle is related to
that of retroviruses, hepadnaviruses
and retrotransposons. Also CaMV ORFs
IV, V and VI show similarities in
organization, function and sequence to
the "gag", "pol" and "env" genes of
retroviruses. On the other hand, there
are ORFs which are peculiar for CaMV
and the expression of which seems to
involve a "polycistronic messenger".
Experiments using SP6-promoted
transcripts from hybrid plasmids grown
in E. coli show that indeed CaMV ORFs
oriented in tandem are recognized by
eukaryotic ribosomes and that the long
and several ATG-codons-containing
leader of the large CaMV transcript is
no hindrance for translation. CaMV
proteins are posttranslationally
modified by protein cleavage and
phosphorylation and the modification
functions might be virus coded. These
and various other aspects of the CaMV
life cycle are compared with other
retroid elements.

INTRODUCTION

Cauliflower mosaic virus (CaMV) belongs to the "retroid elements", a group of viruses and transposable elements that exist in two genomic forms, DNA and RNA. Other members of this group are retroviruses, hepadnaviruses and retrotransposons such as mammalian intracistronic A-particles (IAP), insect copia and copia-like elements and yeast Tyl elements (Table 1). Retroid elements resemble each other in gene organization, sequence and basic steps of their life cycle, indicating a common ancestry and a general importance for the underlying mechanism of reverse transcription, but have adapted to the different host systems and specialized in strategies of persistence: the "DNA" viruses spread systemically within the host organism and the host population, while the transposable elements are maintained and proliferated within a given cell genome. The "RNA" viruses (oncorna- and lentiviruses considered in this contribution) can use both strategies.

In the following we will discuss details of the CaMV strategy. We are aware of the speculative nature of this contribution, however we hope to stimulate some discussion in this field of comparative virology. References are choosen to lead to further citations. For additional information books and reviews by Weiss et al (1), Tiollais (2), Temin (3), Hohn et al (4), Mason et al (5) can be consulted.

Figure 1 Genome organization of retroid elements. Uppermost diagram: CaMV 35S RNA showing the leader sequence, the primer binding site (~) and the individual ORFs. Following is a section of the above diagram in comparison to the gag, pol and env genes (symbolized by different shades) of various retroid elements. Additional reading frames outside the region discussed and split ORFs are omitted for clarity. The position of three conserved aminoacid sequence motives is indicated: ▼ ,cys motif (see text and Fig.2); o, protease domain (VDT/SGA;16); ■, reverse transcriptase (YXDD;17). Genome organizations are deduced from published DNA sequences (legend of Fig.2).

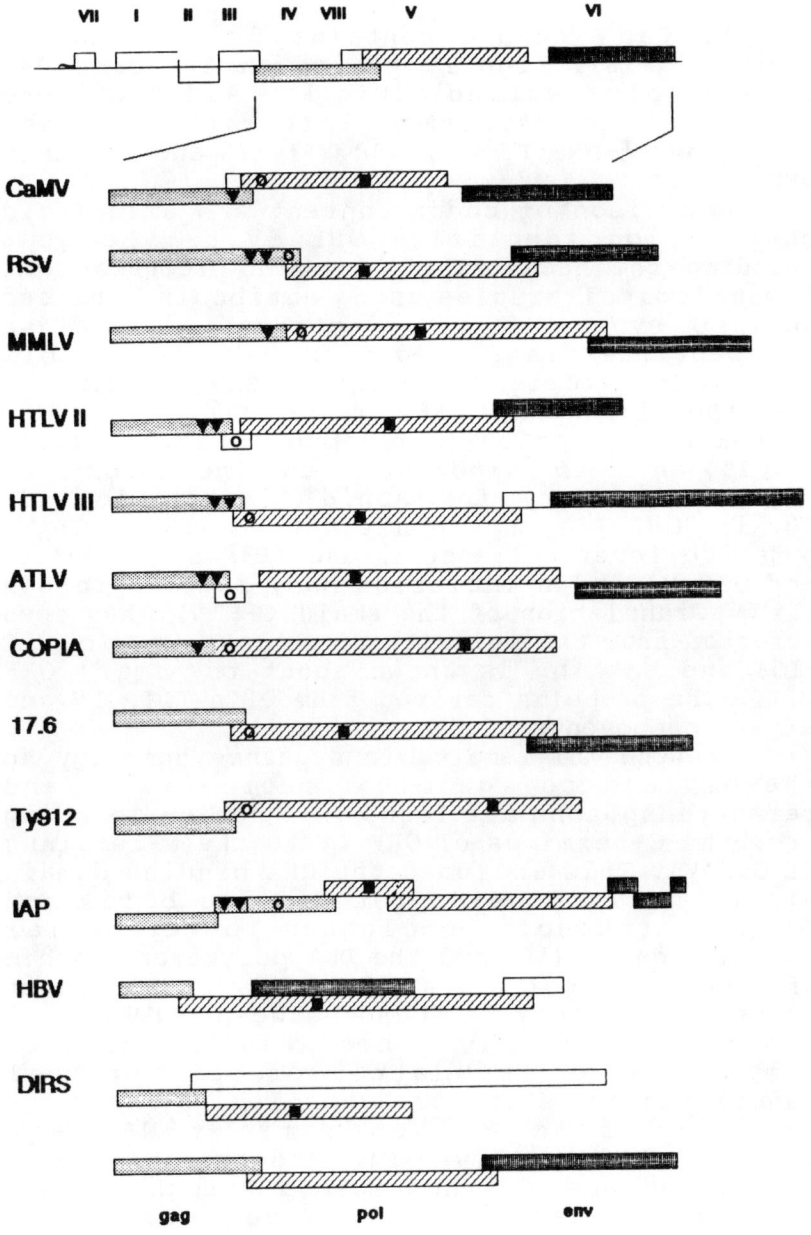

Figure 1.

GENE ORGANIZATION

The CaMV genome contains 8 open reading frames (ORFs), tightly packed on the same DNA strand (plus strand) (Fig.1). All ORFs are essential for systemic infection with the exception of ORF II (6), ORF VII (7) and possibly ORF VIII.

Comparison of codon content and amino acid composition identified ORF IV as the gene encoding the nucleocapsid protein precursor (8). Immunological studies using antibodies, raised against synthetic oligopeptides and against polypeptides translated from ORFs in E. coli expression vectors, identified gene products of the other large CaMV ORFs in infected plants: ORF I, unknown function (our unpublished results); ORF II, an 18 kD component of the inclusion body matrix necessary for aphid transmission (9, 10,11); ORF III, unknown function (12), ORF V, putative reverse transcriptase (13; our results) and ORF VI, main inclusion body matrix protein (14). Translation of the small ORF VII has been inferred from the polarity of mutants in this ORF (15) and nothing is known about the small ORF VIII. The proteins derived from ORFs III, IV and V are components of the virion.

Stretches of remarkable sequence homology to the gag and pol genes of retroviruses and retrotransposons are found in a region spanning from the C-terminus of ORF IV to the N-terminus of ORF VI. This includes the RNA binding domain of the nucleocapsid protein (see below and Fig.2), sequences homologous to retroviral protease genes (16) and the DNA polymerase domain of retrovirus pol genes (17,18,19). Only limited sequence homology is found between CaMV ORF VI and retrovirus env genes (Fig.2), however comparison of the latter proteins reveal functional and structural homologies: both are produced from a subgenomic mRNA, are glycosylated, involved in the determination of the host range (20) and form a structure that surrounds virus particles: env protein in association with a membrane forms an envelope around single virions, the CaMV protein VI is the

major component of the inclusion body matrix
containing many viral particles. ORFs I, II and
III from CaMV have no counterpart in other
retroid elements and are probably specifically
required to infect plant cells, e.g. ORF II
protein for the aphid transmission typical for
many plant viruses. ORF I protein shares
homologies with the 30kD protein involved in
systemic spreading of tobacco mosaic virus (21),
with mitochondrial intron protein and with the
ATP binding sites of various protein kinases (own
observations, Fig.2).

GENOMIC DNA AND ITS TRANSCRIPTION

Transcription of retroid elements, to our
knowledge, always takes place in the nucleus.
While in retrotransposons and most retroviruses
integrated DNA is the usual template for
transcription, unintegrated linear DNA is
considered to act as the template in
lentiviruses, like visna, FeLV or HTLV III (22).
Retrotransposon and retrovirus DNAs are flanked
by long terminal repeats (LTRs). A more complex
system of redundancies including inverted
terminal and internal repeat units has been
reported for DIRS1, a Dictyostelium transposable
element (23). In contrast, the transcribed form
of CaMV and hepadnavirus DNA is a circular
minichromosome with no repeated sequence. Either
the repeats or the circularity of the DNA
template of retroid elements allow synthesis of a
genomic RNA with short terminal repeats, a
feature that will permit RNA template switch
during synthesis of the minus strand DNA.
A peculiarity of CaMV (and hepadnaviruses) is
the presence of two promoters, while other
retroid elements have usually only one. The CaMV
"19S promoter" produces a messenger RNA covering
ORF VI and the "35S promoter" produces terminally
redundant genomic RNA, which may also direct
translation of the remaining ORFs. Retroviruses
create their mRNA diversity by splicing of the
major transcript. Splicing in CaMV has been
deduced from the discovery of a deleted version

a

	Nt.no.													
CaMV	(1)	T	G	G	T	A	T	C	A	G	A	G	C	C	A	T	G
RSV	(370)	T	G	G	-	A	T	C	A	-	A	G	C	-	A	T	G

b

	AA.no.										
CaMV pI	(125)	SMV	H	LGA	V	K	I	L	L	K	A
cAMP-dep. PK	(65)	G	N	H	Y	A	M	K	I	L	D	K	Q
Mu-vMos	(115)	G	V	P	V	A	I	K	Q	V	N	K	C
Hu-cMos	(81)	G	V	P	V	A	I	K	Q	V	N	K	C
vSrc	(289)	T	T	R	V	A	I	K	T	L	K	P	G
gag-Yes	(574)	T	T	K	V	A	I	K	T	L	K	L	G
gag-Fgr	(426)	S	T	K	V	A	V	K	T	L	K	P	G
gag-Abl	(386)	S	L	T	V	A	V	K	T	L	K	E	D
gag-Fps	(944)	N	T	P	V	A	V	K	-	S	C	R	E
gag-Fes	(535)	N	T	L	V	A	V	K	-	S	C	R	E

```
          :            :   *   :   *        :   (:)
```

c

```
      AA.no.
TMV 3OK (106) SYYTAAAKKRPQFKVVPNYAITTQD-AMKNVWQVLVNIRNVKMSAGFLPLSLEFVSVCIVYRNN--------IKLG
              : :: ** * * : / :::::: :* : ** * *              * ***; : :    ***
Mit.Int.(191) DKFGGSVKLRSGVKTIRY-RLQNKEGIIKLINAVWGNIRNSKRLVQF-------WKVCILLNIDFKE----PIKLT
              **;:: :     : * ** :** * * ::::::::*:: :*:!    * ***; *:: ***;
CaMV pI (89)  -----SVDIHDATGKV-YLRLITKEEINKRLSSLKPEVRKTKSMVBL-------GAVKILLKAQFRNGIDTPIKIA
```

d

		Nt.no.																		
CaMV		(3425)	K	K	D	C	R	C	W	I	C	N	I	E	G	H	Y	A	N	E	C	P	N	R	Q	S	S
RSV	a)	(1889)	R	A	R	G	L	C	Y	T	C	G	S	P	G	H	W	Q	A	Q	C	P	K	K	R	S	K
	b)	(1967)	N	S	R	E	R	C	Q	L	C	N	G	M	G	H	N	A	K	Q	C	R	K	R	D	G	N
MMLV		(2115)	L	D	R	D	Q	C	A	Y	C	K	E	K	G	H	W	A	K	D	C	P	K	K	P	R	G
HTLV II	a)	(1878)	P	P	T	Q	P	C	F	R	C	G	K	V	G	H	W	S	R	D	C	T	Q	P	R	P	P
	b)	(1950)	P	P	P	G	P	C	P	L	C	Q	D	P	S	H	W	K	R	D	C	P	Q	L	K	P	P
HTLV IIIa	a)	(1491)	R	K	M	V	K	C	P	N	C	G	K	E	G	H	T	A	R	N	C	R	A	P	R	K	K
	b)	(1554)	P	R	K	K	G	C	W	K	C	G	K	E	G	H	Q	M	K	D	C	T	E	R	Q	A	N
ATLV	a)	(1856)	P	P	N	Q	P	C	F	R	C	G	K	A	G	H	W	S	R	D	C	T	Q	P	R	P	P
	b)	(1925)	P	P	P	G	P	C	P	L	C	Q	D	P	T	H	W	K	R	D	C	P	R	L	K	P	T
Copia		(1109)	K	Y	K	V	K	C	B	H	C	G	R	E	G	H	I	K	K	D	C	F	H	Y	K	R	I
IAP	a)	(2238)	S	N	R	K	A	C	F	N	C	G	R	M	G	H	L	K	K	D	C	Q	A	P	E	R	T
	b)	(2313)	R	E	S	K	L	C	Y	R	C	G	R	G	Y	H	R	A	S	E	C	R	S	V	R	D	V

```
                       *        *         : *          : *
```

e

	AA.no.	
CaMV pVI	(1)	MENIEKLLMQEKILMLELDLVRAKISLARANGSSQ
RSV env	(35)	PTRVNYILIIGVLVLCEVTGVRADVHLLEQPGNLW

```
          ::: :*:    ::: *:  ***:: * :   *
```

Figure 2.

of the CaMV genome,which occurs as satellite of full length viral DNA in infected plants, and is likely to arise by reverse transcription of spliced CaMV RNA (24). However this RNA species is only typical for one strain (CaMV.S(Japan)), it appears only late in systemic infections and might have arisen by accidental use of a pseudosplice site.

While transcription in most retroviruses and retrotransposons is strongly regulated,involving pecific transactivation factors (25,26,27), CaMV promoters are considered to be constitutive. They were therefore used as strong unregulated plant promoters for a variety of hybrid genes in a wide range of plant species (e.g.: 28,29,30).

Figure 2 Sequence homologies in retroid elements.
a) Translational control region: The sequences upstream and including the startcodon (underlined) of the first ORF of CaMV and RSV are compared. In CaMV this sequence is the primer binding site for reverse transcription, in RSV this sequence is involved in control of translation (56). b) ATP binding site: A region within CaMV ORF I is aligned with ATP binding sites of various kinases. Pairwise comparisons (e.g. betweenCaMV pI and ATP-dep. PK reveal higher degrees of homology. c) Homologies between plant virus genes and a mitochondrial intron protein: Pairwise computer comparison between CaMV ORF I derived protein, TMV 30kD protein and the hypothetical cob intron 4 protein of baker's yeast mitochondrion (85) show considerable homologies (see also 86), only a part of which is shown. d) Comparison of aminoacid sequences in an RNA binding domain of gag- and gag-like proteins. e) Alignment of the N-terminus of CaMV pVI (as determined from the DNA sequence) with a region in the RSV pr95 env protein. Aminoacids are shown in single letter code; *, identical aminoacid; :, similar aminoacid; (:), similar aminoacid at either of two positions. The nucleotide number (Nt.no.) or aminoacid number (AA.no.) of the start of the shown sequences are indicated. Additional homologies between CaMV and other retroid elements in the proposed protease and reverse transcriptase region (see Fig.1 and text) are not shown here. Published DNA sequences were used as followed: CaMV(8); RSV(77); MMLV(78); HTLV II(79); HTLV III(80); ATLV(81); Copia(82); 17.6(83); Ty912(67); IAP(49); HBV(84); DIRS-1(23).

REVERSE TRANSCRIPTION, GENERAL PATHWAY

Reverse transcription of CaMV has recently been reviewed in detail (4,31,3;Fig.3). In this contribution we would like to repeat that the reverse transcription process primarily leads to open circular double stranded DNA, the form of the CaMV genome found in mature virions, and that

Figure 3.   Generalised life cycle of retroid elements: RNA with terminal repeats is either produced from an integrated or not integrated DNA form with long terminal repeats (step 1A) or from a supercoiled circular form (step 1B). A primer binds to a primer binding site (MPBS) on the RNA and initiates (-)strand DNA synthesis by reverse transcription from the RNA template. Template RNA is consecutively digested by an RNAse H activity (step 2). When the active reverse transcription complex reaches the 5' end of the RNA template, the nascent DNA strand switches to the sister RNA terminal repeat and continues elongation (steps 3,4). A G-rich RNA sequence known to be relatively stable to RNAse is spared and can now serve as a primer at the primer site (PPBS) to initiate the synthesis of the (+) strand DNA complementary to the newly synthesized (-) DNA strand (steps 4,5). Nascent (+) and (-) strands will pass the site where (-) strand DNA synthesis was initiated (step 6). RNA remnants are digested and the hydrogen bonds that held the ends together are now replaced by ones from the newly synthesized DNA sequences (step 6). Short DNA overhangs are produced by peeling off limited streches of the 3' ends of the two strands, respectively (step 7). Plus and minus strand DNA synthesis can continue in some cases peeling off more and more of the 3' terminal sequences until the origins of plus and minus strand DNA synthesis are reached again (step 8A). The resulting DNA with long terminal repeats can in some cases circularize or integrate into the host chromosome (step 9A). Alternatively the form with the short DNA overhangs can be converted into a supercoil with the help of "repair enzymes" and ligase (step 8b). Either the integrated form (step 1A), the double stranded form with the long terminal repeats or the supercoiled form (step 1B) act as a template for the synthesis of the terminally redundant RNA, allowing the replication cycle to continue for further rounds. (From (4)).

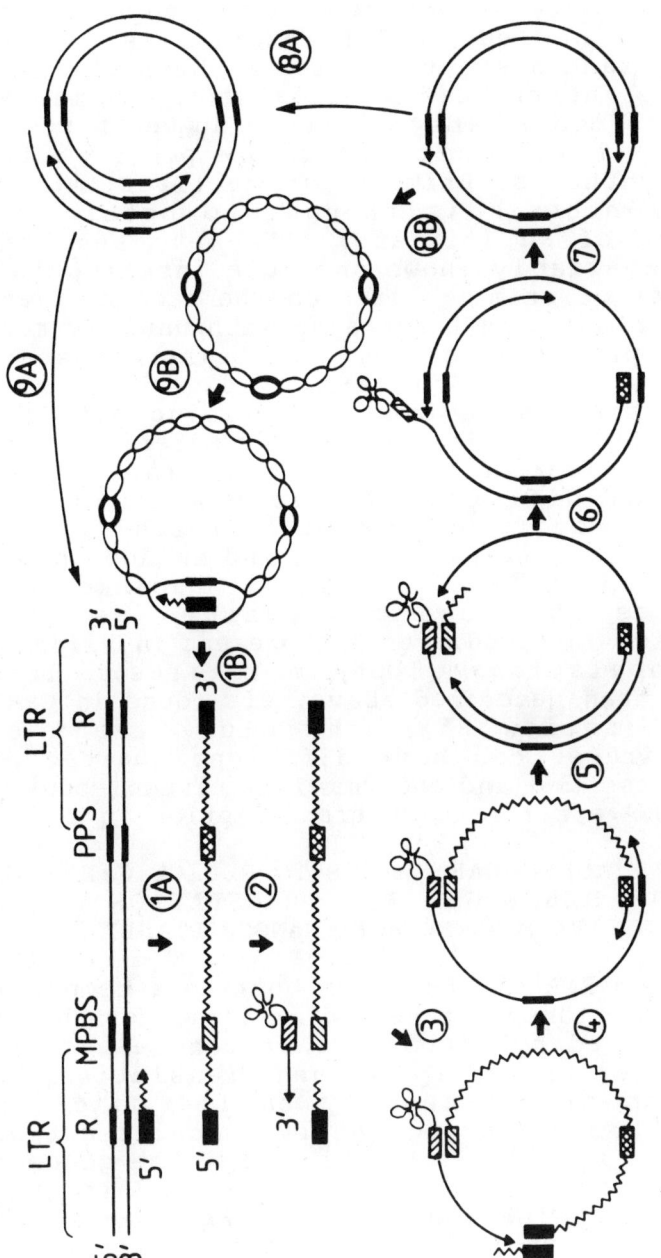

Figure 3.

this form is later converted to the supercoiled form in the nucleus. DNA relaxation is due to single strand discontinuities on both strands, where the interrupted strands show a somewhat variable 3'end briefly overlapping the fixed 5' end; these discontinuities mark the priming sites of DNA synthesis. Priming of the first (minus) strand synthesis is thought to occur upon annealing of an initiator tRNA-met, and this strand invariably shows only one interruption. The number of priming sites on the second strand differs from strain to strain, with usually more than one present, but only one being necessary (32). Also some retroviruses (lentiviruses; 33,34) possess more than one priming site for plus-strand synthesis. The genomic material found in free virions of other retroid viruses represents different intermediates of the replication cycle: RNA in retroviruses, and DNA consisting of a complete minus and an incomplete plus strand in HBV. Additional nucleic acid molecules that can be regarded as true replication intermediates are present in virions: minor populations of DNA molecules, less complete than described above, are found in CaMV virions (review 35); the plus strand of hepadnaviruses can have different degrees of completeness (36) and oncornavirus particles will contain DNA after reverse transcription in vitro.

IS THERE A MOLECULAR SWITCH TO SELECT CaMV RNA EITHER AS MESSENGER FOR TRANSLATION OR AS TEMPLATE FOR REVERSE TRANSCRIPTASE?

Retroid RNA has to serve both as a template for reverse transcriptase and as messenger RNA to be translated by ribosomes. It can be assumed that reverse transcription and translation can not occur simultaneously on the same RNA molecule, and that regulatory circuits act as "molecular switches". In this case, the genomic RNA has to make a crucial choice, and in this context encapsidation may be a key event: in retroviruses translation occurs outside the viral particles and DNA synthesis takes place within upon infecting a new host cell. Packaging is governed by protein-RNA interactions involving one (37) or more (38) protein binding sites on the RNA and one RNA binding domain on the nucleocapsid protein. This last domain includes streches rich in basic amino acids, a "cys" motif "CXXCXXXGHXXXXC" (C,cys; H,his; G,gly; X,any amino acid) (39) and frequently ser or thr residues that can be phosphorylated. Phosphorylation of avian retrovirus nucleocapsid protein has been shown to increase drastically its affinity for RNA (40). Proteins with cys and his clusters are generally considered to complex metal ions (Zn++) and to bind specifically to nucleic acid sequences (41). Thus in a naive view we assume that the "cys" motif folds the RNA binding domain of the nucleocapsid protein, the interaction of the basic amino acids with the phosphate backbone of the RNA provide the energy and the ultimate control of the interaction relies on protein phosphorylation and dephosphorylation.

We have conducted a computer assisted search for the "cys" motif and found it consistently present in retroviruses, copia, IAP and caulimoviruses (but not in HBV, Ty1, DIRS1, and copia like element 17.6) (Fig.2). It is always located close to the end of the gag or corresponding gene - ORF IV in the case of CaMV - (Fig. 1) and in the neighborhood of, or overlapping with a polypeptide region rich in basic amino acids and of ser residues that are in good context for kinasing. Experiments with CaMV nucleocapsid proteins showed that some can be phosphorylated in vitro, and that some can bind CaMV RNA (Fig.4). We therefore assume that also in CaMV the nucleocapsid proteins bind the genomic RNA and package it into particles. Since the final content of CaMV viral particles is double stranded DNA, reverse transcription may occur within the particle and, in contrast to oncornaviruses, in the original host cell. Several observations could be explained by this model: first, in addition to the complete double stranded DNA molecules, CaMV virions contain "strong stop DNA" , which is a short incomplete minus strand DNA molecule still covalently

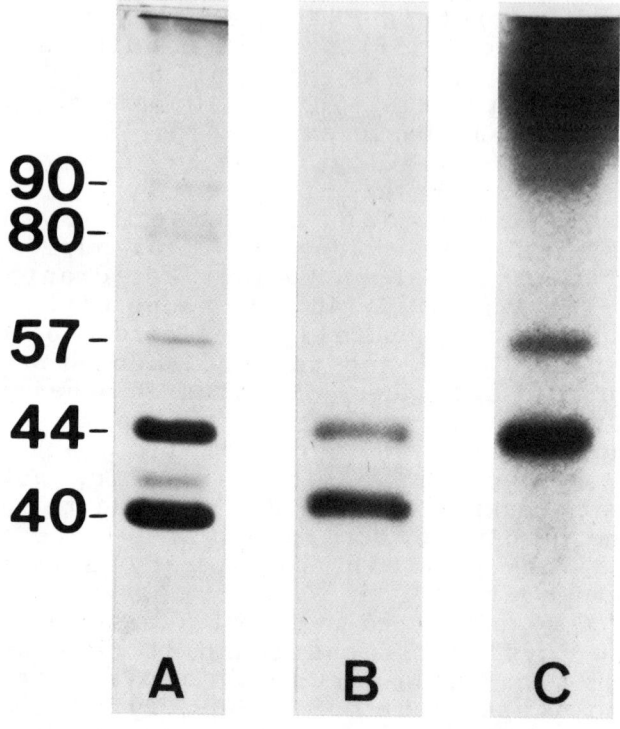

Figure 4 Nucleocapsid proteins of CaMV.
a) Immunoblotting: Viral particles were dissolved in hot
SDS, and the proteins separated on SDS PAGE, transferred to
a nitrocellulose filter and reacted with rabbit antibodies
against the 40 Kd nucleocapsid species. The immunoreactive
proteins were detected by using goat anti-rabbit IgG coupled
with peroxidase. b) RNA-binding properties of nucleocapsid
proteins: After blotting and washing as above the
nitrocellulose filter was incubated with labelled CaMV RNA,
washed again, dried and autoradiographed. c) Kinase activity
in viral particles: Viral particles were incubated in a
kinase standard buffer with $\{\gamma-^{32}P\}ATP$. After the reaction,
the viral proteins were separated by SDS PAGE. The gel was
dried and the radioactive bands were detected by
autoradiography. Molecular weights (kD) were determined by
comparison with protein standards.

connected to primer tRNA nucleotides (42) and
thought to have missed the first template switch;
second, endogenous CaMV DNA synthesis still
occurs in a population of purified virus
particles or viroplasms (43,44 and ourselves,
unpublished); third, ORF V product antigenicity
copurifies with virus particles (own
observations; 45).

Formation of nucleocapsids is not restricted
to viruses: retrotransposons, although not
infectious, have genes related to the retrovirus
nucleocapsid gene and produce virus like
particles (46,47,48,49) and nucleocapsid bound
reverse transcription is not restricted to the
cases discussed above: also in HBV reverse
transcription occurs in viral particles (36).
Perhaps specific packaging of retroid RNA and of
reverse transcriptase into nucleocapsids is a
general mechanism in retroid elements which
ensures that reverse transcription occurs
preferentially on the natural template and
thereby avoids unnecessary scrambling of the host
genome due to reintroduction of cDNA copies from
host messenger RNA into the host genome. In this
context it is interesting to note that some
tumors caused by HBV contain integrated copies of
HBV DNA that produce reverse transcriptase, but
not nucleocapsid protein, since the corresponding
gene became interrupted during the integration
event (50).

If nucleocapsid protein binds to genomic RNA
and thereby removes it from the pool of
translatable molecules, how can translation be
still assured? For MoMuLV it was shown that in a
subpopulation of viral RNA molecules the site
specific for nucleocapsid binding (psi-site) is
spliced out; the resulting shortened RNA is not
packageable and remains available as messenger
(37). Splicing has not been observed in
significant subpopulations of CaMV RNAs in most
strains, however the production of the most
abundant CaMV protein, the gene VI product, is
ensured by the subgenomic 19S RNA messenger, and
it can be assumed that any specific packaging
site on genomic RNA is located outside the
sequence covered by this message. Also the

leaderless version of 35S RNA, a minor component
of CaMV RNA (51), might be devoid of the
packaging signal.

TRANSLATIONAL REGULATION

CaMV 35S RNA could not yet be translated _in
vitro_, neither in reticulocyte lysates, nor in
wheat germ systems. This suggests the operation
of translational control mechanisms in addition
to the one excerted by RNA packaging. To study
these mechanisms experiments were performed using
RNAs produced _in vitro_ under the control of the
bacteriophage SP6 promoter. Genes downstream of
the long natural 5'-leader of 600 nucleotides
could be translated, although less efficiently
than controls lacking this leader sequence. Genes
downstream of the short ORF VII were translated
very poorly but mutations of this ORF in the
coding region restored translation of the
downstream gene. Translation of ORF VII had also
inhibitory effects on translation of other ORFs
if positioned after long leaders. This effect was
observed both with CaMV sequences and with those
of other viruses as shown in Fig.5. An
explanation of this result would be that ORF VII
codes for an RNA-binding translational repressor
protein acting on non coding sequences. CaMV ORF
VII as well as ORF VIII code for aproximately 100
amino acids long polypeptides rich in basic amino
acids (23 and 28%, respectively) typical for
nucleic acid binding polypeptides. Translational
control, albeit positive, was also observed in
RSV (38) and HTLV III (52). The regulator
proposed for the latter, encoded by the bipartite
gene _tat III_, (52) is a polypeptide of similar
length and similar content in basic amino acids
as the proposed CaMV regulator. It would not
surprise us if the transacting regulators of
other viruses (visna: 27; RSV: 26) also exhibit
translational control.
        The target for translational control is
likely to reside within the RNA leader sequences.
Leader sequences are especially long in RSV and
CaMV and in both cases these leaders contain

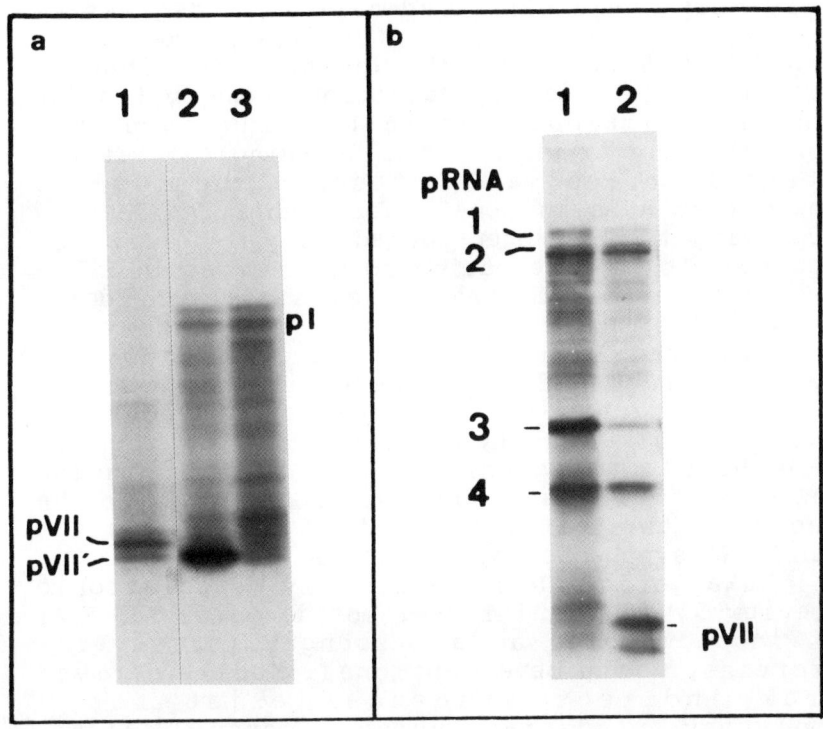

Figure 5 Effect of CaMV ORF VII on translation in vitro.
a) CaMV specific RNAs, covering ORFs VII and I with and without deletions in ORF VII, have been produced with SP6 RNA polymerase and translated in a wheat germ extract. The translation products were analyzed on 8 to 18 % SDS-polyacrylamide gels. 1: complete ORF VII (79AA); only a very small amount of ORF I translation product can be detected. 2: partially truncated ORF VII (49AA). 3: largely truncated ORF VII (12AA). In 2 and 3 ORF I translation (yielding proteins of 37 to 42 kD) is increased. **b)** AlMV RNAs have been translated _in vitro_ in absence (lane 1) or presence (lane 2) of ORF VII RNA. Translation of RNAs with long leader sequences (1,3) is inhibited.

several initiation codons (which however open only very short ORFs) and have high degrees of secondary structure (53, 54 and pers. comm.; our unpublished obsevation). Whether the ribosome scanning rule is still valid and these initiation codons are skipped due to their unfavourable context (55), or whether the scanning rule is not strictly obeyed and ribosomes can move to downstream AUGs in some other manner, perhaps by bypassing the hydrogen-bonded regions, is still debated. It is interesting to note, that in RSV a sequence further upstream the usualy considered eukaryotic initiation context is necessary for optimal translation (56) and that in CaMV this sequence is also found, albeit with three single base insertions, just upstream of CaMV ORF VII (Fig.2); this sequence coincides with the met tRNA homology region, i.e. the primer binding site. It still is an open question whether the choice of the initiator met tRNA as a primer for CaMV DNA synthesis is fortuitous.

Eukaryotic mRNAs are usually monocistronic (review 55) and cauliflower mosaic virus 35S RNA, if indeed acting as a messenger for several proteins, would be exceptional. Models derived from indirect evidences exist for a "polycistronic" translation process (7,15,57, 54). A similar mechanism may be active in translation of RSV mRNAs (58,59). According to the relay race model of Sieg and Gronenborn, ribosomes translating the large transcript reinitiate protein synthesis after passing a termination codon. Studies of CaMV mutants produced in vitro and of their revertants suggest that translation starts at the initiation codon of ORF VII located just adjacent to the tRNA primer binding site (7,15) and then proceeds in a polar fashion through ORFs I, II and III. It is possible that ORFs IV and V are not connected to the "relay race" but translated by a different mechanism since much more ORF IV product (viral nucleocapsid protein) is found in infected tissue than ORF II and ORF III products (11,12). Also ribosomes would have to "back-track" 18 and 37 bases, respectively, if relay racing from ORF III through ORF IV and V occurs. Therefore either some additional factors would be needed to quantitate individual initiation events or an as yet undetected mRNA, initiated from a hypothetical promoter within ORF III (60,61), could be postulated to code for ORF IV and V products. Plant et al (62) reported results suggesting a separate mRNA for ORF V.

POST-TRANSLATIONAL REGULATION. ASSEMBLYAND PROTEIN PROCESSING

Retroviral genes are generally translated into polyproteins that are post-translationally processed to yield the proteins found in mature virions. The env-polyprotein is the precursor of virus envelope proteins and the gag-polyprotein of the structural proteins of the nucleocapsid. The pol gene, which codes for enzymatic functions (protease, reverse transcriptase, integrase) is in all known cases expressed as a gag-pol-fusion protein. This fusion is accomplished by a rare stop codon suppression event with or without frameshift (63,64,65,66,67). This expression mechanism may have several purposes: first, expression of pol is limited and correlated in amount and time to the expression of gag proteins; second, the gag part of the fusion protein directs its incorporation into viral nucleocapsids , where the protease and reverse transcriptase are activated by proteolytic processing; third, the untranslated RNA template is brought in close contact with the enzyme. By combination of these mechanisms no enzymatically active reverse transcriptase with possible deleterious effects for the host cell can be present in a free form at any time of the infection cycle while the replication of the retroid genome is highly effective. According to the general idea retrovirus precursor proteins assemble before proteolytic processing. Studies with protease-minus mutants have shown, that protein processing of gag-precursors is not a precondition for assembly of nucleocapsids (68) and a low amount of unprocessed precursors can be found in mature virions of many retroviruses.

Likewise CaMV capsids contain ORF IV derived 57 kD protein, and shorter and more abundant proteins of related antigenicity (44, 40 kD; Fig.4; see also ref.69) representing different degrees of processing. It is tempting to speculate (Fig.6) that like unprocessed retrovirus gag and gag/pol proteins (70), CaMV ORF IV and ORF V proteins coassemble with the RNA and that they are processed and modified concommitantly with, or even after the assembly process. This would allow proper control and timing of the enzymatic reactions: e.g. in CaMV, the protease would be activated upon assembly, this would cause structural shifts in the nucleocapsid, which in turn would render the packaged RNA available for reverse transcription by an enzyme that itself became activated by

Figure 6. Model describing coupled assembly and reverse transcription in retroid elements. Individual RNA molecules can either become translated or packaged and reverse transcribed. Packaging is initiated by the interaction of RNA with nucleocapsid protein units using specific interaction sites on the respective molecules. During assembly also pol gene products exerting enzymatic functions become trapped in the nucleocapsid, either because they are bound covalently to some of the nucleocapsid protomers (gag/pol fusion proteins), or because of hydrophobic interactions. Structural shifts in the nucleocapsid protein (exagerated in the drawing) occur during assembly and the enzymatic functions of the pol gene products, i.e. protease and reverse transcriptase, become activated. Additional shifts in structure and function of the nucleocapsid components might be induced by action of the activated proteases and by phosphorylation/dephosphorylation, regulating the reverse transcription process and the interaction with the outer proteins. Timing of the enzymatic events varies in different retroid elements: Reverse transcription occurs in retroviruses after infection of a new host cell and in retrotransposons and retroid DNA viruses immediately upon assembly. Capsid strengths vary also, beeing weak in retrotransposons and strong in viruses. Outer proteins participate in envelope formation (env proteins) of retro- and hepadna viruses and inclusion body matrix protein in CaMV.

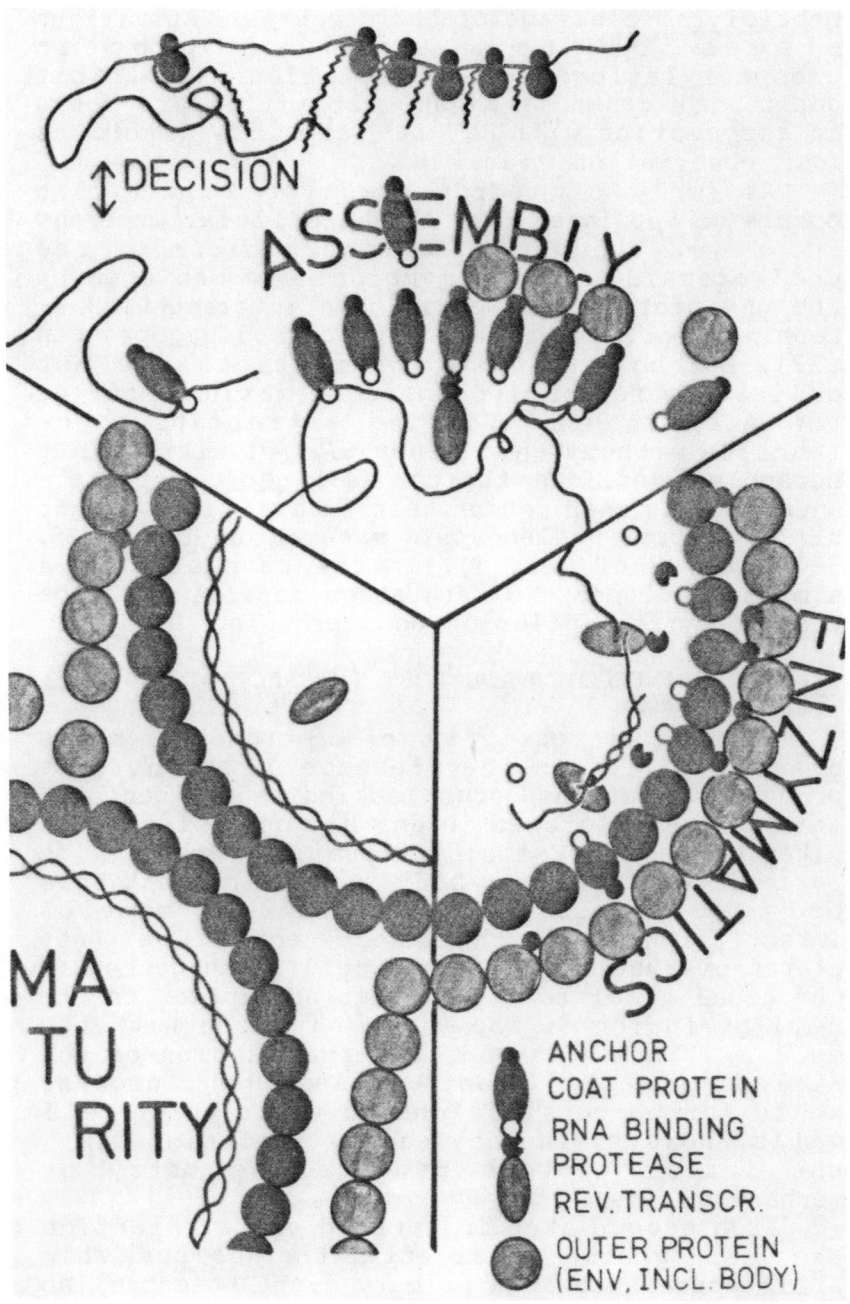

Figure 6.

proteolytic cleavage of the precursor. At various
stages of this pathway protein
phosphorylation/dephosphorylation may exhibit
additional control. Kinase activities are found
in association with HBV (71) and CaMV particles
(our observation; Fig.4).
    Assembly of retrovirus type C particles
occurs on the inner face of the cellular membrane
and it is thought that interactions of the
nucleocapsid with the membrane are mediated by
the env proteins, which in turn interact with N-
terminal portions of the unprocessed gag protein
(72). Although these interactions are not
obligatory for nucleocapsid **formation** (most of
the nucleocapsid forming retrotransposons
function without env genes (73) or with highly
scrambled versions thereof (49) gag/env protein
interactions when env protein is available, might
still control nucleocapsid **maturation** (see p.528,
583 and 707 of ref. 1). In the case of CaMV a
similar nucleocapsid maturation control could be
exerted by the inclusion body protein.

REINFECTION AND HOST INTERACTION

    Natural infectivity of retroid elements
clearly relies on the presence of an env gene
product. Retrotransposons lacking an env gene and
retroviruses mutated in env are not infectious.
Likewise the CaMV aphid transmission protein is
part of the inclusion body indicating that this
body is necessary for the natural mode of
infectivity through the insect vector. The route
of retrovirus entry into the cell is suggested in
the usual model to occur after adsorption to the
cell by fusion of the virus envelope with the
cell membrane (74) and internalization of the
nucleocapsid. For plant viruses such a process
would not be possible due to their thick cell
walls, however in many cases, including CaMV,
the insect vectors mediate the entry by
mechanical wounding.
    As a second step in retroid virus infection
the viral genome has to enter the nucleus. This
will happen both as a primary event to establish
the infection and, at least with caulimo-,

hepadna- and lentiviruses, as a secondary event to increase the number of viral DNA templates in the nucleus. Although CaMV DNA is infectious per se, if mechanically transmitted, natural infection of plant nuclei might involve the whole nucleocapsid. This might explain, why in certain cases nucleocapsids can be observed within the nucleus (75).

In a third step of infection virus has to spread from cell to cell. For this process again different mechanisms must be considered for animal and plant viruses: in the first case release of virus particles from one cell and reinfection of another by the usual entry mechanism will occur, while in plants the likely route is via plasmodesmata, the channels which connect the individual cells (76). For this systemic spreading those CaMV genes might be involved, which do not have homologous counterparts in the other retroid elements.

ACKNOWLEDGEMENTS

We highly ackowledge critical reading of this manuscript by W. Filipowicz and H. Sanfacon. JM.B., K.G. and J.M-I. where supported by EMBO fellowships.

REFERENCES

1. Weiss R, Teich N, Varmus H, Coffin J 1984 RNA tumor viruses. Cold Spring Harbor Lab. Vol.1.
2. Tiollais P, Pourcel C, Dejean A 1985 The hepatitis B virus. Nature 317:489
3. Temin HW 1985 Reverse transcription in the eukaryotic genome:retrovirus,pararetrovirus,retrotransposons and retro- transcripts. Mol.Biol.Evol. 2:455
4. Hohn T, Hohn B, Pfeiffer P 1985 Reverse transcription in a plant virus. TIBS 10:205
5. Mason WS, Taylor JM, Hull R 1986 Retroid virus genome replication. Adv.Virus Res. in press.
6. Howarth AJ, Gardner RC, Messing J, Shepherd RJ 1981 Nucleotide sequence of naturally occurring deletion mutants of cauliflower mosaic virus. Virology 112:678
7. Dixon L, Hohn T 1984 Initiation of translation of the cauliflower mosaic virus genome from a polycistronic mRNA: evidence from deletion mutagenesis.. EMBO J. 3:2731

8. Franck A, Guilley H, Jonard G, Richards K, Hirth L 1980 Nucleotide sequence of cauliflower mosaic virus DNA. Cell 21:285

9. Woolston CJ, Covey SN, Penswick JR, Davies JW 1983 Aphid transmission and a polypeptide are specified by a defined region of the cauliflower mosaic virus genome. Gene 23:15

10. Armour SL, Melcher U, Pirone TP, Lyttle DJ, Essenberg RC 1983 Helper component for aphid transmission encoded by region II of cauliflower mosaic virus DNA. Virology 129:25

11. Givord L, Xiong C, Giband M, Koenig I, Hohn T, Lebeurier G, Hirth L 1984 A second cauliflower mosaic virus gene product influences the structure of the viral inclusion body. EMBO J. 3:1423

12. Xiong C, Lebeurier G, Hirth L 1984 Detection of a new gene product (geneIII) of cauliflower mosaic virus. Proc. Ntl. Acad. Sci. USA 81:6608

13. Ziegler V, Laquel P, Guilley H, Richards K, Jonard G 1985 Immunological detection of CaMV gene V protein produced in engineered bacteria or infected plants.Gene 36:271

14. Xiong C, Muller S, Lebeurier G, Hirth L 1982 Identification by immunoprecipitation of a cauliflower mosaic virus in vitro major translation product with a specific serum against viroplasm protein. EMBO J. 1:971

15. Dixon LK, Jiricny J, Hohn T 1985 Oligonucleotide directed mutagenesis of cauliflower mosaic virus DNA using a repair-resistant nucleoside analogue: identification of an agnogene initiation codon. Gene 41:225

16. Toh H, Ono M, Saigo K, Miyata T 1985 Retroviral protease like sequence in the yeast transposon Tyl. Nature 315:691

17. Toh H, Hayashida H, Miyata T 1983 Sequence homology between retroviral reverse transcriptase and putative polymerases of hepatitis B virus and cauliflower mosaic virus. Nature 305:827

18. Toh H, Kikuno R, Hayashida H, Miyata T, Kugimiya W, Inouye S, Yuki S, Saigo K 1985 Close structural resemblance between putative polymerase of a Drosophila transposable element 17.6 and pol gene product of MoMuLV. EMBO J. 4:1267

19. Volovitch M, Modjtahedi N, Yot P, Brun G 1984 RNA-dependent DNA polymerase activity in cauliflower mosaic virus infected plant leaves. EMBO J. 3:309

20.Schoelz J, Shepherd RJ, Daubert SD 1986 Gene VI of CaMV encodes a hostrange determinant. J.Mol.Cell.Biol. in press.
21.Hull R, Covey SN 1985 CaMV:pathways of infection. BioEssays 3:160
22.Mullins JI, Chen CS, Hoover EA 1986 Desease specific and tissue specific production of unintegrated feline leukemia virus variant DNA in feline AIDS. Nature 319:333
23.Capello J, Handelsman K, Lodish HF 1985 Sequence of dyctostelium DIRS-1: an apparent retrotransposon with inverted terminal repeats and an internal junction sequence. Cell 43:105
24.Hirochika H, Takatsuji H, Ubasawa A, Ikeda JE 1985 Site specific deletion in CaMV DNA: possible involvement of RNA splicing and reverse transcription. EMBO J. 4:1673
25.Arya SK, Guo C, Josephs SF, Wong-Staal F 1985 Trans activator gene of human T-lymphotropic virus type III. Science 229:69
26.Broome S, Gilbert W 1985 Rous sarcoma virus encodes a transcriptional activator. Cell 40:537
27.Hess JL, Clements JE, Naryan P 1985 Cis and trans acting transcriptional regulation of visna virus. Science 229:482
28.Paszkowski J, Shillito RD, Saul M, Mandak V, Hohn T, Hohn B, Potrykus I 1984 Direct gene transfer to plants. EMBO J. 3:2717
29.Odell JT, Nagy F, Chua NH 1985 Identification of DNA sequences required for activity of the CaMV 35S promoter. Nature 313:810
30.DeBlock M, Herrera-Estrella L, VanMontagu M, Schell J, Zambryski P 1984 Expression of foreign genes in regenerated plants and their progeny. EMBO J. 3:1681
31.Maule AJ 1985 Replication of Caulimoviruses in plants and protoplasts. Mol.Plant Virol. (Davies,ed.) CRC Press 2:161
32.Hull R 1984 Caulimovirus group. CMI/AAB descriptions of plant viruses 295:
33.Harris JD, Scott JV, Traynor B, Brahic M, Stowring L, Ventura P, Haase AT, Peluso R 1981 Visna virus DNA: discovery of a novel gapped structure. Virology 113:573
34.Stephens RM, Casey JW, Rice NP 1986 Equine infectious anemia virus gag and pol genes:relatedness to visna and AIDS. Science 231:589
35.Covey SN 1985 Organization and expression of the CaMV genome. Mol.Plant Virol. (Davies;ed.) CRC Press 2:121

36.Summers J, Mason WS 1982 Replication of the genome of a HBV like virus by reverse transcription of an RNA intermediate. Cell 29:403
37.Mann R, Baltimore D 1985 Varying the position of a retro virus packaging sequence results inthe encapsidation of both unspliced and spliced RNAs. J.Virology 54:401
38.Darlix JL, Meric C, Spahr PF 1985 The interaction of viral proteins with Rous sarcoma virus RNA and possible control of reverse transcription, translation and virion assembly. in: Viral messenger RNA (Becker ed).
39.Copeland TD, Oroszlan S, Kalanaraman VS, Sarngadharan MG, Gallo RC 1983 Complete amino acid sequence of human T cell leukemia virus structural protein p15. FEBS letters 162:390
40.Fu X, Phillips N, Jentoft J, Tuazon PT, Traugh JA, Leis J 1985 Site-specific phosphorylation of avian retrovirus nucleocapsid protein pp12 regulates binding to viral RNA. J.Biol.Chem. 260:9941
41.Berg JM 1986 More metal binding fingers. Nature 319:264
42.Turner DS, Covey SN 1984 A putative primer for the replication of CaMV by reverse transcription is virion associated. FEBS Letters 165:285
43.Marsh L, Kuzj A, Guilfoyle T 1985 Identification and characterization of CaMV replication complexes - analogy to HBV. Virology 143:212
44.Mazzolini L, Bonneville JM, Volovitch M, Magazin M, Yot P 1985 Strand-specific viral DNA synthesis in purified viroplasm isolated from turnip leaves infected with CaMV. Virology 145:293
45.Menissier J, Laquel P, Lebeurier G, Hirth L 1984 A DNA polymerase activity is associated with CaMV. Nucleic Acids Res. 12:8769
46.Mellor J, Fulton SM, Dobson MJ, Wilson W, Kingsman SM, Kingsman AJ 1985 A retrovirus like strategy for expression of a fusion protein encoded by yeast transposon Ty1. Nature 313:243
47.Garfinkel DJ, Boeke JD, Fink GR 1985 Ty element transposition: reverse transcriptase and virus like particles. Cell 42:507
48.Shiba T, Saigo K 1983 Retrovirus like particles containing RNA homologous to the transposable element copia in drosophila. Nature 302:119
49.Ono M, Toh H, Miyata T, Awaya T 1985 Nucleotide sequence of the syrian hamster intracisternal A-particle gene. J. Virol. 55:387</cb>

50.Will H, Salfeld J, Pfaff E, Manso C, Theilmann L, Schaller H 1986 Putative reverse transcriptase intermediates of human HBV in primary liver carcinomas. Science 231:594
51.Guilley H, Dudley RK, Jonard G, Balazs E, Richards K 1982 Transcription of cauliflower mosaic virus DNA: detection of promoter sequences, and characterization of transcripts. Cell 30:763
52.Rosen CA, Sodroski JG, Goh WC, Dayton AI, Lippke J, Haseltine WA 1986 Post-transcriptional regulation accounts for the trans-activation of the human T-lymphotropic virus type III. Nature 319:555
53.Darlix JL, Spahr PF 1982 Binding sites of viral protein p19 onto Rous sarcoma virus RNA and possible control of viral functions. J.Mol.Biol. 160:147
54.Hull R 1984 A model for expression of CaMV nucleic acid. Plant Mol.Biol. 3:121
55.Kozak N 1986 Regulation of protein synthesis in virus infected animal cells. Adv.Virus Res. in press:
56.Katz RA, Cullen BR, Malavarca R, Skalka AM 1986 Role of avian retrovirus mRNA leader in expression: evidence for novel translation control. Mol.Cell.Biol. 6:372
57.Sieg K, Gronenborn B 1982 Evidence for polycistronic messenger RNA encoded by cauliflower mosaic virus. Abstract NATO Advanced Studies Inst. Advanced Course 1982:154
58.Hughes S, Mellstrom K, Kosik E, Tamanoi F, Brugge J 1984 Mutation of a terination codon affects src initiation. Mol.Cell.Biol. 4:1738
59.Petersen RB, Hackett PB 1985 Characterization of ribosome binding on Rous sarcoma virus RNA in vitro. J.Virol. 56:683
60.McKnight T, Meagher RB 1981 Isolation and mapping of small cauliflower mosaic virus DNA fragments active as promoters in Escherichia coli. J.Virology 37:673
61.Hohn T, Richards K, Lebeurier G 1982 CaMV on its way to becoming a usefull plant vector. Cur.Top.Microbiol.Immunol. 96:194
62.Plant AL, Covey SN, Grierson D 1985 Detection of a subgenomic mRNA for gene V, the putative reverse transcriptase gene of cauliflower mosaic virus. Nucl. Acids Research 13:8305
63.Yoshinaka Y, Katoh I, Copeland T, Oroszlan S 1985 Translational readthrough of an amber termination codon during synthesis of feline leukemia virus protease. J.Virology 55:870

64. Yoshinaka Y, Katoh I, Copeland TD, Oroszlan S 1985 MuLV protease is encoded by the gag/pol gene and is synthesized through suppression of an amber termination codon. Proc. Ntl. Acad. Sci. USA 82:1618
65. Jacks T, Varmus HE 1985 Expression of the Rous sarcoma pol gene by ribosomal frameshifting. Science 230:1237
66. Mellor J, Malim MH, Gull K, Tuite MF, McCready S, Dibbayawan T, Kingsman SM, Kingsman AJ 1985 Reverse transcriptase activity and Ty RNA are associated with viruslike particles in yeast. Nature 318:583
67. Clare J, Farabaugh P 1985 Nucleotide sequence of a yeast Ty element: evidence for an unusual mechanism of gene expression. Proc. Ntl. Acad. Sci. USA 82:2829
68. Yoshinaka Y, Luftig RB 1982 P65 of Gazdar murine sarcoma viruses contains antigenic determinants from all four of the murine leukemia virus gag polypeptides (p15,p12,p30,p10)and can be cleaved in vitro by the MuLV proteolytic activity. Virology 118:38
69. Al Ani R, Pfeiffer P, Lebeurier G 1979 The structure of cauliflower mosaic virus. II: Identity and location of the viral polypeptides. Virology 93:188
70. Katoh I, Yoshinaka Y, Rein A, Shibuya M, Odaka T, Oroszlan S 1985 Murine leukemia virus maturation: protease region required for conversion from immature to mature core form of virus infection. Virology 145:280
71. Gerlich WH, Goldmann U, Mueller R, Stibbe W, Wolff W 1982 Specificity and localization of the HBV associated protein kinase. J.Virology 42:761
72. Bolognesi DP, Montelaro RC, Frank H, Schaefer W 1978 Assembly of type C Oncorna viruses, a model. Science 199:183
73. Mount SM, Rubin GM 1985 Complete nucleotide sequence of the drosophila transposable element copia: homology between copia and retroviral proteins. Mol.Cell.Biol. 5:1630
74. Anderson KB, Nexo BA 1983 Entry of murine retrovirus into mouse fibroblasts. Virology 125:85
75. Gracia O, Shepherd RJ 1985 CaMV in the nucleus of nicotiana. Virology 146:141
76. Shepherd RJ 1976 DNA viruses of higher plants. Adv.Virus Res. 20:305
77. Schwartz DE, Tizard R, Gilbert W 1983 Nucleotide sequence of Rous sarcoma virus. Cell 32:853
78. Shinnick TM, Lerner RA, Sutcliffe JG 1981 Nucleotide sequence of MoMuLV. Nature 293:543

79. Shimotohno K, Takahashi Y, Shimizu N, Gojobori T, Golde DW, Chen ISY, Miwa M, Sugimura T 1985) Complete nucleotide sequence of an infectious clone of human T-cell leukemia virus type II: an open reading frame for the protease gene. Proc.Natl.Acad.Sci.USA 82:3101

80. Ratner L, Haseltine W, Patarca R, Livak KJ, Starcich B, Josephs SF, Doran ER, Rafalski JA, Whitehorn EA, Baumeister K, Iwanoff L, Petteway SRJr, Pearson ML, Lautenberger JA, Papas TS, Ghrayeb J, Chang NT, Gallo RC, Wong-Staal F 1985 Complete nucleotide sequence of the AIDS virus HTLV-III. Nature 313:277

81. Seiki M, Hattori S, Hirayama Y, Yoshida M 1983 Human adult T cell leukemia virus: complete nucleotide sequence of the provirus genome integrated in leukemia cell DNA. Proc.Natl.Acad.Sci.USA 80:3618

82. Emori Y, Shiba T, Kanaya S, Inouye S, Yuki S, Saigo K 1985 The nucleotide sequence of copia related RNA in Drosophila virus like particles. Nature 315:773

83. Saigo K, Kugimiya W, Matsuo Y, Inouye S, Yoshioka M, Yuki S 1984 Identification of the coding sequence for a reverse transcriptase like enzyme in a transposable genetic element in drosophila menalogaster. Nature 312:659

84. Galibert F, Mandart E, Fitoussi F, Tyiollais P, Charnay P 1979 Nucleotide sequence of the HBV genome (subtype ayw) in E.coli. Nature 281:646

85. Nobrega FG, Tzagoloff A 1980 Assembly of the mitochondrial membrane system. J.Biol.Chem. 255:9828

86. Zimmern D 1983 Homologous proteins encoded by yeast mitochondrial introns and by a group of RNA viruses from plants. J.Mol.Biol. 171:345

Molecular Strategies for Crop Protection, pages 295–305
© 1987 Alan R. Liss, Inc.

REPLICASE AND REPLICATION: STRATEGIES FOR
BROME MOSAIC VIRUS[1]

Timothy C. Hall, Theo W. Dreher and Loren E. Marsh

Biology Department, Texas A & M University
College Station, Texas 77843-3258

ABSTRACT Brome mosaic virus is a positive (+) strand
RNA virus with three genome segments; it infects
grasses. An RNA-dependent RNA polymerase (replicase)
activity which specifically copies RNAs of brome
mosaic virus can be extracted from brome mosaic
virus-infected barley leaves.   The replicase is
template dependent, synthesizing product strands
after de novo initiation opposite a unique C residue
of the template RNA.   Virion RNAs are used as
templates for the accurate synthesis of full length
(-) strand.   Template selection depends on the
promoter activity of the tRNA-like structure present
at the 3' end of each viral RNA; this same structure
is also responsible for specific aminoacylation at
the 3' end with tyrosine. Minus strand RNA3 is used
as a template by the replicase for the synthesis of
the subgenomic RNA4, by a transcriptional mechanism
after initiation internally on (-) RNA3. Mutants in
the tRNA-like structure have been constructed in
order to study the interaction of the replicase and
aminoacyl tRNA synthetase with the RNA.   The
replicative fitness of such mutants in vivo has also
been studied.

---

[1]This work was supported by NIH Grants AI 11572 and
22354 and by Texas A&M University.

INTRODUCTION

Crop productivity is undoubtedly reduced by viral infection, although the financial losses are often hard to determine when the infection is not severe enough to kill the host.   Indeed, viral infections often pass unidentified, and symptomless infections are well known in carrier species (1).   Considerable progress has been made in advancing technology for detecting viral infections of crops, particularly in the use of specific probes for genomic nucleic acids, many of which are commercially available (2).   It is likely that increased awareness of viral infections afforded by modern detection procedures will stimulate demands that new ways to combat viral infections be developed.   The use of biotechnological approaches is attractive, and protection strategies involving expression in plants of products which interfere with viral replication are being considered.   One such approach may involve the expression of sequences complementary to the genome of plant RNA viruses in order to prevent the onset of replication by hybrid-arresting either the translation of viral genes or the promotion of replication itself.   A knowledge of the biochemistry of viral replication is certainly a key element in designing such plant protection strategies.   This laboratory has been engaged in characterizing the replication of the positive-strand RNA virus brome mosaic virus (BMV).   Since the vast majority of the viruses infecting plants have an RNA genome (3), these studies are pertinent to a wide range of plant viruses.

REPLICATION OF BROME MOSAIC VIRUS

The BMV genome comprises RNA1 (3234 nt), RNA2 (2865 nt) and RNA3 (2117 nt).   These are actively translated, capped mRNAs that are separately encapsidated in icosahedra consisting of 180 identical coat protein molecules (4).   The coat protein is translated from RNA4, a subgenomic mRNA that is colinear with the 3' region of RNA3, and arises from that genomic RNA.   RNA4 is co-packaged with RNA3.   All four BMV RNAs have a region of nearly identical sequence at their 3' ends (5), including the region which is folded into a tRNA-like structure (6), and which is responsible for interaction with several host proteins normally associated with tRNAs (7).   BMV RNA can be charged with tyrosine by host tyrosyl tRNA synthetase with high efficiency, and viral RNAs are tyrosylated in vivo (8).

Viral replication is dependent on initial translation of inoculum genomic RNAs, which subsequently serve also as templates for replication. Replication initiates with (-) strand synthesis, in which all four RNAs participate, as evidenced by the appearance of full-length double stranded forms of each RNA in extracts of infected barley protoplasts (9). Antigenome (-) strands do not accumulate to high levels in normal infections, but appear to be hyperactive templates for the asymmetric overproduction of (+) strands, which can then be either encapsidated or enter further cycles of replication. It is not known whether the double stranded RNAs participate actively in replication, or are merely end products in metabolically exhausted cells, or artifacts of extraction, as is the case in poliovirus-infected cells (10). The subgenomic RNA4 appears somewhat late in infection (9), and experiments with pseudorecombinant infections have shown that product RNA4 is derived from input RNA3 and that RNA4 in the inoculum is not replicated (11). Three modes of BMV RNA replication can thus be discerned:

(1) (-) strand copying from (+) strand templates,
(2) hyperactive (+) strand synthesis from (-) strand (or possibly double stranded) templates, and
(3) synthesis of subgenomic RNA4.

Each of these replicative steps has been studied in vitro, as discussed below.

BMV REPLICASE

An RNA-dependent RNA polymerase activity specific for copying BMV RNAs (BMV replicase) appears in barley plants on infection with BMV. We have described the preparation of replicase extracts from infected leaves (12,13). In common with replicases from other eukaryotic positive strand RNA viruses, the activity is associated with the membrane fraction; mild detergent treatment does not solubilize the replicase, and detergent treatments that do completely solubilize the membrane result in the loss of template specificity. Rather than purify an uncoupled polymerase, our studies with BMV replicase have focussed on the characterization of the BMV RNA-specific RNA polymerase in a partially purified membrane fraction (13).

Treatment of extracts with micrococcal nuclease results in a dependence on exogenous template, with only very low levels of activity resulting from endogenous RNA (14). To our knowledge, this is the only eukaryotic viral replicase preparation showing dependence and specificity for the cognate RNA. As will be discussed below, many unique opportunities exist for using this enzyme to characterize features of the viral RNA that contribute to replication and infection.

Synthesis of (-) strand by replicase.

The use of BMV virion RNAs as templates for (-) strand synthesis by BMV replicase is the reaction we have studied in most detail. The typical products are full-length double stranded RNAs which are resistant to ribonuclease digestion. Each of the four virion RNAs serve as template, although RNA3 and RNA4, being significantly shorter, show better yields. RNAs from closely related viruses, such as cowpea chlorotic mottle virus (CCMV), are copied less efficiently, and tRNA or RNAs from unrelated viruses, such as CPMV, a plant picornavirus, are not detectably copied. Product strands are initiated de novo, as shown by incorporation of γGTP but of no other γ-labeled nucleotides (15). Initiation of (-) strand synthesis most likely occurs opposite the penultimate (C2) nucleotide from the $3'$-CCA_{OH} terminus of

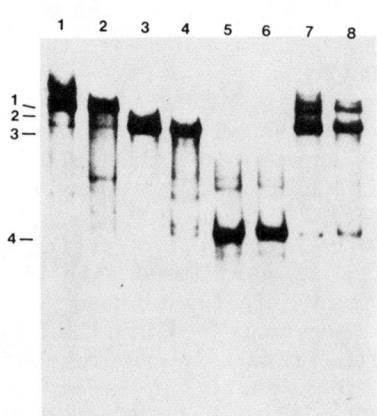

Fig. 1. Fluorograph of ^{3}H-labeled RNA products of BMV replicase treated with micrococcal nuclease (from ref. 15). Bands marked 1→4 label BMV RNAs 1→4. BMV RNAs used as templates for the replicase were: unfractionated RNA (lanes 1&2), RNA 1&2 (lanes 3&4), RNA3 (lanes 5&6), RNA4 (lanes 7&8). All products in even-numbered lanes were ribonuclease treated, and are therefore double-stranded.

the (+) strand. Faithful copying of the (+) strand template was established by direct sequencing of the (-) strand, using a chain-termination method (15). This event results in the (-) strand template, used for synthesis of the new (+) strand, lacking a 5' terminal U residue. The last (Al) residue of the -CCA$_{OH}$ terminus must therefore finally be added in a non-template dependent manner, presumably by host CTP:ATP tRNA nucleotidyl transferase (NTase) in a post-replicational step. BMV RNA is an active substrate for this enzyme. Analysis of double-stranded BMV RNA from infected barley plants by 3'-labeling and nearest neighbor testing revealed the absence of A residues at the 3' ends of these presumed replicative form RNAs (15). Similar results are reported for cucumber mosaic virus double stranded RNAs (16).

It appears that our present replicase preparations are unable to accomplish an entire cycle of replication, giving rise to new (+) RNA from input virion (+) templates. Positive sense products have not been observed in such reactions, and the double stranded nature of the products may represent an inactive form. The proportion of supplied templates used by the replicase is much lower than that prevailing in vivo, and any strand displacement mechanism for keeping template and product strands from forming stable hybrids may be inoperative until we are able to load more replicase molecules on each template.

Synthesis of (+) strand by replicase.

In studying (+) strand synthesis, we have produced (-) sense BMV RNAs in vitro by transcription from cDNA clones using SP6 or T7 RNA polymerases. Minus sense RNA3 was initially used because it was a potential template for synthesis of both (+) RNA3 and the subgenomic RNA4. The cDNA construction used resulted in transcripts with extra nucleotides at the 3' end of the (-) strand, and perhaps for this reason they did not yield full-length (+) RNA3 products (17). It was, however, an active template for subgenomic RNA synthesis by the replicase, discussed below, demonstrating the ability of the preparation to use (-) sense templates.

Various approaches have been used in our attempts to demonstrate full length (+) strand synthesis by the replicase, so far without success (T.W. Dreher, unpublished). It is important to achieve (-) strand

copying _in vitro_, since this may allow direct
demonstration of the inability of (-) RNA4 to replicate.
Correctly terminated (-) RNAs 2, 3 or 4 not are copied by
replicase _in vitro_. This is also true for (-) RNA3 with a
6 base pair deletion serving to remove competition from
the subgenomic promoter (see below). It is possible that
an extra, non-genome-encoded G is required at the 3' end
for activity (16), and this is being investigated. It is
of course possible that factors necessary for (-) strand
copying are absent from the replicase preparation.

Synthesis of subgenomic RNA4 by replicase

    Transcripts encoding (-) polarity BMV RNA3 sequences
of various lengths can be used as templates _in vitro_ for
subgenomic RNA synthesis by replicase. Product synthesis
clearly involves _de novo_ initiation of the RNA chain with
a G residue at the expected internal location of RNA3
(17). Only 20 bases 5' to the site of initiation appear
to be required for subgenomic synthesis; this promoter,
which is responsible for RNA 4 synthesis is being further
characterized (L.E. Marsh, unpublished).
    The above observations have demonstrated that the
subgenomic RNA is produced by a transcriptional mechanism,
rather than by a cleavage event as proposed for cucumber
mosaic virus (18) or by direct replication of RNA4. The
latter would contradict genetic data (11), but it remains
important to prove that replicase is fundamentally unable
to copy (-) RNA4, while copying the genomic (-) strands
very efficiently.

    Because the replicase preparation we employ is only
partially purified, we cannot be sure that one enzyme is
responsible for all the replicative activities we have
observed _in vitro_. It is certainly feasible that addition
or loss of a factor accompanies a switch in the type of
template used.

          CHARACTERIZATION OF THE (+) STRAND PROMOTER
               USING SYNTHETIC MUTANT RNAS

    The template specificity of BMV replicase suggested
that each virion RNA must encode a recognition sequence,
and the conserved sequences of over 200 nt at the 3' end

of each BMV RNA component (5) were obvious candidates for such a recognition locus. Localization of the replicase promoter to this region was accomplished with a number of approaches. The template activity of 3' fragments of the genomic RNAs, produced either by partial ribonuclease digestion or by deoxyoligonucleotide-directed ribonuclease H cleavage indicated that the 3' tRNA-like structure was the apparent promoter. This has been verified using synthetic RNA transcripts produced from cloned cDNA (15).

Another approach to defining the importance of the 3' region for (-) strand initiation involved experiments in which cDNA sequences complementary to the BMV RNA 3' terminus were hybridized to BMV RNA. Such sequences were able to arrest replication, presumably by blocking recognition and initiation on the RNA template (19). Melting of the RNA-DNA hybrid restored template activity. Attempts to arrest replication by sequences complementary to internal regions of BMV RNA were generally unsuccessful, indicating that replication was able to proceed through double-stranded structures.

An important development in our studies of the properties of the tRNA-like structure as replicase promoter was the design of a method for generating mutant RNAs at will in vitro. Wild type RNAs with correct 3' termini were transcribed by SP6 RNA polymerase, and these could be directly tyrosylated without further modification, and were active templates for replicase (20). The specific mutagenization of the BMV cDNA clone has thus made it possible to probe the 3' structure of BMV RNAs by studying the activity of a range of mutants (20, 21). Two locations were initially chosen for mutagenesis; the 3' terminus and the putative tyrosine anticodon (AUA) located 65-67 nt from the 3' end. These experiments demonstrated the power of this approach to studying the role of the multifunctional 3' end of the RNA. Sequences necessary for aminoacylation and replicase template activity could be readily separated. A combination of oligonucleotide-directed mutations and deletions generated from restriction sites present in the cDNA sequence has now been used to comprehensively map functional regions of the 3' sequence (20,21; T.W. Dreher, in preparation). Figure 2 shows the folding of the 3' tRNA-like structure (14) and the location of some of the mutations we have introduced into the RNA. Arm A is the aminoacyl acceptor stem and the -AUA- anticodon appears within the loop of arm C. Apart from arm D, which is absent in the RNA of broad bean mottle virus, the structure (but not necessarily the sequence) is conserved in the RNAs of all bromo- and cucumoviruses (compiled in ref. 22). Tyrosyl tRNA synthetase apparently interacts with only a limited part of the structure: the aminoacyl acceptor stem (arm A), arm B (principally the basal region), and the sequence immediately 5' to the tRNA-like structure shown in Fig. 2. There appears to be no significant interaction with arm C or the anticodon, nor with arm D. Replicase apparently makes more extensive contacts with the RNA, since mutations in several regions result in loss of template activity. The most significant determinant for recognition appears to be the anticodon loop; even single base changes cause a major loss of template activity (20). Arm B and the 3' end are other regions where mutations decrease template activity. Removal of Arm D results in increased template activity. Two loci have now been found in which subtle sequence changes (one or three base substitutions) cause a large decrease in aminoacylation activity without significantly affecting the _in vitro_ replicase template activity (Dreher and Hall, in preparation). Such mutants are valuable in assessing the importance of efficient tyrosylation of viral RNAs during infection.

The effect that mutations in the tRNA-like structure have on replicatability _in vivo_ can be studied. This can be accomplished by incorporating the mutated 3' region into cDNA clones that can be transcribed to give full length, infectious transcripts (23). Some of the deletions described previously (21) have been spliced into the RNA3 transcript. Initial tests confirm that mutations

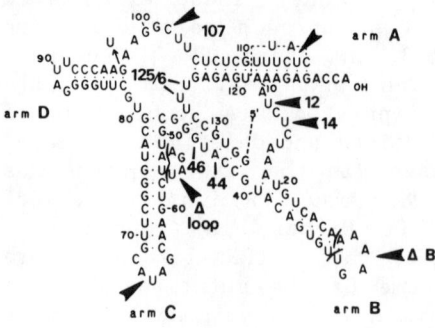

Fig. 2.   Sites within the tRNA-like structure of BMV RNA3 mutated and tested for altered functions.

that are detrimental to the activities tested <u>in vitro</u> yield transcripts that are non-infectious (Bujarski <u>et al</u>., in preparation). Deletion of arm D, which did not decrease activity in the <u>in vitro</u> assays, also did not prevent replication in either barley plants or protoplasts. Replicase template activities observed <u>in vitro</u> thus appear to reflect the replicatability of those RNAs <u>in vivo</u>.

CONCLUSIONS

BMV replicase preparations are able to complete accurately <u>in vitro</u> several aspects of the replicative cycle of BMV RNA. In those cases tested so far, the replication activities of mutants tested <u>in vitro</u> has appropriately reflected their ability to replicate <u>in vivo</u>. These characteristics support our belief that the replicase activity we extract from BMV-infected barley is that specifically involved in the replication of BMV RNA. It is unfortunate that the activities of other replicase systems have not been characterized in similar detail. Uncritical reliance on levels of nucleotide incorporation has led to many spurious conclusions in regard to processes involved in viral replication. Major challenges in understanding BMV replicase are the identification of the polypeptides in the active enzyme complex, and their role in promoter selection and chain elongation. Elucidation of the nature of the replicative systems will be important in defining possible ways in which viral infections can be impeded or prevented. Our studies have demonstrated that the ability to specifically modify RNA sequences provides an entirely new approach to solving the problems of pathogenesis of the RNA viruses.

REFERENCES

1.   Garrett RG, Cooper JA, Smith PR (1985). Virus epidemiology and control. In Francki RIB (ed): "The Plant Viruses Volume I", New York: Plenum Press, p 269.
2.   Garger SJ <u>et al.</u> (1983). Rapid detection of plant RNA viruses by dot blot hybridization. Plant Molecular Biology Reporter 1:21.

3. Francki RIB, Milne RG, Hatta T (1985). "An Atlas of Plant Viruses." Boca Raton: CRC Press.
4. Francki RIB (ed; 1985). "The Plant Viruses Volume I", New York: Plenum Press.
5. Ahlquist P, Dasgupta R, Kaesberg P (1981). Near identity of 3' RNA secondary structure in bromoviruses and cucumber mosaic virus. Cell 23:183.
6. Rietveld K, Pleij WA, Bosch L (1983). Three-dimensional models of the tRNA-like 3' termini of some plant viral RNAs. EMBO J. 2:1079.
7. Hall TC (1979). Transfer RNA-like structures in viral genomes. Intl. Rev. Cytology 60:1.
8. Loesch-Fries LS, Hall TC (1982). In vivo aminoacylation of brome mosaic and barley stripe mosaic virus RNAs. Nature 298:771.
9. Loesch-Fries LS, Hall TC (1980). Synthesis, accumulation and encapsidation of individual brome mosaic virus RNA components in barley protoplasts. J. gen. Virol. 47:323.
10. Baltimore D (1968). Structure of the poliovirus replicative intermediate RNA. J. molec. Biol. 32:359.
11. Lane LC, The Bromoviruses. Advances in Virus Research 19:151.
12. Hardy SF, German TL, Loesch-Fries LS, Hall TC (1979). Highly active template-specific RNA-dependent RNA polymerase from barley leaves infected with brome mosaic virus. PNAS (USA) 76:4956.
13. Bujarski JJ, Hardy SF, Miller WA, Hall TC (1982). Use of dodecyl-β-D-maltoside in the purification and stabilization of RNA polymerase from brome mosaic virus-infected barley. Virology 119:465.
14. Miller WA, Hall TC (1984). Use of micrococcal nuclease in the purification of highly template dependent RNA-dependent RNA polymerase from brome mosaic virus-infected barley.
15. Miller WA, Bujarski JJ, Dreher TW, Hall TC (1986). Minus-strand initiation by brome mosaic virus replicase within the 3' tRNA-like structure of native and modified RNA templates. J. molec. Biol. 187:537.
16. Collmer CW, Kaper JM (1985). Terminal sequences ofthe double-stranded RNAs of cucumber mosaic virus and its satellite: implications for replication. Virology 145:249.

17. Miller WA, Dreher TW, Hall, TC (1985). Synthesis of brome mosaic virus subgenomic RNA *in vitro* by internal initiation on (-)-sense genomic RNA. Nature 313:68.

18. Gould AR, Symons RH (1982). Cucumber mosaic virus RNA3: Determination of the nucleotide sequence provides amino acid sequences of protein 3A and viral coat protein. Eur. J. Biochem. 126:217.

19. Ahlquist P, Bujarski JJ, Kaesberg P, Hall TC (1984). Localization of the replicaserecognition site within brome mosaic virus RNA by hybrid-arrested RNA synthesis. Plant molec. Biol. 3:37.

20. Dreher TW, Bujarski JJ, Hall TC (1984). Mutant viral RNAs synthesized *in vitro* show altered aminoacylation and replicase template activities. Nature 311:171.

21. Bujarski JJ, Dreher TW, Hall TC (1985). Deletions in the 3' terminal tRNA-like structure of viral RNA differentially affect aminoacylation and replication *in vitro*. PNAS (USA) 82:5636.

22. Joshi RL, Joshi S, Chapeville F, Haenni A-L (1983). tRNA-like structures of plant viral RNAs: conformational requirements for adenylation and aminoacylation. EMBO J. 2:1123.

23. Ahlquist P, French R, Janda M, Loesch-Fries LS (1984). Multicomponent RNA plant virus infection derived from cloned viral cDNA. PNAS (USA) 81:7066.

Molecular Strategies for Crop Protection, pages 307–318
© 1987 Alan R. Liss, Inc.

STUDIES ON THE REPLICATION OF TOMATO GOLDEN MOSAIC VIRUS AND THE CONSTRUCTION OF GENE VECTORS [1]

Robert H.A. Coutts, Kenneth W. Buck, Eulian J.F. Roberts, Clare L. Brough, Robert J. Hayes, Heather Macdonald, Samuel W. MacDowell, Ian T. D. Petty, Marek J. Slomka, William D.O. Hamilton [2] and Michael W. Bevan [2]

Department of Pure and Applied Biology, Imperial College of Science and Technology, Prince Consort Road, London SW7 2BB, United Kingdom

ABSTRACT Investigations on the replication of tomato golden mosaic virus (TGMV) DNA in whole plants, protoplasts, tissue culture and isolated nuclei indicate that a number of approaches are available to study gene expression in plants. Two types of high copy number transient expression vector, based on TGMV replicons, are being constructed for the rapid assay of promotor function and other factors affecting plant gene expression; firstly vectors replicating as viruses for whole plants and secondly vectors replicating as plasmids for transformation of plant protoplasts. Viral genes essential for the replication of TGMV replicons are to be incorporated into the plant genome using Ti based plasmid vectors and constitutively expressed. Our progress in these studies and the general applicability of geminiviral vectors for such investigations are discussed.

[1] This work was supported by the Science and Engineering Research and Medical Research Councils of the United Kingdom, the Potato Marketing Board, the Nuffield Foundation, The British Petroleum Company PLC, and CAPES Brazil.
[2] Plant Breeding Institute, Maris Lane, Trumpington, Cambridge

INTRODUCTION

Considerable progress has been made developing
Agrobacterium Ti plasmid gene vectors and selectable
antibiotic resistance markers for plant cell
transformation. Disarmed vectors are now available which
will allow cloned genes to be inserted into the genome of a
wide range of dicotyledonous plants (1,2). Transformed
plants are capable of normal growth and differentiation and
foreign genes are stably inherited as dominant markers in
heterozygotes (3). Transformation of plants by direct gene
transfer without a vector has now also been achieved (4,
5).

Using Ti plasmid vectors or direct gene transfer it
is now possible to study the mechanisms of plant-specific
gene regulation in whole plant tissues. For this purpose
plant promoter sequences are coupled to "reporter" genes.
However, even using the latest techniques, periods of
several weeks are required after transformation before the
progeny are available for testing.

For investigations of the control of plant gene
transcription, for example using site-specific mutagenesis
to define the importance of 5'-flanking sequences, more
rapid systems would be useful. Several transient
expression assays have been designed for this purpose in
animal cell systems; those involving vectors based on virus
replicons have an additional advantage of high copy number
and therefore a higher level of expression (6,7). We are
now developing similar rapid expression vectors for dicotyledonous plant hosts based on geminivirus replicons.

The geminiviruses are a unique group of plant viruses
characterized by geminate particles measuring ca. 18x30nm
and a genome of covalently closed single-stranded(ss) DNA
(8). Two recent reviews detail the biology (9) and
molecular biology (10) of the geminiviruses. Virus, virion
ssDNA, and virus-specific double-stranded (ds) DNA,
isolated from infected plants, are infectious for whole
plants and for plant protoplasts (10, 11). Electron
microscopy (e.g. 12) and studies with isolated nuclei (13)
from infected plants suggest that virus DNA replicates and
virus particles accumulate in the nuclei of infected cells.

We have been investigating the genome organization

and expression of a number of geminiviruses including
tomato golden mosaic virus (TGMV) and potato yellow mosaic
virus (PYMV) both of which infect a range of solanaceous
plants. We first characterized TGMV and showed it to be an
authentic geminivirus with a genome of ssDNA circles ca.
2500 nucleotides in length (14). For cloning we used the
virus replicative form (RF) dsDNA and developed a new
procedure which facilitated the isolation of the
intracellular forms of TGMV in high yield. Several
virus-specific DNAs were characterized: circular and linear
ssDNA, identical with DNA obtained from virions; open
circular dsDNA of ca. 2 500 base pairs with a single break
in one strand (RFII); covalently closed circular
(supercoiled) dsDNA (RFI); genomic length linear dsDNA; a
subgenomic ssDNA and three dsDNAs of greater than genomic
length, possibly concatemers (15, 16, 17). Similar DNA
species have been described in cassava latent virus (CLV)
(18) and PYMV infected cell extracts.

Restriction endonuclease digestion of TGMV RF DNA
produced fragments, the sum of the sizes of which added up
to twice that expected from its apparent length suggesting
the presence of two different molecules. Using the RF II
DNA we cloned the TGMV genome into Escherichia coli plasmid
pAT 153 (19) and proved conclusively that TGMV DNA
consisted of two components of similar size A and B.
Cloned TGMV dsDNA is fully infective for plants and gives
rise to progeny virus indistinguishable from native TGMV
(20). Whole plant infection was achieved with a mixture of
components A and B, but not with either component
separately thereby providing unequivocal evidence for the
bipartite nature of the TGMV genome. Two other bipartite
geminiviral DNA genomes CLV (21, 22), and bean golden
mosaic virus (BGMV) (23, 24) have been cloned, sequenced
and proven to be infectious for whole plants. A recent
report suggests that the larger of the cloned DNA
components of CLV is independently infectious for tobacco
protoplasts (11). The infectious cloned DNA components of
TGMV have been sequenced in their entirety (25) and an
examination of the sequence revealed the following.

(a) A region of ca. 200 bp which is highly conserved
between the two DNA components and in which a putative
origin of DNA replication has been identified.
(b) Four major open reading frames (ORF), two on DNA
A and two on DNA B, in opposite orientations, suggesting

bidirectional transcription from within, or close to, the 200 bp "common" region. These were designated AR1, AL1, BR1, BR2, with R and L indicating rightward and leftward transcription respectively.

(c) Two smaller overlapping ORFs, AL2 and AL3, one of which (AL2) overlaps the 3' end of AL1. These two ORFs may have a promoter and polyadenylation signal in common. They may also share a polyadenylation signal with AL1, although their promoter appears to be distinct.

(d) All six ORFs are conserved in CLV (22) and BGMV (23).

(e) AR1 encodes the virus coat protein. The major polyadenylated mRNAs of CLV have been mapped (26), and preliminary results obtained from TGMV suggest a similar transcription pattern. Experiments on the construction and analysis of mutants carrying deletions in specific genes are also underway.

In parallel with studies on the full-length clones of the two components of the TGMV genome we have characterised a subgenomic DNA, detected in plants infected with native TGMV (15), but not cloned DNA (20). This subgenomic DNA of ca. 1.2 kb is encapsidated in virions (20) and replicates to a copy number some 30% of that of the genomic DNA. Heteroduplex mapping and the sequencing of a number of clones of the subgenomic DNA has revealed that this subgenomic DNA is derived from DNA B. Subgenomic DNA molecules include the 200 bp "common" region, but lack complete ORFs with ORF BR1 being completely deleted and the C-terminus of ORF BL1 being absent. Similar subgenomic DNAs have been described and sequenced in CLV DNA preparations (18) and identified in extracts of PYMV infected plants.

Sufficient background information is now available for us to develop two types of high copy number expression vector systems i.e. those vectors which replicate as virions and vectors which replicate as plasmids. Our recent results on the construction of such vectors and the relative merits of the two systems are outlined below.

RESULTS AND DISCUSSION

Our first approach to vector construction was based on the naturally occurring subgenomic DNA described above. Since all functions required for virus replication are

provided by the two full length genomic DNAs subgenomic DNA
may only require the putative 200 bp "common" region for
replication. We have thus constructed tripartite vector
systems for plants comprised of the two cloned TGMV DNA
components in combination with a third vector component
based on the subgenomic DNA, into which foreign DNA has
been inserted. This system is analogous to some SV40
helper virus vector systems (27).

We set out initially to confirm the feasibility of
the system described above by inoculating tobacco plants
with the two genomic components of TGMV in combination with
the cloned subgenomic component. Similar experiments with
CLV DNA suggested that in such inoculations the inclusion
of the subgenomic component results in interference with
virus replication and a delay in symptom expression (18)
and we noted similar results with TGMV. Dissimilar from
the inoculation of plants with the cloned genomic
components of the virus alone, which does not generate
subgenomic DNA (20) inclusion of cloned sub-genomic DNA in
the inoculum results in its synthesis and encapsidation in
systemically infected leaves. Subsequently several
potential vectors constructed from clones of naturally
occurring TGMV subgenomic DNA and subgenomic DNA
constructed in vitro have been produced for tripartite
inoculation of plants. In one series of constructs, a
nopaline synthase (NOS) promotor has been coupled to the
neomycin phosphotransferase II (npt II) coding region which
confers resistance to the antibiotics kanamycin and G418 to
which plant cells are normally sensitive (28) and the NOS
3' signal region (29) to express the coding region. This
DNA, inserted into cloned subgenomic DNA, is to be
inoculated onto plants in combination with the two cloned
genomic components in an attempt to demonstrate its
replication and expression of npt-II enzyme activity. It
is not thought that packaging constraints will obviate
against the replication of this DNA since dimeric forms of
either of the genomic components of CLV (18) and TGMV DNA
are apparently encapsidated, albeit at low efficiency.
Additionally, possible naturally occurring unencapsidated
bipartite geminiviruses have been described (30), and once
characterized at the molecular level may be useful for
similar constructions. As an alternative to vectors based
on the naturally occurring subgenomic DNAs artificial
constructs consisting of portions of genomic DNA are also
being investigated. In one series of such constructs a

complementation of the genomic functions of the smaller
TGMV B component DNA is being utilised. In these
constructs the 200 bp region and the complete ORF for BL1
has been retained and has been linked to the npt-Ⅱ gene
and the NOS polyadenlylation signal. The ORF for BR1 in
these constructs has been deleted. Clones of this DNA are
to be inoculated onto plants in combination with TGMV A
component DNA and cloned DNA of the B component which has
been inactivated by insertional mutagenesis within ORF BL1
but retains ORF BR1 intact. If replication of the vector
DNA occurs in these plants symptoms will be apparent and
npt-Ⅱ enzyme activity should be detectable.

It is likely that the TGMV coat protein gene (AR1)
will not be required for dsDNA replication in protoplasts,
and possibly not whole plants. We are concentrating on
further constructs in the tripartite system, deleting the
coat protein coding region and replacing it with selectable
marker genes. The relative efficiency of geminiviral
promoters for gene expression is as yet unknown.
Nevertheless, initial constructs have deleted ORF AR1 from
DNA A and replaced it by the chloramphenicol acetyl
transferase (CAT) coding region, leaving the 3'and 5'
untranslated regions and the rest of the molecule intact.
This construct has been recloned and the DNA inoculated
onto whole plants in combination with clones of both
genomic DNAs. The vector DNA may replicate, become
encapsidated, and produce the CAT enzyme. This particular
vector may replicate as dsDNA in a bipartite infection with
DNA B if TGMV capsid is not required for this process.

A number of other constructions of this type are
planned using different reporter genes and promoters in the
vector DNA including dihydrofolate reductase (DHFR) and npt
Ⅱ. However, packaging constraints may preclude
combinations that include the cauliflower mosaic virus
(CaMV) 35S promoter (31) and the npt Ⅱ gene in tripartite
systems, but this has yet to be investigated. The coat
protein replacement vectors may be of particular use for
the transfection of protoplasts where, if as in the case of
CLV DNA, the smaller of the viral genome components is not
essential for dsDNA replication (11) and high copy number
vector DNA replication is required. However, we can only
speculate at this stage on the behaviour and replication of
cloned TGMV DNA in transfected protoplasts. Our efforts at
transfecting tobacco protoplasts with the viral or cloned

DNA components of TGMV have sporadically given results in terms of the production of ssDNA of genomic length and higher molecular weight, possibly concatemeric DNA species. In most experiments while cloned TGMV DNA was retained in tobacco leaf protoplasts, transfected using polyethylene glycol (PEG) in the presence of $CaCl_2$ and petunia leaf protoplasts transfected by electroporation, only a progressive degradation and not replication of the DNA was noted during subsequent culture in growth media.

Our recent studies on nuclei isolated from TGMV infected plants (13) suggest that host-encoded polymerases are intimately involved in viral DNA replication. Nuclei from protoplasts transfected using PEG and $CaCl_2$ contained a proportion of the viral DNA inoculum. Bearing these results in mind it would appear essential that those enzymes involved in DNA synthesis should be at their most active during transfection and subsequent culture. Such a situation may have been achieved in the recent successful transfection of tobacco protoplasts with cloned CLV DNA virus by isolating the protoplasts, transfecting and culturing them in a rich growth medium (11). However the efficiency of transfection in these studies was not stated. We feel that the use of protoplasts synchronized to the S phase of development by aphidicolin perhaps in combination with a herbicide that inhibits cell-wall formation without affecting nuclear division (e.g. 2, 6 dichlorobenzonitrile) will yield protoplasts more useful for high efficiency transformation as was reported recently in direct gene transfer experiments (32). More hopeful for the prolonged expression of vector DNA in transfected tissue, once successful transfection has been achieved, have been our studies of callus initiated from TGMV infected tobacco plants where we have demonstrated active virus DNA synthesis during a number of successive passages over a six month period. These results are particularly encouraging for the replication of the bipartite vectors with deleted capsid ORF detailed earlier.

The development of vectors replicating as plasmids alone has also been approached in a manner analogous to that used in animal systems (6). Here, genes essential for virus DNA replication are integrated into the host chromosome so that their products are produced constitutively, and the vector contains only the origin of virus replication. Since the vector replicates as a plasmid, the size of any inserted passenger DNA is not limited by packaging constraints. For these vectors genes essential for virus DNA replication will be introduced into the genome of tobacco plants using the binary Ti plasmid vector system (2). We do not know how many genes are essential for TGMV DNA replication as opposed to multiplication of virus particles. In phage ϕX174 several genes are required for virion production but only one, gene A is required for DNA replication (33). In the case of TGMV one or more ORF may be required and we intend to only introduce those genes required for DNA replication and to exclude those genes responsible for symptom development.

In our preliminary experiments we were unsure of the level of viral gene expression in plants when these genes were introduced linked to their own promoters. Thus in recent transformations we have utilised the CaMV 35S promoter to ensure adequate virus gene expression. Transformed leaf discs containing either TGMV A or B DNA have been grown in culture and regenerated callus has produced plants yielding seed for further study. The inoculation of plants raised from these transformed seeds with the second virus DNA component should result in successful infection. By introducing different combinations of the virus genes by similar transformations and ascertaining which genes can be deleted without interfering with virus dsDNA replication we can determine which gene combination, constitutively expressed by the plant genome will actively support the sustained replication of plasmid vectors in protoplasts isolated from these plants. These vectors contain the proposed 200 bp TGMV origin of replication, a col E1 replicon, an ampicillin-resistance gene and a lac polylinker site for insertion of foreign DNA and cloning in E. coli and a NOS gene to act as an internal control for monitoring gene expression levels. Clearly this vector system requires a reliable, high-efficiency transfection system with protoplasts which we are developing, as outlined earlier.

Both virus and plasmid vectors are designed as high copy number rapid expression vectors for studying factors affecting plant gene transcription, particularly promotor function by site-specific mutagenesis and possibly hormonal control of gene expression. These virus vectors will also be useful for the systemic invasion of whole plants and hence studies on tissue-specific gene expression. Plasmid vectors may be used to study RNA splicing mechanisms, or in combination with T-DNA, as a method of inserting multiple copies of a gene into the plant genome.

Since geminiviruses infect both di- and monocotyledonous hosts, the potential for constructing similar vectors for monocots exists and investigations with the cloned DNA of the wheat dwarf geminivirus (34) to this end are currently in progress.

ACKNOWLEDGEMENTS

We thank the Ministry of Agriculture, Fisheries and Food for a licence (PHF 29/171) allowing us to work with TGMV.

REFERENCES

1.   Zambryski P, Joos H, Genetello C, Leemans J, Van Montagu M, Schell J (1983) Ti plasmid vector for the introduction of DNA into plant cells without alteration of their normal regeneration capacity. EMBO J 2:2143.

2.   Bevan MW, (1984) Binary Agrobacterium vectors for plant transformation. Nucleic Acids Res 20:8711.

3.   De Block M, Herrera-Estrella L, Van Montagu M, Schell J, (1984) Expression of foreign genes in plants and their progeny. EMBO J 3:1681.

4.   Paszkowski J, Shillito RD, Saul M, Mandak V, Hohn T, Hohn B, Potrykus I (1984). Direct gene transfer to plants. EMBO J 3:2717.

5.   Hain R, Stabel P, Czernilowsky AP, Steinbiss HH, Herrera-Estrella L, Schell J. (1985). Uptake, integration, expression and genetic transmission of a selectable chimaeric gene by plant protoplasts. Mol Gen Genet 199:161.

6.   Mellon P, Parker V, Gluzman Y, Maniatis T (1981). Identification of DNA sequences required for transcription of the human α1-globin gene in a new SV40 host-vector system. Cell 27 :279.

7.   Di Maio D, Treisman R, Maniatis T (1982). Bovine papilloma virus vector that progagates as a plasmid in both mouse and bacterial cells. Proc Nat Acad Sci USA 79:4030.

8.   Goodman RM (1981). Geminiviruses. J Gen Virol 54:9.

9.   Harrison BD (1985). Advances in geminivirus research.
      Ann Rev Phytopathol 23:55.

10.   Stanley J (1985). The Molecular Biology of
      Geminiviruses. Adv Virus Res 30:139.

11.   Townsend R, Watts J, Stanley J. (1986). Synthesis of
      viral DNA forms in Nicotiana plumbagnifolia
      protoplasts inoculated with cassava latent virus
      (CLV); evidence for the independent replication of one
      component of the CLV genome. Nucleic Acids Res
      14:1253.

12.   Adejare GO, Coutts RHA (1982). Ultrastructural
      studies on Nicotiana benthamiana tissue following
      infection with a virus transmitted from mosaic
      diseased Nigerian cassava. Phytopathol Z 103:87.

13.   Coutts RHA, Buck KW (1985). DNA and RNA polymerase
      activities of nuclei and hypotonic extracts of nuclei
      isolated from tomato golden mosaic virus infected
      leaves. Nucleic Acids Res 13:7881.

14.   Hamilton WDO, Sanders RC, Coutts RHA, Buck KW (1981).
      Characterization of tomato golden mosaic virus as a
      geminivirus. FEMS Microbiol Lett 11:263.

15.   Hamilton WDO, Bisaro DM, Buck, KW (1982).
      Identification of novel DNA forms in tomato golden
      mosaic virus infected tissue. Evidence for a two
      component genome. Nucleic Acids Res 10:4901.

16.   Buck KW, Coutts RHA, Bisaro DM, Hamilton WDO, Stein
      VE, Sunter G (1983). Intracellular DNA forms from
      tomato golden mosaic virus-infected plants. In
      Robertson HD, Howell SH, Zaitlin M, Malmberg RL (eds)
      : "Plant Infectious Agents", Cold Spring Harbor
      Laboratory Publications, p 54.

17.   Sunter G, Coutts RHA, Buck K W (1984). Negatively
      supercoiled DNA from plants infected with a
      single-stranded DNA virus. Biochem Biophys Res Comm
      118:747.

18.   Stanley J, Townsend R (1985). Characterization of DNA
      forms associated with cassava latent virus infection.
      Nucleic Acids Res 13:2189.

19.   Bisaro DM, Hamilton WDO, Coutts RHA, Buck KW (1982).
      Molecular cloning and characterization of the two DNA
      components of tomato golden mosaic virus. Nucleic
      Acids Res 10:4913.

20. Hamilton WDO, Bisaro DM, Coutts RHA, Buck KW (1983). Demonstration of the bipartite nature of the genome of a single-stranded DNA plant virus by infection with the cloned DNA components. Nucleic Acids Res 11:7387.
21. Stanley J, Gay MR (1983). Nucleotide sequence of cassava latent virus DNA. Nature 301:260.
22. Stanley J (1983). Infectivity of the cloned geminivirus genome requires sequences from both DNAs. Nature 305:643.
23. Howarth A J, Caton J, Bossert M, Goodman RM (1985). Nucleotide sequence of bean golden mosaic virus and a model for gene regulation in geminiviruses. ProcNatl Acad Sci USA 82:3572.
24. Morinaga T, Ikegami M, Miura K (1983). Infectivity of the cloned DNAs from multiple genome components of bean golden mosaic virus. Proc Japan Acad B 59:363.
25. Hamilton WDO, Stein VE, Coutts RHA, Buck KW (1984). Complete nucleotide sequence of the infectious cloned DNA components of tomato golden mosaic virus : Potential coding regions and regulatory sequences. EMBO J 3:2197.
26. Townsend R, Stanley J, Curson SJ, Short MN (1985). Major polyadenylated transcripts of cassava latent virus and location of the gene encoding coat protein. EMBO J 4:33.
27. Mulligan RC, Howard, BH, Berg P (1979). Synthesis of rabbit β-globin in cultured monkey kidney cells following infection with a SV40 β-globin recombinant genome. Nature 277:108.
28. Zambryski P, Herrera-Estrella L, De Block M, Van Montagu M, Schell J. (1984). In Hollaender A, Setlow, J. (eds): "Genetic Engineering," Vol 6 New York : Plenum, pp 253.
29. Bevan MW, Flavell RB, Chilton M-D (1983). A chimaeric antibiotic resistance gene as a selectable marker for plant cell transformation. Nature 304:184.
30. Sequeira JC, Harrison BD (1982). Serological studies on cassava latent virus. Ann Appl Biol 101:33.
31. Hohn T, Richards K, Lebeurier G (1982). Cauliflower mosaic virus on its way to becoming a useful plant vector. Curr Topics Microbiol Immun 96:193.
32. Meyer P, Walgenbach E, Bussmann K, Hombrecher G, Saedler H (1985). Synchronised tobacco protoplasts are efficiently transformed by DNA. Mol Gen Genet 201:513.

33.   Kornberg A (1980).   In "DNA replication" New York :
      Freemans, p 508.
34.   MacDowell SW, Macdonald H, Hamilton WDO, Coutts RHA,
      Buck KW (1985).   The nucleotide sequence of cloned
      wheat dwarf virus DNA.   EMBO J 4:2173.

Molecular Strategies for Crop Protection, pages 319–334
© 1987 Alan R. Liss, Inc.

THE ROLE OF RNA PROCESSING IN VIROID REPLICATION[1]

Andrea D. Branch and Hugh D. Robertson

Laboratory of Genetics, The Rockefeller University
New York, New York 10021

ABSTRACT

INTRODUCTION

Viroids are the smallest microbial agents causing stress in plants. They are autonomously replicating, covalently closed circular RNA molecules about 300 nucleotides in length. Unlike all other classes of infectious agents (viruses, mycoplasma, bacteria, fungi and eukaryotic parasites), viroids do not appear to code for any proteins (1), and thus can be used as probes to seek host molecules involved in viroid replication and disease induction. The role of such factors in normal plant physiology and development may provide intriguing insights into the mechanism of eukaryotic gene expression. At the moment, the mechanism of viroid pathogenesis is entirely unknown; however, viroid symptoms have an extensive history, which has been documented for a variety of economically important host plants (1-4).

[1]This work was supported in part by grants from the NSF, the USDA, and the McKnight Foundation.

The first viroid disease, originally described as a
"potato trouble", was detected in the potato fields of South
New Jersey in the fall of 1921 (5). Affected plants had small
vines and elongated tubers with prominent eyes. Their yield
of high grade potatoes was low. Because the symptoms and
route of transmission resembled those of potato viruses, the
infectious agent was first designated potato spindle tuber
virus (PSTV). PSTV was later transmitted to tomato plants, a
more favorable experimental host than potato plants. A
number of field isolates of PSTV were collected which fell
into distinctive groups based on the pattern of symptoms
induced in tomato plants. "Severe" strains caused marked
stunting and leaf curling (epinasty), while "mild" strains
produced few symptoms and were largely detectable because
their presence in a plant conferred resistance to subsequent
inoculation with a severe strain, as shown by Fernow in his
classical cross protection test (6). For many years, bioassay provided the only means of viroid detection; however, in
1972 Diener and colleagues (7) opened the viroid field to
molecular analysis by showing that the peak of viroid infectivity exactly coincided with the position of an RNA gel band
which appeared exclusively in extracts from infected plants.
The term "viroid" was introduced by Diener to emphasize the
novel properties of the agent responsible for spindle tuber
disease.

THE VIROID ·REPLICATION CYCLE

It may not be surprising that disease agents as exotic
as viroids have an unconventional replication cycle. Like the
vast majority of plant viruses, viroids replicate through an
RNA intermediate [i.e., the minus strand is composed of RNA
(8-12)]. However, as mentioned above, unlike viruses, viroids
appear to be unable to supply any of their own proteins and
thus must acquire replication factors from their host plants.

Although various host activities (including RNA polymerases I, II, and III) will copy viroid RNA in vitro (3), no
consensus has yet emerged concerning the identity of the
enzyme(s) responsible for copying viroid RNA in vivo. The
ability of plants to supply viroid-copying activities raises
the intriguing possibility that certain RNA molecules may be
copied in plant cells under normal conditions. In fact, it
has been known for some time that control plants contain
endogenous RNA replicases (13, 14). A search for possible

endogenous RNA templates for this host RNA replicase is
currently underway in a number of laboratories.

A rolling circle replication cycle has been proposed
for viroids and certain other infectious RNA's (satellites of
particular plant viruses) (10, 11, 15, 16). Figure 1A shows
a model which is based on studies of PSTV-specific nucleic
acids extracted from infected tomato plants--the system in
which multimeric viroid RNA's were originally described (10).

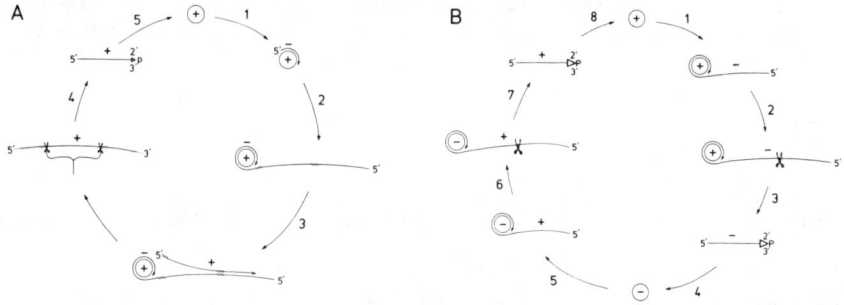

Figure 1. Rolling circle replication

The PSTV replication cycle begins at the top of the
diagram, with an infecting, circular plus strand (arbitrarily
marked "+", although it appears to lack message activity)
which is copied by a rolling circle mechanism into a multimeric minus strand (marked "-"; Steps #1 and #2). In Model A,
the minus strand is directly copied into a long plus strand
(Step #3), which is cleaved to produce unit length molecules
with 2', 3' cyclic phosphate termini (Step #4), in keeping
with experiments showing that unit length linear PSTV molecules
present in plant extracts have a 2', 3' cyclic phosphate
terminal moiety. Circularization produces mature viroid RNA
(Step #5). A second version of this cycle, Model B, illustrates the possibility that multimeric minus strands are cleaved
and ligated to give circular molecules, which serve as plus
strand templates. Circular minus strands have been associated
with avocado sunblotch viroid (ASBV) and certain satellite
RNA's (17); however, at the moment, it is not clear whether
circular minus strands are a general feature of the viroid
replication pathway since ASBV does not appear to be closely

related to any of the other viroids [and yet all known viroids, except ASBV, have sequence homology to each other (18, 19)].

PATHWAYS OF RNA SPLICING

Whatever the exact form the minus strand turns out to have, there is general agreement that plus strand synthesis involves the cleavage and ligation of a multimeric plus strand precursor. Thus, this phase of viroid replication is reminiscent of RNA splicing. The three well-characterized RNA splicing pathways are outlined in Figure 2. It is likely that viroid replication involves steps from one or more of them.

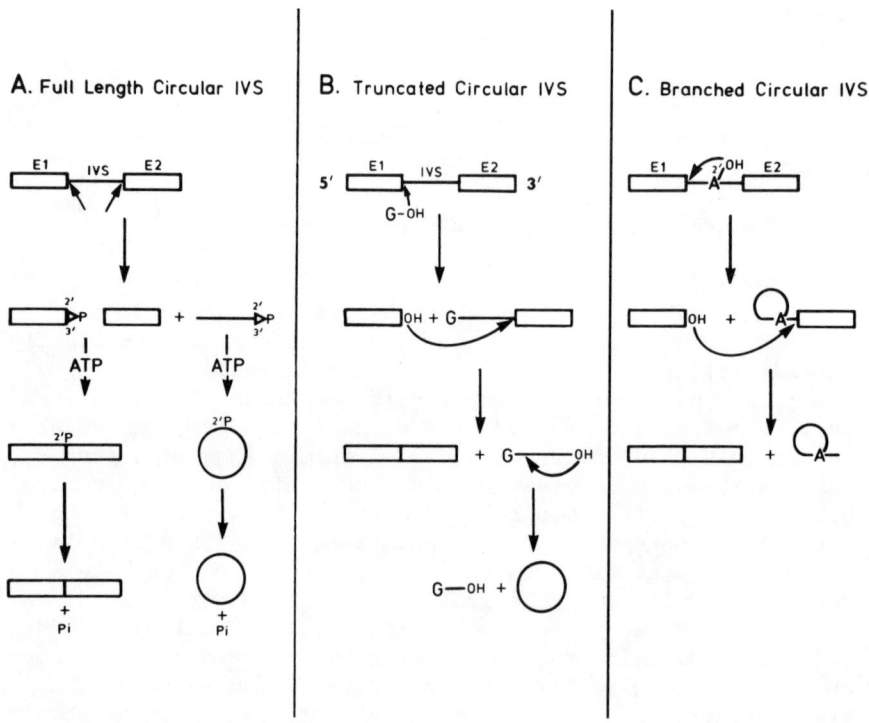

Figure 2. Three pathways of RNA splicing

The process of tRNA splicing is depicted in Pathway A. Transfer RNA genes from a number of eukaryotes have intervening sequences which are released by endonucleolytic cleavage. As shown by studies carried out in yeast extracts, this step is catalyzed by a specific "splicing ribonuclease", which leaves 5' hydroxyl and 2', 3' cyclic phosphate termini. The tRNA exons are then joined by an RNA ligase in a reaction, first worked out in wheat germ extracts (20, 21), in which a 2' phosphate group is created at the junction, in addition to the conventional 3', 5' phosphodiester bond. The 2' phosphate, detectable through in vitro ligation studies because it renders the junction site insensitive to ribonuclease digestion, is thought to be removed in vivo by a phosphatase activity.

In contrast to the splicing of tRNA precursors, which require two separate protein enzymes (a processing enzyme and a ligase), the splicing of Tetrahymena ribosomal RNA (22, 23) can occur in vitro in the absence of any proteins, through a series of autocatalytic transesterification reactions set in motion by the addition of a guanosine cofactor at the junction between the 5' exon and the intervening sequence (see Fig. 2B). This reaction produces a 5' exon with a 3' hydroxyl terminus, releasing the intervening sequence still attached to the 3' exon, but with the non-encoded guanosine present on its 5' end. As this concerted reaction continues, the hydroxyl group at the 3' end of the 5' exon attacks the 3' splice site: the exons are joined and the intervening sequence is released as a linear fragment. The free (linear) intervening sequence may then circularize autocatalytically in a reaction leading to the cleavage and release of 15 nucleotides from its 5' end. In a third cleavage and ligation pathway, typical of mRNA splicing in mammalian cell nuclei (24, 25), the intervening sequence is released in the form of a lariat (see Fig. 2C).

Of the three mechanisms of RNA splicing, only the pathway used for splicing tRNA precursors (Fig. 2A) has the potential for generating a circular form encompassing the entirety of the intervening sequence. Therefore, cleavage and ligation reactions involved in viroid replication might be most closely related to the process of tRNA splicing, since in viroid replication, the circular product must preserve the complete genome. Interestingly, it has been shown that the entire excised tRNA intervening sequence can be circularized by the same RNA ligase which joins exon segments to give mature-sequence tRNA. The RNA ligase from wheat germ which is capable of joining tRNA exons is also able to circularize linear viroid RNA extracted from infected tomato plants (26, 27).

THE ROLE OF RNA PROCESSING IN VIROID REPLICATION

Kiberstis et al. (28) recently provided strong evidence
that replication of one of the circular satellite RNA's
includes a ligation step characteristic of tRNA splicing.
Direct analysis of ^{32}P-labeled circular "virusoid" RNA associated with Solanum nodiflorum mottle virus revealed that a single
oligonucleotide in the fingerprint contained a 2' phosphate
group--molecular evidence of RNA ligation by a tRNA-like
mechanism. Evidence of an analogous 2' phosphate group in
studies of in vivo-labeled circular viroid RNA would provide
additional signs that a unit length linear RNA with a 2', 3'
cyclic phosphate terminus is the immediate precursor to mature
viroid RNA and that the ligation mechanism is similar to the
one described for eukaryotic RNA ligases thought to be part of
the tRNA splicing pathway. Equally important, detection of a
junction site would identify the residues within the circular
viroid RNA which had been the terminal nucleotides on the unit
length linear form. This piece of information would establish
the site at which multimeric plus strands are cleaved to unit
length molecules in vivo.

The position of the in vivo cleavage site has been a
major question ever since the detection of multimeric minus
strands suggested that a multimeric plus strand might exist.
Autocatalytic cleavage and processing reactions relying
heavily on host factors have been considered as possible
mechanisms for the production of unit length plus strands. As
a first step toward understanding the mechanism of cleavage,
we incubated ^{32}P-labeled dimeric transcripts of PSTV [synthe-
sized in vitro and already proven infectious (29)] under a
variety of conditions and found a low level of self-catalyzed
cleavage. A qualitatively identical pattern of cleavage
products was generated when transcripts were incubated in a
variety of buffers containing divalent cations. Reaction
products included three well-defined bands, one of which comigrated with unit length linear PSTV, and two smaller fragments, indicating that cleavage occurred at two points in the
dimeric transcript. Fingerprinting roughly localized a processing site to the region between bases 250 and 270. Studies
with the wheat germ RNA ligase carried out on the unit length
product suggested that this species is terminated by a 2', 3'
cyclic phosphate moiety. Efforts to obtain more efficient
self-cleavage have been carried out in a number of laboratories, including our own, so far with no improved results.
However, Tsagris et al., recently reported promising evidence

that the addition of host extracts may accelerate the processing of PSTV transcripts in vitro (30). In other studies,
the infectivity of plasmids containing slightly longer than
unit length PSTV inserts (31) and of plasmids and citrus
exocortis viroid (CEV) transcripts containing a short terminal
duplication (32) led Symons and colleagues (19, 32) to suggest
that the site of cleavage is within the duplicated sequence
[bases 87 to 98 in PSTV (33)]. Further direct studies are
needed to identify rigorously the site of cleavage and to
determine the relative importance of autocatalysis and more
conventional RNA processing to cleavage of viroids such as
PSTV and CEV. However, at the moment, it appears that a host
factor is required for efficient in vitro cleavage of PSTV and
related viroids. This contrasts with results of studies
carried out on ASBV, the satellite of tobacco ringspot virus,
and the virusoid of lucern transient streak virus (17, 34, 35).
Transcripts of these three agents cleave autocatalytically in
vitro at a high rate. This cleavage almost certainly has
biological relevance since Bruening and colleagues (35) have
shown that the linear products generated from dimeric transcripts of tobacco ringspot virus satellite have the same ends
as the linear molecules found in infected plants [where, unlike
viroid systems, the linear form predominates (16)]. When
examined, the autocatalytic products of these in vitro
reactions have been found to have 2', 3' cyclic phosphate
termini (35). In the case of the unit length minus strand of
the tobacco ringspot virus satellite, self-circularization was
demonstrated in vitro (36), sparking a search for circular
minus strands in plants replicating this pathogen.

At the moment, it is not at all clear why the transcripts
of some agents carry out efficient autocatalytic cleavage in
vitro and others at a low rate. It is possible that mechanis-
tically similar processing reactions take place during the
replication of all of these infectious RNA's and that they
differ only in the relative effect of host factors on the rate
of RNA-catalyzed cleavage. Similar ideas are now being tested
in mRNA splicing and in other systems. Furthermore, it may be
significant that PSTV and all other viroids except ASBV
[which, T. J. Morris (personal communication) has suggested,
should be tested for satellite characteristics] share a region
of highly conserved sequence. This region is composed of two
interacting, but non-contiguous, segments located near the
center of the rodlike viroid structure (33). Its function
is a topic of heated debate, but some vital role in replication is generally conceded, due to its high degree of

evolutionary conservation. Keese and Symons (19) have
suggested that the central conserved region can assume two
alternative configurations: one important for cleavage, the
other for copying. Diener has explored similar ideas (37).
As discussed below, this region contains a novel element of
RNA tertiary structure, which may be related to conformational
switching. Perhaps the form of the molecule required for
cleavage of PSTV and CEV is highly unfavorable under the in
vitro conditions tested thus far. Further experimental manipulations might reveal conditions in which the configuration
needed for cleavage is more likely to form, enhancing the rate
of autocatalysis. A maximum change in structure might result
from the use of partial transcripts, for example, or by driving
the transcripts into RNA:RNA duplex structure, following the
lead provided by studies of processing in the cucumber mosaic
virus satellite system (38).

A NOVEL STRUCTURAL ELEMENT IN VIROID RNA

The novel element of local tertiary structure in the
central conserved region of PSTV was detected because it
contains two bases which become covalently crosslinked in
response to irradiation with ultraviolet light (39, 40). Thus,
either purified viroid RNA or RNA of PSTV-infected plants was
exposed to ultraviolet light, and then fractionated by two-
dimensional gel electrophoresis. Because the crosslinked form
of PSTV has a unique gel mobility, it created a new spot in the
gel and could be readily isolated in pure form. RNA fingerprinting, secondary analysis and primer extension were used to
identify the bases covalently attached by the UV-irradiation.
A diagram of the PSTV crosslinking site appears in Figure 3.

Forty of the 46 bases in and around the crosslinking
site in PSTV also occur in the analogous regions of CEV,
chrysanthemum stunt and coconut cadang viroids (3, 18, 19).
This high degree of conservation implies that the novel
structural element plays a critical role in viroid biology, and
may be required for replication.

The crosslink connects bases G-98 and U-260. As
indicated in Figure 3, these bases occur in regions usually
depicted as single-stranded loops flanked by short helical
segments. However, our data suggest that G-98 and U-260
actually lie in close proximity to each other, perhaps
stabilized by the kinds of tertiary bonds identified in the
crystal structure of tRNA (41). It is interesting to note

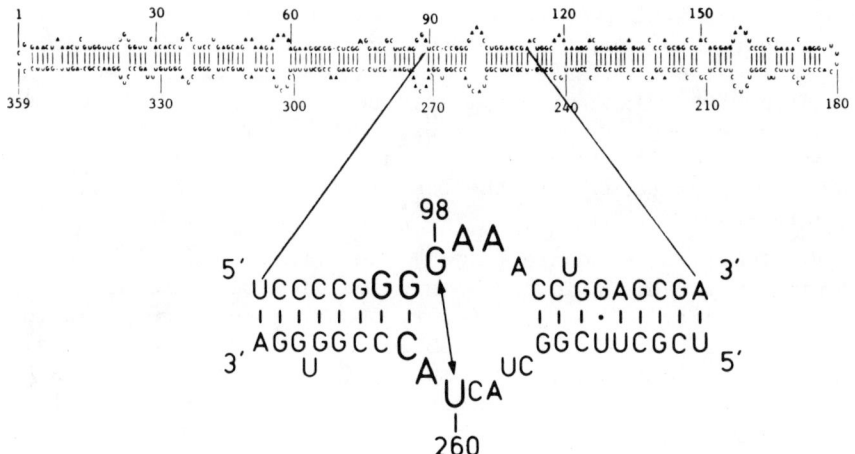

Figure 3. The PSTV crosslinking site

that some of the first evidence suggesting the true three-
dimensional structure of tRNA came from studies of a UV-sensitive crosslinking site present in tRNA (42). In the same
way, detailed knowledge of the chemistry of the crosslinking
site in PSTV should permit a more accurate model of viroid RNA
to be constructed. A specific suggestion concerning the role
of the PSTV crosslinking site derives from studies of a host
RNA with a similar structural feature.

Several lines of evidence suggest that host cells have
molecules with similarities to viroid RNA which may be important to understanding both normal gene expression in plants
and viroid molecular biology. For example, Sänger and
colleagues reported that inoculation with 50 molecules of PSTV
led to infection in 10% of the plants tested (18). Such high
infectivity of a naked RNA molecule may imply that a host
receptor exists which can be used by viroid RNA, but which is
designed for binding to an endogenous factor of unknown
identity and function. Furthermore, as mentioned earlier, it
is likely that the host machinery which copies viroid RNA has
endogenous templates. In addition, stable circular RNA's
have been detected in a number of systems; Halbreich reported
that an excised circular intervening sequence from yeast can be
isolated as part of an RNA:RNA duplex (43)--a structure

suggestive of RNA-level replication events. Recently,
Epstein and Gall (44) have shown that multimeric transcripts
of a newt satellite DNA (a tandemly repeated sequence from the
chromosomal DNA) cleave autocatalytically in vitro, recalling
studies of ASBV and other plant virus satellite RNA molecules.
Finally, we have uncovered evidence that PSTV and 5S rRNA have
a nearly identical element of local tertiary structure and are
now beginning to explore the ramifications of this
finding (39).

The first evidence for a UV-sensitive site in 5S rRNA
came from two dimensional gel studies of nucleic acids from
tomato plants. These preliminary observations were followed
by detailed analysis of ^{32}P-labeled 5S rRNA from HeLa cells,
from which highly purified samples of UV-crosslinked 5S rRNA
were prepared (40). RNA fingerprinting and secondary analysis
were used to map the crosslinking site in 5S rRNA. A detailed
comparison of the crosslinking sites in PSTV and 5S rRNA
appears elsewhere (39).

It is not clear how the unusual bonding pattern present
at the crosslinking site in viroid and 5S rRNA contributes to
RNA function. Various explanations have been considered.
First, the crosslinking site in PSTV may be near the site at
which multimeric plus strand precursors are cleaved. By
analogy to the tertiary structural elements that often comprise
key parts of protein active sites, it is possible that the RNA
structural element revealed by the UV-induced crosslink could
help to form the active site for RNA-catalyzed cleavage or
other RNA catalyzed reactions. Alternatively, the crosslinking
site might provide the binding site for one or more proteins.
In this regard, it may be significant that the UV-sensitive
site in HeLa 5S rRNA occurs in the analogous region of the
molecule to the TFIIIA binding site in Xenopus 5S rRNA
(45, 46).

5S rRNA binding to TFIIIA occurs in many cell types and
may play a role in the developmental regulation of 5S ribosomal
genes. As 5S genes are transcribed, it appears that the
resulting 5S rRNA binds to TFIIIA [which has a higher affinity
for 5S rRNA than for 5S rDNA (47)] and transcription stops.
We are interested in the possibility that an in vivo interaction between viroid RNA and TFIIIA may be an important
component of viroid replication and, perhaps, viroid disease
induction as well. TFIIIA is a positive transcription factor
for RNA polymerase III, the enzyme responsible for the
synthesis of tRNA, in addition to 5S rRNA. Given the

similarities between tRNA splicing and viroid cleavage and
ligation reactions (discussed above) and the possibility
that transcription and processing are coordinated events
within the cell, it may be worth looking closely for a direct
role of RNA polymerase III in viroid replication. Furthermore, with the continuing interest in transcription factors
like TFIIIA in the regulation not only of 5S rRNA gene
expression (48, 49), but possibly in the action of other
genes as well (50), it would not be surprising to find that
viroids may contain "escaped" binding sites for host transcription components which they exploit both for their own
proliferation and to the detriment of the host organism.

ACKNOWLEDGMENTS

We thank Drs. Olke C. Uhlenbeck, Milton Zaitlin, George
Bruening, and Robert H. Symons for lively discussions of RNA
structure on one hand, and viroid replication on the other.

REFERENCES

1. Diener TO (1979). "Viroids and Viroid Diseases." New
    York: J. Wiley & Sons.
2. Dickson, E (1979). Viroids: infectious RNA in plants.
    In Hall, TC, Davies, JW (eds): "Nucleic Acids in Plants
    Vol. II," Boca Raton: CRC Press, p 153.
3. Sänger, HL (1986). Minimal infectious agents: the vir-
    oids. In Mahy, BWJ, Pattison, JR (eds): "The Microbe
    1984: Part I, Viruses," Cambridge: Cambridge Univer-
    sity Press, p 281.
4. Branch, AD, Willis, KK, Davatelis, G, Robertson, HD
    (1984). In vivo intermediates and the rolling circle
    mechanism in viroid replication. In Maramorosch, K,
    McKelvey, JJ (eds): "Subviral Pathogens of Plants and
    Animals: Viroids and Prions," New York: Academic
    Press, p 201.
5. Martin, WH (1922). Spindle tuber, a new potato trouble.
    Hints to Potato Growers of the NJ State Potato Assoc.
    3:8.
6. Fernow, KH (1967). Tomato as a test plant for detect-
    ing mild strains of potato spindle tuber virus. Phy-
    topathology 57:1347.
7. Diener, TO (1972). Potato spindle tuber viroid, VIII.
    Correlation of infectivity with a UV-absorbing compo-
    nent and thermal denaturation properties of the RNA.
    Virology 50:606.
8. Grill, LK, Semancik, JS (1978). RNA sequences complemen-
    tary to citrus exocortis viroid in nucleic acid prepa-
    rations from infected Gynura aurantiaca. Proc. Nat.
    Acad. Sci., USA, 75:896.
9. Branch, AD, Dickson, E (1980). Tomato DNA contains no
    detectable regions complementary to potato spindle
    tuber viroid as assayed by Southern hybridization.
    Virology 104:10.
10. Branch, AD, Robertson, HD, Dickson, E (1981). Longer
    than unit length viroid minus strands are present in RNA
    from infected plants. Proc. Nat. Acad. Sci., USA
    78:6381.
11. Owens, RA, Diener, TO (1982). RNA intermediates in pota-
    to spindle tuber viroid replication. Proc. Nat. Acad.
    Sci., USA 79:113.
12. Rohde, W, Sänger, HL (1981). Detection of complementary
    RNA intermediates of viroid replication by Northern blot
    hybridization. Biosci. Rep. 1:327.

13. Duda, CT, Zaitlin, M, Siegel, A (1973). In vitro synthesis of double-stranded RNA by an enzyme system isolated from tobacco leaves. Biochim. Biophys. Acta 319:62.
14. Duda, CT (1976). Plant RNA polymerases. Ann. Rev. Plant Physiol. 27:119.
15. Branch, AD, Robertson, HD (1984). A replication cycle for viroids and other small infectious RNA's. Science 223:450.
16. Kiefer, MC, Daubert, SD, Schneider, IR, Bruening, G (1982). Multimeric forms of tobacco ringspot virus RNA. Virology 121:262.
17. Hutchins, CJ, Keese, P, Visvader, JE, Rathjen, PD, McInnes, JL, Symons, RH (1985). Comparison of multimeric plus and minus forms of viroids and virusoids. Plant Mol. Biol. 4:293.
18. Sänger, HL (1982). Biology, structure, functions and possible origin of viroids. In Parthier, B, Boulter, D (eds): "Nucleic Acids and Proteins in Plants II," Vol. 14B of the Encyclopedia of Plant Physiology, New Series, Berlin, Springer-Verlag, p 368.
19. Keese, P, Symons, RH (1985). Domains in viroids: evidence of intermolecular RNA rearrangements and their contribution to viroid evolution. Proc. Nat. Acad. Sci. USA 82:4582.
20. Konarska, M, Filipowicz, W, Domdey, H, Gross, HJ (1981). Formation of a 2' phosphomonoester, 3',5'-phosphodiester linkage by a novel RNA ligase in wheat germ. Nature 293:112.
21. Konarska, M, Filipowicz, W, Gross, HF (1982). RNA ligation via 2' phosphomonoester, 3',5'-phosphodiester linkage: requirement of 2',3'-cyclic phosphate termini and involvement of a 5' hydroxyl polynucleotide kinase. Proc. Nat. Acad. Sci. USA 79:1474.
22. Kruger, K, Grabowski, PJ, Zaug, AJ, Sands, J, Gottschling, DE, Cech, TR (1982). Self-splicing RNA: autoexcision and autocyclization of the ribosomal RNA intervening sequence of Tetrahymena. Cell 31:147.
23. Cech, TR (1986). The generality of self-splicing RNA: relation to nuclear mRNA splicing. Cell 44:207.
24. Ruskin, B, Krainer, AR, Maniatis, T, Green, MR (1984). Excision of an intact intron as a novel lariat structure during pre-mRNA splicing in vitro. Cell 38:317.
25. Grabowski, PJ, Padgett, RA, Sharp, PA (1984). Messenger RNA splicing in vitro: an excised intervening sequence and a potential intermediate. Cell 37:415.

26. Branch AD, Robertson HD, Greer C, Gegenheimer P, Peebles C, Abelson J (1982). Cell-free circularization of viroid progeny RNA by an RNA ligase from wheat germ. Science 217:1147.

27. Kikuchi Y, Tyc K, Filipowicz W, Sänger HL, Gross HJ (1982). Circularization of linear viroid RNA via 2'-phosphomonoester, 3',5'-phosphodiester bonds by a novel type of RNA ligase from wheat germ and Chlamydomonas. Nucleic Acids Res. 10:7251.

28. Kiberstis PA, Haseloff J, Zimmern D (1985). 2'-phosphomonoester, 3',5'-phosphodiester bond at a unique site in circular viral RNA. EMBO J. 4:817.

29. Robertson HD, Rosen DL, Branch, AD (1985). Cell-free synthesis and processing of an infectious dimeric transcript of potato spindle tuber viroid RNA. Virology 142:441.

30. Tsagris M, Tabler M, Eisenrieth U, Sänger HL (1986). Processing of multimeric potato spindle tuber viroid (PSTV) RNAs in vitro. Abstr. 1986 Cold Spr. Harb. Mtg. "RNA Processing," p 164.

31. Tabler M, Sänger HL (1985). Infectivity studies on different potato spindle tuber viroid (PSTV) RNAs synthesized in vitro with the SP6 transcription system. EMBO J. 4:2191.

32. Visvader JE, Forster AC, Symons RH (1985). Infectivity and in vitro mutagenesis of monomeric cDNA clones of citrus exocortis viroid indicates the site of processing of viroid precursors. Nucleic Acids Res. 13:5843.

33. Gross HJ, Domdey H, Lossow C, Jank P, Raba M, Alberty H, Sänger HL (1978). Nucleotide sequence and secondary structure of potato spindle tuber viroid. Nature 273:203.

34. Hutchins CJ, Rathjen PJ, Forster AJ, Symons RH (1985). Replication of viroids: an in vitro system to characterize the self-processing of oligomeric RNA intermediates of avocado sunblotch viroid to plus and minus monomers. Abstr. OR-13-01, 1st Int. Congr. Plant Mol. Biol., Savannah, p 30.

35. Prody GA, Bakos JT, Buzayan JM, Schneider IR, Bruening G (1986). Autolytic processing of dimeric plant virus satellite RNA. Science 231:1577.

36. Buzayan JM, Hampel A, Gerlach WL, Bruening G (1986). Autolytic processing and subsequent spontaneous ligation of RNAs that are complementary to a plant virus satellite RNA. Abstr. 1986 Cold Spr. Harb. Mtg. "RNA Processing," p 192.

37. Diener TO (1986). Viroid processing: a model involving the central conserved region and hairpin I. Proc. Nat. Acad. Sci. USA 83:58.
38. Young ND, Palukaitis P, Zaitlin M (1986). Replication of cucumber mosaic virus satellite-RNA in vitro. J. Cel. Biochem., Supp. 10C, p43.
39. Branch AD, Benenfeld BJ, Robertson HD (1985). Ultraviolet light-induced crosslinking reveals a unique region of local tertiary structure in potato spindle tuber viroid and HeLa 5S RNA. Proc. Nat. Acad. Sci. USA 82:6590.
40. Branch AD, Benenfeld BJ, Robertson HD (1985). Unusual properties of two branched RNA's with circular and linear components. Nucleic Acids Res. 13:4889.
41. Ladner JE, Jack A, Robertus JD, Brown RS, Rhodes D, Clark BFC, Klug A (1975). The structure of yeast phenylalanine transfer RNA at 2.5 Å resolution. Proc. Nat. Acad. Sci. USA 72:4418.
42. Yaniv M, Favre A, Barrell BG (1969). Evidence for interaction between two non-adjacent nucleotide residues in tRNA$_{\mathrm{Y}}^{\mathrm{val}}$ from E. coli. Nature 223:1331.
43. Halbreich A (1984). Yeast mitochondria contain a linear RNA strand complementary to the circular intronic bI1 RNA of cytochrome b. Abstr. 1984 Cold Spr. Harb. Mtg. "RNA Processing," p 126.
44. Epstein LM, Gall JG (1986). Self-cleaving transcripts of satellite DNA. Abstr. 1986 Cold Spr. Harb. Mtg. "RNA Processing," p 195.
45. Andersen J, Delihas N, Hanas JS, Wu C-W (1984). 5S RNA structure and interaction with transcription factor A. 2. Ribonuclease probe of the 7S particle from Xenopus laevis immature oocytes and RNA exchange properties of the 7S particle. Biochemistry 23:5759.
46. Pieler T, Erdmann VA (1983). Isolation and characterization of a 7S RNA particle from mature Xenopus laevis oocytes. FEBS Letters 157:283.
47. Hanas JS, Bogenhagen DF, Wu, C-W (1984). Binding of Xenopus transcription factor A to 5S RNA and to single-stranded DNA. Nucleic Acids Res. 12:2745.
48. Pelham HRB, Brown DD (1980). A specific transcription factor that can bind either the 5S RNA gene or 5S RNA. Proc. Nat. Acad. Sci. USA 77:4170.
49. Honda BM, Roeder RG (1980). Association of a 5S gene transcription factor with 5S RNA and altered levels of factor during cell differentiation. Cell 22:119.

50.   Rosenberg UB, Schröder C, Preiss A, Kienlin A, Coté S, Riede I, Jäckle M (1986). Structural homology of the product of the Drosophila Krüppel gene with Xenopus transcription factor IIIA. Nature 319:336.

Molecular Strategies for Crop Protection, pages 335–338
© 1987 Alan R. Liss, Inc.

WORKSHOP SUMMARY: MOLECULAR BIOLOGY OF PLANT VIRUSES

Paul Ahlquist

Biophysics Laboratory and Department of Plant Pathology
University of Wisconsin-Madison
1525 Linden Drive
Madison, Wisconsin 53706, U.S.A.

The increasingly common and powerful application of molecular biology and recombinant DNA technology has been a major driving force for recent advances in plant virology. One important revelation of recent virus studies from many laboratories is the recognition, from sequence comparisons and other data, of family relationships among a wide range of distinct virus groups. Viruses within each such family are related by nonstructural protein homologies, and a variety of evidence suggests that these homologies reflect similar nucleic acid replication mechanisms. Three such families have been recognized to date. One includes the plant caulimoviruses, the animal retroviruses and hepatitis B viruses, and the retrotransposons. Another relates the animal alphaviruses to a broad collection of superficially dissimilar single-stranded RNA plant viruses including tobacco mosaic, tobacco rattle, alfalfa mosaic, cucumber mosaic, and brome mosaic viruses, as well as others. A third relates cowpea mosaic virus and its relatives to the animal picornaviruses, such as poliovirus.

These results suggest that despite their exceedingly great outward diversity, it may be possible to group most plant viruses in just a few categories, according to common nucleic acid replication mechanisms which have been stable over long evolutionary periods. One possibility that arises from this view is that, with better understandings of viral replication mechanisms, it may be possible to develop broad spectrum control strategies that will inhibit central replication steps in not just one, but many viruses.

Another important trend in plant virus studies is the extended use of cloned genomes to genetically manipulate not only DNA viruses, but also RNA viruses, which comprise the majority of known plant viruses, and subviral elements such as viroids and virus satellites. Such infectious nucleic acid clones provide direct control of virus genetics, and thus important avenues for studying virus replication and identifying the determinants and mechanisms of virus-host interactions which control host defense responses and pathogenicity.

Discussions in this workshop covered a variety of topics relating to the genetic organization, expression, and replication of both RNA and DNA plant viruses, as well as virus-host interactions. More information on individual presentations can be found in the symposium abstracts (Journal of Cellular Biochemistry, Supplement 10C, pp. 37-43, 1986).

David Baulcombe (Plant Breeding Institute, Cambridge, U.K.) and Nevin Young (Cornell University) both described experiments with the satellite (Sat) RNA of cucumber mosaic virus (CMV). Sat RNA is a capped ssRNA around 335 bases long which is found in some natural isolates of CMV. Sat RNA is dependent on CMV for replication and encapsidation, but is itself unnecessary for CMV replication. Presence and packaging of Sat RNA normally reduces the yield of virions containing CMV RNAs and, depending on the Sat RNA strain and host, the presence of Sat RNA may either attenuate or exacerbate symptoms of CMV infection. Both the Cambridge and Cornell groups find not only linear monomeric but also various multimeric forms of (+) and (-) strand Sat RNA in infected plants. The Cornell group has isolated a subcellular fraction from CMV/Sat infected plants which synthesizes predominantly new (+) strands of Sat RNA from an endogenous template, yielding largely double stranded RNA products. Preliminary experiments with isolated double stranded product RNA suggests that it is capable of limited self processing.

In a different approach, the Cambridge group has constructed concatomeric Sat cDNA clones of around 1.5 or 2.5 unit lengths, from which infectious transcripts can be made in vitro. These Sat cDNA concatomers have been inserted between the cauliflower mosaic virus 35S promoter and a nopaline synthase polyadenylation signal, and transformed into plants using the Agrobacterium/Ti plasmid system. Regenerated transformed plants express polyadenylated, Sat-homologous RNA of the sizes predicted for primary transcripts from these constructions. Processed, monomeric Sat RNA appears, however, only upon infection of such plants with CMV, suggesting that CMV may induce a processing factor, or that only double-stranded Sat RNA, made in the presence of CMV, can self process. CMV infection is attenuated in plants transformed with such Sat RNA constructs, suggesting a possible control strategy once the integrated Sat sequences are disabled to prevent transmission to untransformed plants, or possible mutation into a pathogenic Sat RNA strain. In the poster session Baulcombe also presented further sequencing results which showed close homology between the two component tobacco rattle virus and the single component tobacco mosaic virus.

Roy French (University of Wisconsin-Madison) discussed the use of infectious in vitro transcripts from cloned brome mosaic virus (BMV) cDNA in molecular genetic studies of viral replication and gene expression. Construction of a number of designed BMV mutants was described, as well as the uses of such mutants in mapping and characterizing regulatory sequences which direct productive viral RNA replication. Gene replacement experiments leading to the construction of BMV-based gene expression vectors were also reviewed. Substitution of a foreign gene for the natural BMV coat protein gene in a number of related constructs produced novel viruses able to both replicate and efficiently express the foreign gene in transfected protoplasts.

Most plant viruses depend on biological vectors, often insects, for transmission. Donald Nuss (Roche Institute of Molecular Biology) described the interaction of wound tumor virus (WTV) with its leafhopper vector. In natural transmission, WTV actively replicates in the leafhopper vector but causes no detectable pathogenicity. Similarly, the virus replicates persistantly in cultured leafhopper cells, but causes no detectable cytopathology. In such cells, virus production initially passes through an acute phase of active replication over the first five days of infection, but then subsides by 90-95% to a stable, persistant level maintained in subsequent passages. Reduced virus production appears mediated by a change not in the level but the translational activity of the viral mRNAs synthesized, possibly due to a reduction in viral mRNA capping. This reduced virus production may represent an adaptation to insure the survival and transmission efficiency of the insect vector.

Eulian Roberts (Imperial College, London) described work on the replication of geminivirus DNA and its potential use for gene vector constructions, focusing on the related geminiviruses tomato golden mosaic virus (TGMV) and potato yellow mosaic virus (PYMV). The genome of each virus consists of two 2.5 kilobase, single-stranded, circular DNAs designated components A and B. After observing and characterizing naturally occurring deletion mutants of virus component B which replicated in the presence of wild type components A and B, a potential virus vector was constructed from a clone of wild type TGMV component B by replacing the region lost in the natural deletions with a selectable NPTII gene. However, after the hybrid coponent was inoculated on plants with wild type components A and B, no NPTII sequences could be detected in the progeny virus. Rather, only wild type A and B DNAs and a deleted version of B lacking NPTII sequences could be recovered. This suggests that, as with early attempts at cauliflower mosaic virus vectors, the hybrid B component may have been unstable towards recombination. Alternative approaches to geminivirus vectors are being explored.

James Schoelz (University of Kentucky) reported a molecular genetic analysis of the determinants of host defense responses to cauliflower mosaic virus (CaMV) infection. In solanaceous hosts, CaMV strains CM1841 and Cabb-B both induce necrotic local lesions, while CaMV strain D4 replicates systemically. Recombinant hybrid viruses were constructed in which various segments of CaMV D4 DNA were replaced by the corresponding regions of either CM1841 or Cabb-B DNA. The hybrid viruses were tested for host responses on suitable solanaceous plants, with results indicating that the determinant(s) for CaMV induction of solanaceous hypersensitivity response map within a single 469 bp fragment in the 5' half of CaMV gene VI, which encodes the major viral inclusion body protein. Sequence comparisons in this region show 17 amino acid changes between the gene VI product of compatible and incompatible viruses. One of these changes creates a new glycosylation site in the gene VI protein of the compatible D4 strain, and preliminary studies (M. Young; Davis, CA) indicate that this new glycosylation site is used in vivo. Along with further fine structure mapping of the hypersensitivity determinant(s), the relevance of the altered gene VI glycosylation pattern to host reponse is under further investigation.

Molecular Strategies for Crop Protection, pages 339–341
© 1987 Alan R. Liss, Inc.

WORKSHOP SUMMARY: VIRAL REPLICATION
AND PLANT RESPONSES

Milton Zaitlin

Department of Plant Pathology
Cornell University
Ithaca, New York 14853

This workshop considered two general topics, each
relating to how crop plants might resist the deleterious
effects of plant virus disease. The first -- a continu-
ation of the class of studies described in lectures given
earlier at the Symposium by M. Bevan and by R. N. Beachy
(published in this volume) -- dealt with the integration
of portions of plant viral genomes into host genomes to
induce resistance. The second topic concerned efforts to
clone resistance genes from virus-resistant plants. In
addition, a paper was presented in which the bacteriophage Qß was used as a model to test whether introduction
of a viral gene into its host's genome would produce
"pathogen-derived resistance". Abstracts of all of the
workshop papers, except for the one given by J. Kridl of
Calgene, appear in the Journal of Cellular Biochemistry,
Supplement 10C, 1986.

Three papers described the introduction of viral
sequences into the plant genome in attempts to generate
virus resistant plants. The theoretical base for these
studies derives from the phenomenon known as "cross
protection". Cross protection was first described in
1929 and has seen limited use in commercial practice to
protect plants from viral disease. Generally, plants
infected with a mild strain of a given virus do not show
additional symptoms when subsequently challenged by a
severe symptom-inducing virus "related" to the original
one; "unrelated" viruses are usually not inhibited. The
mechanism of this phenomenon is not understood; currently
the two most popular hypotheses are: 1) inhibition of
the challenge virus by the coat protein of the resident
virus, and 2) the annealing of incoming nucleic acids to
those of the resident virus in the cell. In the latter
case, this inhibition could result from annealing of
challenge RNA to "plus sense" RNA (represented by the
resident viral RNA itself) or "antisense" RNA (represented by viral RNA sequences generated during replication).

J. Kridl and her associates have introduced cloned DNA sequences derived from tobacco mosaic virus into the tobacco genome; they have used 5' sequences (1-452) and 3' sequences (5705-6394). The latter contain the coat protein gene. Both sets of sequences were introduced in both the "sense" and "anti-sense" orientations. She presented preliminary data on the potential for viral inhibition of these inserted sequences, but concluded that without further study it was not possible to assess the potential for virus resistance of these transformed plants.

S. Loesch-Fries and her colleagues have similar objectives, but have utilized alfalfa mosaic virus and the tobacco plant. They have introduced sequences coding for the coat protein into plants and have shown that the coat protein was expressed. Expression was very dependent on the position of the leaf on the plant; young leaves contained more coat protein than did older ones. The RNA transcript was unaffected by leaf position suggesting there was either differential degradation of the coat protein or differential association of mRNA with ribosomes in the plant. Infection of protoplasts prepared from these transformed plants with virions of alfalfa mosaic indicated they were more resistant than the controls. However, the protoplasts were infectible when viral RNA rather than the virion was used as the inoculum, suggesting a role for coat protein in the phenomenon. They also inserted "antisense" constructs in plants, but felt they needed higher expression before they could test their potential for viral protection.

R. Grumet gave the results of a series of experiments in which Escherichia coli were transformed to express the coat protein gene of the bacteriophage Qß. Her group found that these bacteria were about 1000-fold more resistant to Qß infection than were bacteria without the coat protein gene. This gene insertion could also convey resistance to a lesser extent to another related bacteriophage, but not to several unrelated ones. It was shown that for resistance to occur, a functional coat protein had to be generated in the host since transformants containing deleted or frame-shifted coat sequences were ineffective. Coat protein is known to normally have a regulatory effect in phage replication and it remains to be determined if transformation with other viral genes can protect the E. coli from infection.

The final two papers discussed progress in isolating
genes from tobacco cultivars which display a hypersensitive (resistance) reaction when infected with
tobacco mosaic virus. The gene of interest (N) is
inherited as a single dominant Mendelian trait; it was
originally transferred to certain <u>Nicotiana</u> <u>tabacum</u>
cultivars from <u>N. glutinosa</u>. Infected plants of the N
genotype also display an "acquired resistance" to
subsequent infection of uninoculated leaves. D. Dunigan
reported that he had isolated mRNAs from N gene-containing plants by a temperature shift from a high, nonpermissive temperature to a low, permissive one. He
reported at least 5 unique proteins by <u>in</u> <u>vitro</u>
translation of these mRNAs; he and his colleagues are
currently isolating cDNA clones representing these
mRNAs. J. Horowitz has isolated cDNA clones from TMV
infected, uninfected and mock-infected N gene-containing
plants. By differential colony screening she has
isolated several clones representing mRNAs induced by the
TMV infection. The mRNAs represented by these clones
show differential expression in various plant parts. She
is now in the process of trying to identify what genes
these clones represent and what their relationship might
be to the N gene and to acquired resistance.

The discussion that followed the workshop was lively, indicating the degree of interest in the two topics
considered. In a sense, however, most of the papers were
progress reports; another 6 to 12 months will undoubtedly
tell whether either or both of these approaches will lead
to a better understanding of the mechanism(s) of virus
resistance and if either concept might provide the
technology to produce virus resistant plants.

III. Protection Against Insects and Environmental Stress

Molecular Strategies for Crop Protection, pages 345–353
© 1987 Alan R. Liss, Inc.

EXPRESSION OF A BACILLUS THURINGIENSIS INSECTICIDAL
CRYSTAL PROTEIN GENE IN TOBACCO PLANTS

M.J. Adang, E. Firoozabady, J. Klein, D. DeBoer,
V. Sekar, J.D. Kemp[1], E. Murray, T.A. Rocheleau,
K. Rashka, G. Staffeld, C. Stock, D. Sutton[1],
and D.J. Merlo

Agrigenetics Advanced Science Company
5649 East Buckeye Road
Madison, Wisconsin 53716

Agrigenetics Adv. Sci. Co. Publication No. 60

ABSTRACT   A major strategy for controlling pest
insects is the development of plants resistant to
insect attack. Introduction of proteins toxic to
insects into plants via genetic engineering is one
approach to developing such plants. We have previously cloned and characterized the DNA encoding
the insecticidal crystal protein gene of Bacillus
thuringiensis strain HD-73. To transfer the toxic
crystal protein gene into plants, the HD-73 gene was
engineered 3' to the (ORF 24) promoter of pTi15955.
This promoter/gene cassette was cloned into a binary
micro T-DNA vector containing a plant selectable
marker and delivered into tobacco cells. In regenerated transgenic tobacco plants, we have detected
B. thuringiensis peptides by immunoassay. Crystal
protein-specific transcripts in these tissues are
truncated at the 3' end of the gene. Nonetheless,
leaves of some transgenic plants kill tobacco horn-
worms in feeding bioassays.

[1] Present address: Plant Genetic Engineering
Laboratory, Box 3GL, New Mexico State University,
Las Cruces, New Mexico 88003

INTRODUCTION

A major problem of crop production world-wide is damage caused by insect feeding. Consequently, efforts of applied entomologists and plant breeders have focused on developing insect resistant crop plants. Success depends on the identification of resistant germplasm and the ability to breed these resistance factors into related cultivars. Genetic engineering and plant tissue culture can also be used to accomplish this objective with one additional advantage. Genes contributing to insect resistance can be transferred from other cultivars or even non-related species into plants. Crop plants transformed with genes encoding insecticidal proteins, for example, may resist insect feeding.

A necessity for engineering plants with novel traits is the availability of genes encoding proteins with the desired qualities. For insect resistance, a well-characterized source of genes is the insect pathogen, Bacillus thuringiensis (Bt). This bacterium produces a proteinaceous crystal that is lethal to insects. Feeding insects ingest crystals which are solubilized in the midgut. Released subunit polypeptides act on the midgut cells of susceptible insects, resulting in a cessation of feeding, followed by death in several days. Various strains are active against the larvae of Lepidoptera (butterflies and moths), while other strains produce crystal proteins toxic to Diptera (black flies and mosquitos) or Coleoptera (beetles). We have used the insecticidal crystal protein (ICP) gene of Bt subspecies kurstaki HD-73. The HD-73 gene encodes a M_r=133,000 peptide that is lethal to lepidopteran larvae (1).

This report describes the transfer of the ICP gene to tobacco plants, its expression, and the insect resistance observed in transgenic plants.

RESULTS

To introduce a Bt ICP gene into tobacco plants, we used a binary vector system and Agrobacterium tumefaciens. One of the vectors, the micro Ti plasmid pH400 has the following features. All of the oncogenic T-DNA functions have been deleted, while the genes encoding octopine synthase (OCS) and the 25 base pair border signals (A and B) were retained. A neomycin phospho-

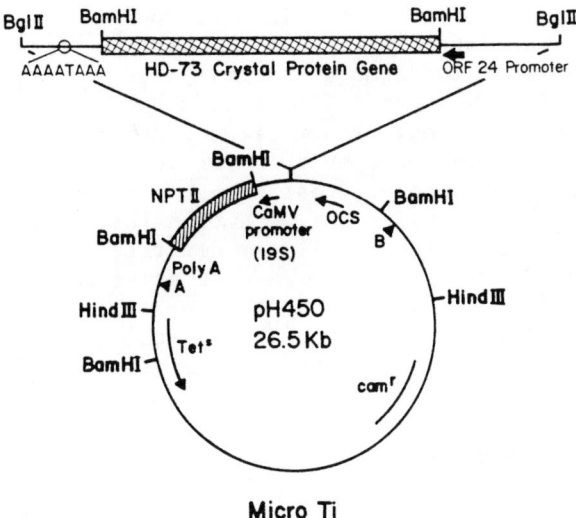

FIGURE 1. Micro T-DNA construction containing Bt
ICP coding sequence in ORF 24 cassette. Plasmid pH450
was mated into A. tumefaciens strain LBA4404 and used to
transform N. tabacum cv. Xanthi.

transferase II (NPTII) gene under the control of the CaMV
19S promoter was inserted for selection of transformed
plants. The replicon is plasmid pSUP106, an 11 kb broad
host range vector (2). Plasmid pH400 replicates in
Escherichia coli and A. tumefaciens, and is readily
mobilized between these bacteria. The DNA between the
two 25 bp repeats is transferred to plants by the trans-
acting functions of Ti plasmid vir regions.
      The DNA for transfer was constructed in the
following manner. The full-length HD-73 ICP gene was
engineered to introduce BamHI sites prior to the initial
methionine codon, and after the terminal translation stop
codon. The Bt ICP open reading frame (ORF) contained in
this 3.7 kb BamHI fragment was subcloned into an
expression cassette. This cassette contains a T-DNA
ORF 24 promoter and poly(A) addition signals (3, 4) previously shown to be functional in plants. Figure 1 shows
plasmid pH450, which is the micro Ti vector after inser-

tion of the ORF 24 promoter, Bt ICP ORF, and polyadenylation signals.

    Plasmid pH450 was mated into A. tumefaciens strain
LBA4404 to generate a binary vector system. This Agrobacterium strain contains a severely deleted Ti plasmid,
which no longer harbors a T-region, but contains a fully
functional vir region. Plant cells transformed by the
binary vector, mobilized by this vir region, are toti-
potent and non-tumorous. Leaf segments of tobacco,
Nicotiana tabacum cv. Xanthi were transformed and regenerated into plants essentially as described (5). Plants
selected on 300 µg/ml kanamycin were screened for OCS
activity and Bt ICP antigen by enzyme-linked immunoassay
(ELISA). Approximately 75% of the tested plants had
detectable OCS activity. After optimizing the ELISA (6),
antigenic peptides could be detected in leaf extracts of
some transgenic plants. Subsequently, ten OCS positive
plants were chosen for molecular analysis. The presence
of Bt ICP DNA in these plants was confirmed by Southern
blot analysis. Blots of BamHI cut tobacco DNA were
probed with the ^{32}P-labelled Bt ICP gene. Figure 2 shows
that eight of the ten plants contain the expected 3.7 kb

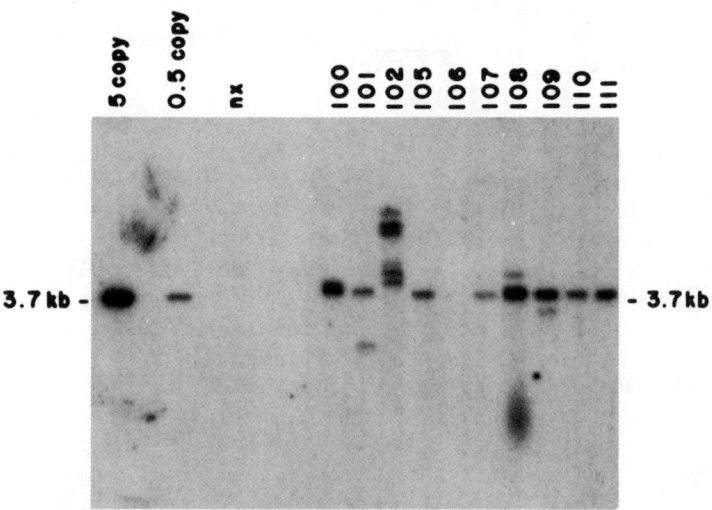

    FIGURE 2. Southern hybridization analysis of DNA
from transgenic tobacco plants. DNA was digested with
BamHI, fractionated on an agarose gel, transferred to
nitrocellulose and hybridized with ^{32}P-labeled BamHI Bt
fragment. nx indicates non-transformed tobacco control.

hybridizing band. Plant 102 has multiple hybridizing
bands, none of which are 3.7 kb, suggesting multiple
rearranged copies of the Bt gene. The plant 106 lane in
Figure 2 has a band at 3.7 kb visible after a long exposure of the film. These transgenic plants were vegetatively propagated and analyzed further.

Leaf tissues from the above plants were analyzed for
Bt ICP specific mRNA. Northern blots were prepared using
poly(A)-enriched RNA. Hybridization was carried out with
a message-specific ^{32}P-labeled RNA probe prepared using
the Bt BamHI fragment in an SP6 vector from Promega
Biotec (Madison, WI). Figure 3 shows the 1.7 kb RNA
present in plants containing the Bt crystal protein
gene. The 3.7 kb RNA, also seen in non-transgenic

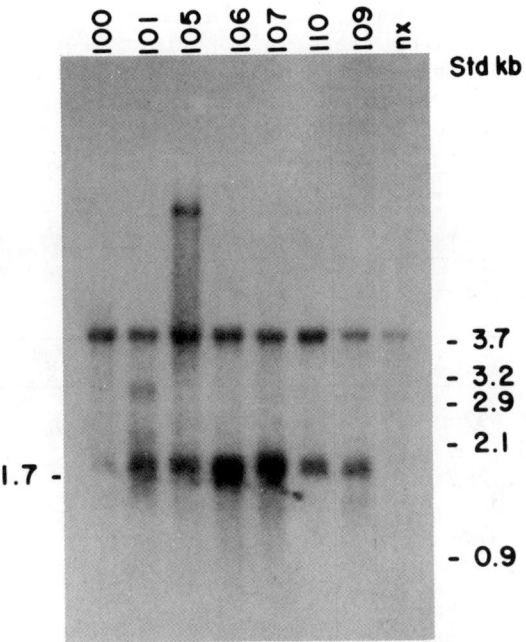

FIGURE 3. Northern blot hybridization analysis of
RNA. Poly(A)-enriched RNA was prepared from transgenic
or non-transformed (nx) leaf tissues, subjected to electrophoresis through a 1.5% formaldehyde agarose gel,
transferred to nitrocellulose, hybridized to a
^{32}P-labeled complementary strand SP6 Bt RNA probe and
autoradiographed.

control plants, is not specific for the Bt probe. Using
a probe from the 5' region of the ICP gene we obtained
similar results, whereas a 3' probe did not hybridize to
any plant RNAs. Blots of the same RNAs done in parallel,
but hybridized with NPTII or OCS specific probes showed a
much stronger signal to the respective 1.25 and 1.2 kb
RNAs, indicating steady state amounts of ICP RNA are low
relative to the NPTII and OCS RNAs in these plants. The
large RNA in plant 105 (Figure 3) also hybridizes to
NPTII and OCS probes. These Northern blot results indicate that a 1.7 kb RNA, which is present in transgenic
plants, is truncated from the 3' end of the ICP gene.

   To detect Bt protein in plants, we used an ELISA
against the $M_r = 133,000$ ICP. Table 1 shows the results of
two assays on aqueous leaf extracts. Prior to sample
application, Assay I samples were standardized for equal
tissue weight. Samples in Assay II had equal amounts of
protein per microtiter plate well. Plant 100 was positive for ICP in both assays, while other plants gave
variable results. By comparison of experimental values
to a standard curve, consisting of plant extract and
$M_r = 133,000$ ICP, we estimate that leaves from plant 100

TABLE 1

ELISA FOR BT ICP IN TRANSGENIC LEAF TISSUE

	Assay I[a]		Assay II[b]	
Plant	A(492nm)	ICP	A(492nm)	ICP
100	0.17 ± 0.03	+	0.16	+
101	0.02 ± 0.01	−	0.07	+−
102	0.06 ± 0.02	+−	0.21	+
105	0	−	0.10	+
106	0.10 ± 0.02	+	0.01	−
107	0.06 ± 0.03	+−	0.06	+−
109	0.11 ± 0.05	+	0.03	−
110	0.06 ± 0.01	+−	0.13	+
111	0.12 ± 0.02	+	0	−
nx[c]	0(n=18)		0(n=19)	

[a]Standardized for equal tissue mass.
[b]Standardized for equal protein amounts.
[c]nx indicates non-transformed Xanthi tissue.

cuttings contained approximately 2 ng cross-reacting
protein per mg leaf protein.

Transgenic plants were screened for insect resistance against newly hatched tobacco hornworm (THW)
(Manduca sexta) larvae. Excised leaves were placed on
moistened filter paper in Petri plates. Three THW larvae
were applied per plate and scored for mortality after
three days. Leaves from number 100 plants killed 53% of
the larvae in five feeding trials. Three other clones of
transgenic plants killed approximately 25% of the worms
tested, while mortality was 7% on leaves from non-transformed plants. These trials demonstrated the toxicity of
the transgenic plants to THW larvae.

DISCUSSION

This report describes the expression of an insecticidal gene in tobacco plants. We have engineered the Bt
HD-73 ICP gene to be expressed in plants. This gene was
inserted between the promoter and poly(A) addition
signals of the T-DNA gene encoding the first step in
mannopine synthesis (ORF 24). Via a micro T-DNA vector
and A. tumefaciens, this Bt gene was introduced into
tobacco plants. Southern blots confirmed that plants
regenerated on media containing kanamycin were transgenic. Northern RNA blots indicated that the major Bt
transcript in leaves is 1.7 kb and not the expected
3.9 kb (Figure 3). Hybridization with 5' and 3' probes
show this 1.7 kb RNA to contain sequences of the 5' portion of the ORF. S_1 nuclease protection mapping is
necessary to delimit the 5' and 3' ends of this transcript. Since the N-terminal 612 amino acid residues of
the 1178 residue HD-73 ICP are sufficient to kill insects
(1), a 3' truncated mRNA may be sufficient to encode an
insecticidal polypeptide.

It is surprising that the truncated Bt mRNA appears
as a discrete species. One explanation for this is that
sequences within the coding region are functioning as
regulatory signals for transcription termination or
cleavage/polyadenylation. While the Bt ORF does not
contain an AATAAA sequence for poly(A) processing,
numerous hexanucleotide sequences are present which
differ from the consensus sequence by only a single
nucleotide. One of these may function in plants as a
polyadenylation signal. Detailed experiments are

necessary to determine the factors which contribute to
the expression of low level, 3' truncated Bt transcripts
in tobacco leaves.

Using an ELISA, we have detected Bt ICP in leaves of
transgenic plants. The estimated amount of cross-
reacting protein for plant 100 was 2 ng/mg plant protein. It is possible that the amount of antigen is
underestimated by this assay. Both the primary and
secondary antisera used in the double antibody sandwich
ELISA were raised against the complete peptide, and not
to a truncated peptide. These sera react less strongly
with truncated peptides than with full-length peptides.
Several attempts have been made to detect ICP using
Western blots. While no peptides can be detected in leaf
tissues, this may not be surprising, since the amount of
antigen present as determined by immunoassays is below
the detection limit of our assay. Truncated peptides
have been detected by Western blots of protein from
callus and shoot tissues.

Feeding trials using THW larvae indicated that some
tobacco plants expressing the Bt ICP gene were toxic to
the larvae. However, many plants containing the ICP gene
had no affect on feeding THW larvae. It is important
that we eliminate the possibility that resistance is due
to altered secondary characteristics such as plant morphology or allelochemicals. Nevertheless, these results
are encouraging because of the correlation between resistant plants and the presence of ICP in the leaves.

REFERENCES

1. Adang MJ, Staver MJ, Rocheleau TA, Leighton J, Barker
    RF, Thompson DV (1985). Characterized full-length
    and truncated plasmid clones of the crystal protein
    of Bacillus thuringiensis subsp. kurstaki HD-73 and
    their toxicity to Manduca sexta. Gene 36:289-300.
2. Priefer UB, Simon R, Pühler A (1985). Extension of
    the host range of Escherichia coli vectors by incor-
    poration of RSF1010 replication and mobilization
    functions. J Bacteriol 163:324-330.
3. Barker RF, Idler KB, Thompson DV, Kemp JD (1983).
    Nucleotide sequence of the T-DNA region from the
    Agrobacterium tumefaciens octopine Ti-plasmid
    pTi15955. Plant Mol Biol 2:335-350.

4. Gelvin SB, Karcher SJ, Goldsbrough PB (1985). Use of a T_R T-DNA promoter to express genes in plants and bacteria. Mol Gen Genet 199:240–248.
5. Horsch RB, Fry JE, Hoffmann NL, Eichholtz D, Rogers SG, Fraley RT (1985). A simple and general method for transferring genes into plants. Science 227:1229–1231.
6. Stanley CJ, Johannsson A, Self CH (1985). Enzyme amplification can enhance both the speed and the sensitivity of immunoassays. J Immunol Methods 83:89–95.

Molecular Strategies for Crop Protection, pages 355–366
© 1987 Alan R. Liss, Inc.

ENGINEERING OF INSECT RESISTANT PLANTS
USING A B. THURINGIENSIS GENE

M. Vaeck, H. Höfte, A. Reynaerts, J. Leemans, M. Van
Montagu,M. Zabeau

Plant Genetic Systems N.V., Gent, Belgium

ABSTRACT

A crystal protein gene (bt2) has been cloned from
plasmid DNA of Bacillus thuringiensis (B.t.) berliner 1715
and directs the synthesis of a 130 kd protein (Bt2) in E.
coli which is toxic to larvae of Pieris brassicae and
Manduca sexta. Treatment of the Bt2 protein with trypsin
or chymotrypsin yields a 60 kd protease resistant fragment
which is fully toxic towards insect larvae "in vivo" and
insect cell lines "in vitro". The minimal portion of the
Bt2 protein required for toxicity has been mapped by
deletion analysis and coincides with the 60 kd protease
resistant Bt2-fragment. Tobacco plant cells have been
transformed with chimeric toxin genes using a Ti plasmid
vector. Transformed plants express a functional toxin and
exhibit resistance against insect larvae.

INTRODUCTION

Bacillus thuringiensis (B.t.) is a gram positive
bacterium which produces endogenous crystals upon
sporulation. The crystals are composed of protein and are
specifically toxic against certain insect larvae, mainly
lepidopteran and dipteran species [1]. Upon ingestion by
larvae the crystals disolve in the alkaline conditions of
the insect midgut and release proteins of molecular weight
(M.W.) 130 - 160 kd [2]. These high M.W. protoxins are
proteolytically processed by midgut proteases to yield

smaller toxic fragments [3]. Most crystal protein genes have been localised on large plasmids [4, 5,]. Some genes have recently been cloned and functional toxin was expressed in E.coli [6, 7, 8, 9]
B.t. toxins are highly specific in their activity and therefore represent a perfectly safe insecticide and an interesting alternative for chemical control agents. Commercial preparations based on B.t. have already been used for several years as a biological insecticide, however only with limited success, mainly due to high production costs and instability of B.t. in the field. We have used the Agrobacterium vector system to transfer and express the B.t. toxin gene in tobacco plants. We present data on the resistance of those plants against insect larvae.

RESULTS

Cloning of a B.t. gene encoding a 130 kd crystal protein exhibiting insect toxicity. Since most crystal protein genes in B. thuringiensis strains are plasmid-born [5], we used a purified plasmid preparation from strain B.thuringiensis berliner 1715 for the cloning of crystal protein genes. A library was constructed by cloning size fractionated Sau 3A digested plasmid DNA into an E. coli expression vector. Using a colony immunoblot assay with rabbit anti-crystal protein serum, 4 clones containing overlapping DNA fragments were isolated. A 7.5 kb BamHI-PstI fragment containing the toxin gene was subcloned into pUC8, generating plasmid pGI502.
Total cell extract of E. coli K514 (pGI502) was analysed on SDS-PAGE. An intense protein band, with apparant M.W. of 130 kd, was visible and was not present in K514 containing the pUC8 plasmid without insert. This protein, termed Bt2, comigrated in SDS-PAGE with one of the major crystal proteins of B.t. berliner. Bt2 protein represented between 5 and 10% of the total protein content in K514 (pGI502) and was present as a precipitate in E. coli.
The relationship between Bt2 and B.t. crystal proteins was confirmed by its immunological properties and biological activity. In Western blotting,Bt2 reacted strongly with a rabbit anti-Bt berliner crystal serum and in ELISA, 8 out of 16 monoclonal antibodies, generated

against B.t. berliner crystal proteins, reacted with the purified Bt2. The toxic activity of Bt2 protein was assayed on Pieris brassicae and Manduca sexta larvae. Purified preparations of Bt2 showed toxicity levels comparable to those of solubilized crystals from B.t. berliner (Table 1).

TABLE 1
INSECT TOXICITY OF B.T. CRYSTAL PROTEINS

	P. brassicae (LD50) ng/larva	M. sexta (LD50) ng/cm2)
Bt berliner cry	15	N.T.
Bt berliner cry (solubilized)	0.6 (+0.3)	7.5
Bt2 protein	1.6 (+1.3)	6
Bt2 protein/trypsin	1.5	5

Nucleotide sequence of the gene and N-terminal amino acid sequence of the crystal protein. The bt2 toxin gene was localized by deletion mapping on a 4343 bp HpaI-PstI fragment. The sequence of this fragment showed one large open reading frame encoding a protein of 1155 amino acids with a predicted molecular weight of 130533 d. This agrees well with the molecular weight of Bt2 determined in SDS-PAGE. Bt2 protein from K514 (pGI502) was additionally purified by DEAE-ion exchange chromatography and Sephacryl gel filtration. The amino acid sequence of the N-terminal end of this purified protein was determined by gas-phase sequencing to be X - Asp - Asn - Asn - Pro - Asn - Ile - Asn - Glu - X - Ile - Pro - Tyr - Asn - Leu - X - Asn - Pro. This sequence was identical to the N-terminal amino acid sequence deduced from the nucleotide sequence.

Generation of a 60 kd toxic polypeptide through proteolytic degradation of the 130 kd Bt2 protein. The delta-endotoxins of B.t. are generally believed to be protoxins which, upon ingestion by insects, are degraded by insect gut proteases into smaller active toxin(s) [10]. We therefore investigated whether smaller toxic polypeptides could be generated from Bt2 by proteolytic cleavage. Purified Bt2 protein was digested with either trypsin or chymotrypsin and at defined time intervals, aliquots were analysed on SDS-PAGE. The results demonstrated that the 130 kd Bt2 protein is rapidly degraded by trypsin or chymotrypsin (after 10 min. at 37°C) yielding a predominant polypeptide of 60 kd. This 60 kd polypeptide was relatively resistant to further degradation by both enzymes over a 2 h period, indicating that it constitutes a protease resistant fragment within the Bt2 protein. The 60 kd polypeptide generated by trypsin cleavage was purified by gel filtration and its insect toxicity was determined. It was equally toxic to P. brassicae and M. sexta larvae as intact 130 kd Bt2 (Table 1). To position this peptide within the intact Bt2 protein molecule, its N-terminal amino acid sequence was determined and was identical to a sequence starting with the isoleucine residue at position 29 in the complete sequence. Thus, a polypeptide fragment exhibiting the complete toxic activity of Bt2 was localized in the N-terminal half of the molecule.

To determine whether this 60 kd polypeptide represented an active toxin or was still a precursor which has to be further processed by insect midgut enzymes to yield the final toxic unit, we performed functional assays on insect cells "in vitro". Three different cell lines, derived from Lepidoptera species, known to be susceptible to B.t. toxins, were used, as well as one Diptera cell line. The cells were incubated either with untreated 130 kd Bt2 protein or with a purified preparation of the trypsin generated 60 kd fragment of Bt2. After 2 h at 28°C, the percentages of dead cells were scored. While the intact 130 kd Bt2 protein had no effect on cell viability, the 60 kd fragment induced a significant mortality (>75%) in the three Lepidoptera cell lines, but had no effect on the Diptera cell line (Table 2).

Therefore the 60 kd trypsin fragment represents an active and specific toxic molecule.

Genetic mapping of the minimal gene fragment encoding an active toxin. Different deletions at the 5′ or the 3′ end of the bt2 gene were constructed and expressed in E. coli. We analysed the produced polypeptides to delineate the minimal Bt2 fragment exhibiting insect toxicity. At the 3′end, the gene fragment encoding the active toxin is contained within the first 2170 bp, but extends beyond

TABLE 2
Toxicity of Bt2 protein on insect cells in vitro

Origin of cell line	Intact 130 K	60 K trypsin fragment
Trichoplusia ni (1)	-	+
Spodoptera frugiperda (1)	-	+
Choristoneura fumiferana (2)	-	+
Aedes albopictus	-	-

(1) Cell lines obtained from the N.E.C.R. Institute of Virology, Oxford, (2) cell line from Dr S.S. Sohi, F.P.M.I., Ontario

the HindIII site at position 1692. To precisely determine the 3'end point of the minimal fragment, random deletion mutants encoding N-terminal fragments were constructed using exonuclease Bal31. Total cell extracts from 10 derivatives which have their deletion endpoints in the HindIII-KpnI region were analysed in a toxicity assay on P. brassicae larvae. Extracts containing Bt2-fragments larger than pLB879 were fully toxic whereas extracts containing pLB834 or smaller fragments were completely non-toxic to P. brassicae larvae (Table 3).

The deletion endpoints in pLB879 and pLB834 were determined by DNA sequencing (Fig. 1) and indicated that the critical endpoint of a DNA fragment encoding an active toxin maps between positions 1798 and 1821 on the bt2 gene.
The absence of toxicity in pLB834 could result from a higher susceptibility to degradation of the pLB834 encoded polypeptide in E. coli extracts. To investigate this possibility, we constructed inframe fusions of the Bt2 gene fragments of pLB834 and pLB879 to the 5'end of lacZ to produce pBZ12 and pBZ13 respectively. SDS-PAGE analysis showed that the bt2-lacZ-fusion genes directed the production of high amounts of protein with the predicted molecular weight of the fusion protein. Reactivity with anti-B.t. berliner crystal serum and

TABLE 3

Toxicity of cellextracts of NF1 strains containing different plasmids expressed as % mortality after 4 days; 50 3rd instar P. brassicae were used per dilution of extract.

Type of construct	E. coli clone	Position 3'end Bt2 sequence	1	1/10
Deletions	pLB16	2170	100	100
	pLB820	2031	100	100
	pLB822	1911	100	100
	pLB828	1830	100	100
	pLB826	1827	100	100
	pLB884	1824	100	100
	pLB879	1821	98	50
	pLB834	1798	0	2
	pLB950	1731	6	0
	pLB876	1695	2	0
	pLB12	1692	0	0
Bt2-lacZ	pBZ12	1821	100	74
fusions	pBZ13	1798	4	0

anti-beta-galactosidase serum in Western blotting confirmed the presence of both Bt2 and beta-galactosidase determinants. Tests on P. brassicae larvae revealed that the toxicity of cell extract from pBZ12 was comparable to that of pLB879, whereas the equally stable fusion protein encoded by pBZ13 was completely nontoxic.

To delineate the 5'end of the gene fragment encoding an active toxin, we constructed a 5' deletion in the Bt2 gene up to the ClaI site at the 110th bp, removing the first 36 codons of the bt2 gene. A polypeptide encoded by this deletion derivative starts only 8 amino acids beyond residue 29 which is the N-terminus of the fully active 60 Kd processed toxin. This polypeptide had the expected size and was reactive with anti-Bt2 serum. However it was completely non-toxic to P. brassicae larvae.

Together these data show that the minimal toxic fragment of the Bt2 protein is a 60 kd polypeptide delineated by amino acid residues 29 and 37 at the N-terminus and amino acids 601 and 607 at the C-terminus (Fig. 1).

```
Trypsin cleavage
       site  ↓
 GLY GLU ARG ILE GLU THR GLY TYR THR PRO ILE ASP ILE
...'GGA'GAA'AGA'ATA'GAA'ACT'GGT'TAC'ACC'CCA'ATC'GAT'ATT'..
    80          90          100         ↑110
                                        pRB210


        600                     605
 TYR ILE ASP ARG ILE GLU PHE VAL PRO ALA GLU VAL
...'TAT'ATA'GAT'CGA'ATT'GAA'TTT'GTT'CCG'GCA'GAA'GTA'
     1800↑          1810        1820 ↑
     pLB834                      pLB879
```

FIGURE 1. Mapping of the 5' and 3'end for the minimal toxic fragment.

Transfer and expression of chimeric B.t. genes in tobacco plants. Recent progress in plant genetic engineering techniques made it possible to transfer foreign DNA into many plant species using vectors derived from the Ti-plasmid of Agrobacterium tumefaciens. Using such vectors, several genes of bacterial origin - mostly antibiotic resistance genes - have been expressed in transgenic plants using plant specific promotor sequences. The latter have been obtained from the Ti-plasmid T-DNA genes [11, 12,], from plant viruses [13] or from nuclear plant genes [14, 15, 16]. We have used Ti-plasmid derived vectors [17] to transfer the bt2 toxin gene into N. tabacum plants using a leaf disc transformation procedure [18]. Details on these chimeric gene constructs will be published elsewhere.

Transformed tobacco plants containing a chimeric bt2 gene were analysed for production of Bt2 related polypeptides in an ELISA assay using a rabbit anti-Bt crystal serum, as well as monoclonal antibodies generated against the Bt2 protein. In reconstruction experiments, the detection limit of this assay was shown to be around 1 ng/ml. Variable expression levels of B.t. toxin were recorded in independant transformants. Quantification of the amount of Bt2 related polypeptide present in plant cell extracts and toxicity assays on these extracts showed that the specific toxicity of the plant produced B.t. toxin was comparable to that of the bacterial gene product.

Insect resistance of transformed tobacco plants. We evaluated the toxic activity exhibited by transformed tobacco plants on insect larvae. Twenty first instar larvae of M. sexta were placed on leaf discs of 4 cm diameter in petri-dishes. Fresh leaf material was given each day, growth rate and mortality of the larvae feeding on the leaves were scored over a 6 day period. Larvae feeding on Bt positive plants rapidly showed marked growth inhibition, which retarded their transition to the second larval stage. Most importantly, significant mortality among the larvae became apparent at day 3 and reached 100% at day 6 for several of the independent transformed plants tested (Table 4).

TABLE 4

Mortality rates in M. sexta larvae feeding on transformed tobacco plants.

	% Mortality		
	Day 3	Day 4	Day 6
Bt± plants			
1	90	100	100
2	10	45	65
3	80	95	100
4	5	60	95
5	5	35	85
6	85	90	100
7	65	75	95
Control plants			
8	0	0	0
9	0	0	0
10	0	0	5
11	5	5	5

In contrast, larvae growing on untransformed control plants developed normally and their mortality rate was only 0-5%. Thus, the effect on insect larvae of B.t. toxin expressed in plant cells was comparable to the effect of B.t. toxin of bacterial origin, namely growth inhibition in the initial stage followed by death.
As expected from the observed variability in expression levels of B.t. protein, insecticidal activity exhibited by independant transformants was also variable (Table 4). However for one and the same transformant the degree of insect resistance was highly reproducible when tested in independant experiments over a period of time.
    In a next set of experiments, we investigated whether the levels of B.t. toxin produced in transformed plants would protect them from significant insect damage. Fifteen freshly hatched larvae were placed on the leaves of 40 cm high tobacco plants in the greenhouse. On control plants, considerable damage was obvious after 4-7 days and they were completely consumed after 10-15 days. In contrast, all larvae died within 4 days when feeding on plants expressing high levels of Bt2 protein, and leaf damage was restricted to holes of a few mm^2 caused by each insect (Fig. 2).

    Conclusion and prospects. A functional analysis has been performed on Bt2, a cloned Bacillus thuringiensis crystal protein. The Bt2 protein is highly toxic to larvae of Pieris brassicae and Manduca sexta. This 130 kd polypeptide is a protoxin which is not toxic in an in vitro assay on insect cell lines. However, treatment of Bt2 with proteolytic enzymes results in a 60 kd protease resistant and fully toxic polypeptide.
    We have precisely determined, through genetic mapping, the minimal fragment of Bt2 which still exhibits full toxicity. This fragment was found to coincide with the 60 kd polypeptide generated by proteolytic degradation and located in the NH$_2$-terminal half of the protein. According to our data, it is very unlikely that toxic fragments, smaller than this 60 kd can be obtained. A detailed analysis to identify the functional domains of Bt2 is presently in progress, using monoclonal antibodies specific for well defined regions on the molecule.
The feasibility of using plant genetic engineering techniques to obtain insect resistant plants is clearly exemplified here. Fully normal plants were obtained,

expressing sufficiently high levels of <u>B.t.</u> toxin to give resistance against insect damage under greenhouse conditions. Further experiments will be directed to fully evaluate this new approach under natural conditions of stress caused by insect pests and to investigate the inheritance of the newly acquired defense mechanisms.

A large number of natural <u>B.t.</u> strains have been isolated, covering a wide spectrum of insecticidal activities and directed against Lepidoptera, Coleoptera and Diptera species.

Analysis of the molecular basis of the specificity will allow to design new types of toxins with the desired activity against agriculturally important insect pests. This technology, in combination with new developments in the field of plant engineering will enable us to engineer genetically modified varieties of commercially important crops, expressing resistance against major insect pests.

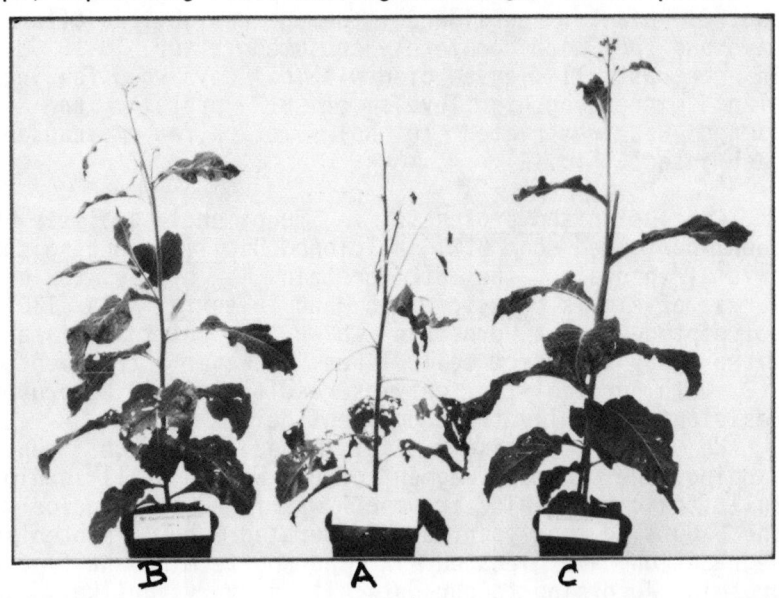

FIGURE 2. Insect assay on transformed tobacco plants showing a control plant (A), a transformed plant expressing a low level (B) or a transformed plant expressing a high level of <u>B.t.</u> (C), 13 days after 15 first instar <u>M</u>. <u>sexta</u> larvae were placed on each plant.

ACKNOWLEDGEMENTS

We thank J. Seurinck, C. Ampe, J. Vandekerckhove, S. Jansens, H. Vanderbruggen, C. Matthijs, A. De Sonneville, H. Van De Wiele for valuable contributions, and K. Tenning for typing.

This work was financed in part by the Rohm and Haas Company, Philadelphia, U.S.A.

REFERENCES

1. Dulmage HT (1980). In Burges HD (ed): Insecticidal activity of isolates of Bacillus thuringiensis and their potential for pest control. "Microbial control of pests and plant diesease 1970-1980," London: Academic Press, p 193.
2. Huber HE, Luthy P, Ebershold H-R (1981). The subunits of the parasporal crystal of Bacillus thuringiensis: size, linkage and toxicity. Arch Microbiol 129:14.
3. Lilley M, Ruffel RN, Sommerville HJ (1980). Purification of the insecticidal toxin in crystals of Bacillus thuringiensis. Gen Microbiol 118:1.
4. Gonzalez JM, Brown BJ, Carlton BC (1982). Transfer of Bacillus thuringiensis plasmids coding for delta-endotoxin among strains of B. thuringiensis and B. cerem. Proc Natl Acad Sci USA 79:6951.
5. Kronstad JW, Schnepf HE, Whiteley HR (1983). Diversity of locations for Bacillus thuringiensis crystal protein genes. Bact 154:419.
6. Schnepf HE, Whiteley (1981). Cloning and expression of the Bacillus thuringiensis crystal protein in Escherichia coli. PNAS 78:2893.
7. Klier A, Fargette F, Ribier J, Rapoport G (1982). Cloning and expression of the crystal protein genes from Bacillus thuringiensis strain berliner 1715. Embo J 1:791.
8. Shibano Y, Yagamata A, Nakamara N, Iizuka T, Sugisaki H, Takanami M (1985). Nucleotide sequence coding for the insecticidal fragment of the Bacillus thuringiensis crystal protein. Gene 34:243.

9. Adang MJ, Staver MJ, Rochelean TA, Leighton J, Barker RF, Thompson DV (1985). Characterized full-length and truncated plasmid clones of the crystal protein of Bacillus thuringiensis subsp. kurstaki HD-73 and their toxicity to Manduca sexta. Gene 36:289.
10. Thomas WE, Ellar DJ (1983). Bacillus thuringiensis var. israeliensis crystal delta-endotoxin: effects on insect and mammalian cells in vitro and in vivo. Cell Sci 60:181.
11. Herrera-Estrella L, De Block M, Messens E, Hernalsteens J-P, Van Montagu M, Schell J (1983). Chimeric genes as dominant selectable markers in plant cells. Embo J 2:987.
12. Velten J, Velten L, Hain R, Schell J (1984). Isolation of a dual plant promoter fragment from the Ti plasmid of Agrobacterium tumefaciens. Embo J 3:2723.
13. Odell JT, Nagy F, Nam-Hai C (1985). Identification of DNA sequences required for activity of the cauliflower mosaic virus 35S promoter. Nature 313:810.
14. Herrera-Estrella L,Van Den Broeck G, Maenhaut R, Van Montagu M, Schell J, Timko M, Cashmore A (1984). Light-inducible and chloroplast-associated expression of a chimaeric gene introduced into Nicotiana tabacum using a Ti plasmid vector. Nature 310:115.
15. Simpson J, Timko M, Casmore AR, Schell J, Van Montagu M, Herrera-Estrella L (1985). Light-inducible and tissue-specific expression of a chimaeric gene under control of the 5'-flanking sequence of a pea chlorophyll a/b-binding protein gene. Embo J 11:2723.
16. Kaulen H, Schell J, Kreuzaler F (1986). Light-induced expression of the chimeric chalcone synthase-NPTII gene in tobacco cells. Embo J 5:1.
17. Deblaere R, Bytebier B, De Greve H, Deboeck F, Schell J, Van Montagu M, Leemans J (1985). Efficient octopine Ti plasmid-derived vectors for Agrobacterium-mediated gene transfer to plants. Nucl Acid Res 13:4777.
18. Horsch RB, Fry JE, Hoffman NL, Eichholtz D, Rogers SG, Fraley RT (1985). A simple and general method for transferring genes into plants. Science 227-1229.

Molecular Strategies for Crop Protection, pages 367–373
© 1987 Alan R. Liss, Inc.

A NOVEL STRATEGY FOR THE DEVELOPMENT
OF NEMATODE RESISTANT TOMATOES

by W.H.-T. Loh, S.A. Kut, and D.A. Evans
DNA Plant Technology Corporation
Cinnaminson, New Jersey 08077

ABSTRACT Previous attempts to break the linkage between
nematode resistance (Mi) and associated deleterious
agronomic traits in tomato have been prevented by
reduced meiotic recombination near the centromere. An
effort was made to break the linkage with Mi through
tissue culture induced mitotic crossing-over. Although
the linkage was not broken, the results suggest that the
map distances between Mi and other centromeric genes are
greatly underestimated by restriction of meiotic recombination, and that tissue culture induced mitotic
crossing-over may be uniquely suited to break such
linkages.

INTRODUCTION

In tomatoes, nematode infestation is manifested by the
development of shallow and knotted root systems. The
infected plants wilt despite adequate irrigation and suffer
from mineral deficiency, leading to reduced top growth and
decreased fruit yield. Two species of root knot nematodes,
Meloidogyne incognita and M. javanica, are prevalent in the
Sacramento and San Joaquin Valleys where most processing
tomatoes are grown (1). Chemical control is costly and has
limited efficacy. Instead, efforts to transfer natural
resistance from wild tomato species have been far more
successful.
Nematode resistance was originally introduced into Lycopersicon esculentum via a cross with L. peruvianum var.
dentatum (2). The trait has been characterized as a single
dominant gene (Mi) at position 35 cM on chromosome 6 (3) and
confers resistance to M. incognita, M. javanica, M. acrita,
and M. arenaraia (4). Rick and Fobes (5) subsequently
observed that the Mi locus was tightly linked to an isozyme

marker, acid phosphatase 1 (Aps-1). Thus, nematode resistance can be determined by staining for Aps-1 (6). Plants which synthesize the fast migrating Aps-1^1 polypeptide, otherwise restricted to accessions of L. peruvianum var. dentatum, have functional nematode resistance (7).

While the incorporation of two copies of the resistance allele would greatly simplify commercial tomato breeding schemes, this gene combination also results in soft fruit with high stem retension and reduced yield. Considering the source of the resistance, such effects have been attributed to tightly linked deleterious gene combinations rather than a pleiotropic effect of the Mi locus. Extensive efforts to break this linkage have not been successful, possibly due to the proximity of Mi to the centromere, where meiotic recombination is suppressed.

We proposed to break this linkage by inducing mitotic crossover events in tissue culture. The results obtained suggest that the map distances of the genes produced by mitotic crossing-over are not the same as meiotic recombination frequencies.

METHODS

MCO stock, heterozygous for the following traits, yv, Mi, Aps-1 , coa, c, were obtained from Dr. H.P. Medina-Filho, Instituto Agronomico, Sao Paulo, Brazil. A brief description of each trait, all located on the long arm of chromosome 6, is provided in Table 1. Young leaf tissues were placed into culture according to Evans and Sharp (8). Plants regenerated

TABLE 1
SUMMARY OF GENETIC MARKERS ON CHROMOSOME 6

Symbol	Description of trait	Chromosome position
yv	yellow virescence	34
Mi	nematode resistance	35
Aps-1	acid phosphatase	35
coa	corrotundata	64
c	potato leaf	93

from tissue culture, the Ro generation, were then scored for
changes at each locus.

Nematode resistance was determined by staining for Aps-1
according to the procedure of Medina-Filho and Tanksley (9)
with minor modifications. Isozymes were separated on hori-
zontal 12.5% hydrolyzed starch (Sigma) gels run at 5°C with
the following buffers: electrodes - 0.3M boric acid, pH 8.6;
gel - 0.076M Tris, 0.006M citric acid, pH 8.6. The other
traits were determined by scoring for segregation in the
selfed progeny.

RESULTS

A total of 51 regenerates, Ro plants, could be
unambiguously characterized. The resulting genotypes,
grouped into five classes, are summarized in Table 2. The
corrotundata trait, coa, which leads to reduced plant height
was excluded from the analysis, since the trait is often
encountered in tissue culture derived plants. Isozyme
analysis showed no recombination involving the Aps-1 locus of
any of the regenerates.

TABLE 2
SUMMARY OF SOMACLONAL VARIANTS
RESULTING FROM MITOTIC CROSSING-OVER

Type	Number of plants		Genotype	
A	42	Yv/yv	Aps-1(+/1)	C/c
B	2	Yv/yv	Aps-1(+/1)	c/c
C	2	Yv/yv	Aps-1(+/1)	C/C
D	4	Yv/Yv	Aps-1(+/1)	C/c
E	11	Yv/Yv	Aps-1(+/1)	C/C

Based on segregation patterns in the subsequent (R_1)
generation, 42 regenerates (type A) were unchanged with
respect to the three marker loci, yv, Aps-1, and c. Fifteen
regenerated plants (D and E types) contained two copies of
the wild type allele for yellow virescence. However, no

yv/yv combinations were found among the Ro plants. Two
regenerates (type B) were homozygous for the recessive potato
leaf trait (c/c), while thirteen plants were homozygous for
the dominant wild type allele (type C). In addition, four
regenerates had changes at both the yv and c loci (type E).

DISCUSSION

Mitotic crossing-over has been suggested as a source of
tissue culture derived somaclonal variation (8,10). Since
organogenesis originates from a single cell, a somatic recombination event will change the genetic make-up of plants
which originate from that cell. Therefore, it may be possible to use tissue culture to generate a mitotic crossing-over
event to produce a tomato plant homozygous for nematode
resistance. Such an approach may break the linkage between
the Mi locus and closely linked undesirable traits, e.g.
reduced yield and soft fruit, that have apparently been
cotransferred from L. peruvianum.

Of the 51 plants examined, none exhibited the Aps-1$^{(1/1)}$
gene combination which would indicate the presence of both
nematode resistant alleles. However, a crossover event
between Aps-1 and c produced two regenerates (type B) homozygous for the potato leaf allele, c/c. Figure 1 diagrams the
mitotic crossing-over event leading to the production of such
plants. Loss of the c locus, associated with traits derived
from the wild species, may lead to improved agronomic habit
for these plants and their progeny.

The frequency of recombination between any two markers
on the same arm of the chromosome should reflect the map
distance between these two markers. Thus, the ratio of the
observed changes were unexpected. Although a single map unit
separates yv from Aps-1, it was possible to identify 15
mitotic crossing over events between these two traits in 51
plants. In fact, the number of recombinational events
observed between yv and Aps-1 is equal to the number between
Aps-1 and c, which is 69 map units away (11). Such a high
frequency of recombination suggests that the actual chromosome position of these two traits is much farther than a
single map unit. The absence of yellow virescent (yv/yv)
regenerates may result from in vitro selection in favor of
healthier, wild type shoot primordia (unpublished data, D.A.
Evans, 1985).

The frequency of meiotic recombination is suppressed in
regions adjacent to the centromere. Crossovers near the

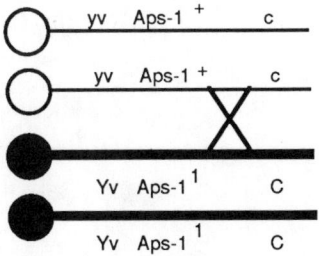

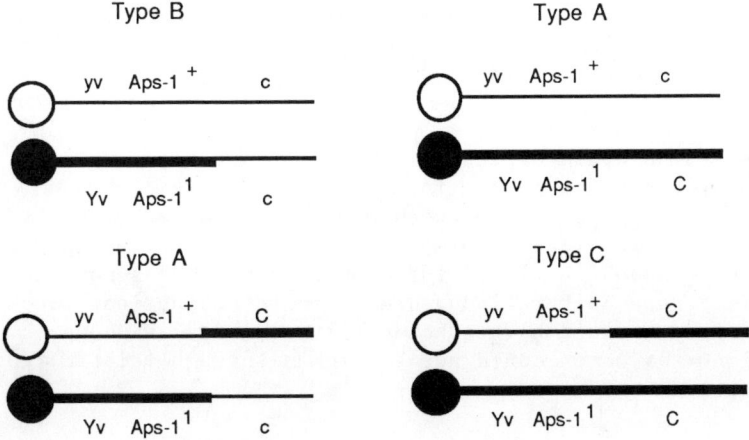

FIGURE 1. Mechanism of a single mitotic crossing-over event leading to production of type A, B, and C plants.

centromere leading to distorted spindle formation and suppression of recombination in this region would be evolutionarily advantageous. As shown in Figure 2, the yv, Mi, and Aps-1 loci are located near both the centromere and the proximal heterochromatin on chromosome 6L (12). Thus, calculations of the map distances between these loci near the centromere based on meiotic recombination rates are likely to be abnormally low.

Mitotic crossing-over takes place during chromosome pairing prior to DNA replication, and random pairing of the chromosomes would tend to occur at regions of homology. Since the heterochromatin surrounding the centromere contains mostly repetitive sequences, pairing would be favored in this region and other heterochromatic portions of the chromosome. Map positions calculated through estimates of mitotic recom-

Chromosome 6

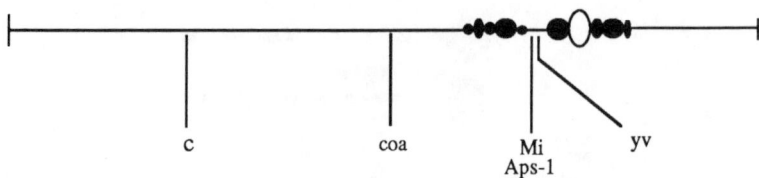

FIGURE 2. Position of marker loci on chromosome 6 in
relation to the centromere (open circle) and heterochromatin (closed circles).

bination would overestimate distances between markers located
near the heterochromatin.

Discrepancies between map distances generated by meiotic
versus mitotic recombination have been reported in yeast (13)
and Drosophila (14). The same phenomenon has now been found
in tissue culture induced mitotic recombination. Such an
approach would have great utility in breaking genetic
linkages which could not be broken through meiotic recombination.

REFERENCES

1. Siddiqui IA, Sher SA, French AM (1973). "Distribution of
    plant parasitic nematodes in California." Department of
    Food and Agriculture, Division of Plant Industry, Sacra-
    mento, California.
2. Bailey DM (1941). The seedling test method for root-knot
    nematode resistance. Proc Amer Soc Hort Sci 38:573.
3. Gilbert JC, Macguirre DC (1955). One major gene for
    resistance to severe galling from Meloidogyne incognita.
    Tomato Genet Coop Rep 5:15.
4. Barham WS, Winstead NN (1957). Inheritance of resistance
    to root knot nematodes. Tomato Genet Coop Rep 7:3.
5. Rick CM, Fobes J (1974). Association of an allozyme with
    nematode resistance. Tomato Genet Coop Rep 24:25.
6. Medina-Filho HP (1980). Linkage of Aps-1, Mi and other
    markers on chromosome 6. Tomato Genet Coop Rep 30:26.
7. Medina-Filho HP, Stevens MA (1980). Tomato breeding for
    nematode resistance: Survey of resistant varieties for
    horticultural characteristics and genotype of acid phos-

phatase. Acta Hort 100:383.

8. Evans DA, Sharp WR (1983). Single gene mutations in tomato plants regenerated from tissue culture. Science 221:949.

9. Medina-Filho HP, Tanksley SD (1983). Breeding for nematode resistance. In Evans DA, Sharp WR, Ammirato PV, Yamada Y (eds): "Handbook of Plant Cell Culture, Vol. I." New York: Macmillan Publ Co, p 904.

10. Evans DA, Sharp WR, Medina-Filho HP (1984). Somaclonal and gametoclonal variation. Amer J Bot 71:759.

11. Rick CM (1982). Linkage map of the tomato (Lycopersicon esculentum). Genetic Maps 2:360.

12. Khush GS, Rick CM (1968). Cytogenetic analysis of the tomato genome by means of induced deficiencies. Chromosoma 23:452.

13. Zimmermann FK, Vig BK (1975). Mutagen specificity in the induction of mitotic crossing-over in Saccharomyces cerevisiae. Mut Res 35:255.

14. Mather K (1939). Crossing over and heterochromatin in the X chromosome of Drosophila melanogaster. Genetics 24:413.

Molecular Strategies for Crop Protection, pages 375-392
© 1987 Alan R. Liss, Inc.

MOLECULAR DETAILS OF PLANT CELL WALL HYDROXYPROLINE-RICH GLYCOPROTEIN EXPRESSION DURING WOUNDING AND INFECTION[1]

Allan M. Showalter and Joseph E. Varner

Department of Biology, Washington University, St. Louis, MO   63130

ABSTRACT   Hydroxyproline-rich glycoproteins (HRGPs) are major structural components of plant cell walls, especially in dicotyledonous species. These glycoproteins accumulate in the cell wall in response to wounding and infection. HRGP cDNA and genomic clones from carrot and tomato were isolated in our laboratory and found to encode the repetitive pentapeptide sequence Ser-(Pro)$_4$. These repeat units are posttranslationally modified to Ser-(Hyp)$_4$ repeats which are characteristic of these wall-bound glycoproteins. The molecular characterization of the tomato cDNA and genomic HRGP clones is described here; the characterization of the carrot HRGP clones has been published previously. The carrot and tomato HRGP sequences were used as probes in RNA blot hybridization experiments to monitor HRGP mRNA levels in wounded carrot roots and tomato plants. These experiments demonstrated that wounding results in the accumulation of HRGP mRNA. Similar experiments were carried out in elicitor-treated bean cell suspension cultures and in race:cultivar-specific interactions between bean hypocotyls and the partially biotrophic fungus

[1]This work was supported by research grants from the National Science Foundation (PCM8104516), the Department of Energy (DE-FG 02-84ER13255), and by an unrestricted grant from the Monsanto Co.

Colletotrichum lindemuthianum. In these experiments, HRGP mRNA levels increased in response to
fungal elicitor and infection. Moreover, HRGP
mRNA accumulated earlier in an incompatible
interaction (host resistant) than in a compatible
interaction (host susceptible), and this accumulation was correlated with the expression of
hypersensitive resistance, indicating a role for
HRGPs in plant disease resistance. HRGPs may
function in plant defense by allowing for the
formation of a dense, impenetrable cell wall
barrier and/or by the immobilization of pathogens
in the cell wall by virtue of their behavior as
microbial agglutinins.

INTRODUCTION

Plants respond to stress in a variety of ways,
and these responses are crucial for their survival.
Wounding and pathogen infection are two of the many
stresses which plants encounter in the field. The
similarity of the responses which these two stresses
induce indicates that they are related to one
another. Specifically, mechanical wounding and
infection result in the accumulation of phytoalexins, deposition of lignin or lignin-like material, deposition of callose, accumulation of cell
wall hydroxyproline-rich glycoproteins (HRGPs),
increases in the activity of certain hydrolytic
enzymes such as chitinase and glucanase, accumulation of proteinase inhibitors, and alterations in
a number of other cellular metabolic pathways
(1,2). *A priori,* one might expect to find similarities in plant responses to these two stresses,
because several plant pathogens require wounding
before entry and because mechanical wounding of
plant tissue occurs during infection.
   During the past six years, our laboratory has
focused on the biochemistry and molecular biology of
plant cell wall HRGPs, also known as "extensins".
These glycoproteins have molecular weights of
approximately 100 kilodaltons and constitute a major
structural component of the wall, particularly in
dicotylendous plants (3,4). The abundance of
hydroxyproline (Hyp), serine, valine, tyrosine,

lysine, and histidine and the repeated occurrence of the pentapeptide Ser-(Hyp)$_4$ further characterize these glycoproteins. Arabinose (largely in the form of tri- and tetra-arabinosides) and galactose are O-glycosidically linked to hydroxyproline and serine respectively and account for about 50-65% of the molecular weight.

Several investigations (5-8) have demonstrated an accumulation of HRGPs in response to wounding and infection and support a role for HRGPs in plant defense. In addition, artificial enhancement or suppression of HRGP levels correlates with increased and decreased levels of disease resistance respectively (9). Recently, our group has isolated and characterized cDNA and genomic clones from carrot (10,11) and tomato (12) and has used these clones to monitor HRGP mRNA levels following wounding and pathogen infection. These studies are presented below and discussed with respect to the role of HRGPs in plant defense.

RESULTS

Effect of wounding on HRGP mRNA levels

A carrot genomic clone for a cell wall HRGP has recently been isolated and characterized in our laboratory (11). This gene encodes a 306 amino acid polypeptide sequence which includes a putative signal peptide and 25 Ser-(Pro)$_4$ pentapeptide repeats. In order to monitor changes in HRGP mRNA levels in response to wounding, the carrot genomic clone was used as a hybridization probe in RNA blot hybridizations to unwounded and wounded carrot root RNA (Fig. 1). In this experiment, HRGP mRNA accumulated dramatically in response to wounding; this finding is consistent with the reported accumulation of HRGP in the cell walls of wounded carrot roots (5). Moreover, three HRGP mRNA species accumulate; while two of them (the 1.4 kb and 1.8 kb mRNA species) apparently represent transcriptional products from one gene, the identity of the large size transcript remains unknown.

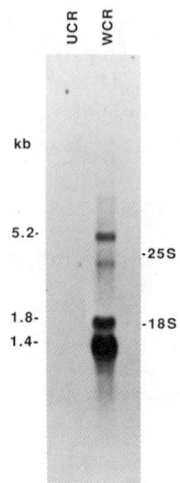

Figure 1.   RNA blot hybridization of unwounded
and wounded carrot root RNA to radiolabeled carrot
genomic HRGP sequences.   Lanes UCR and WCR contain 4
μg of poly A+ RNA isolated from unwounded and
wounded carrot roots respectively.   Wounding was
done by slicing the roots into 1 mm thick disks and
incubating these disks in a humid environment at
25°C for 24 hours.   The carrot HRGP probe used here
was the 4 kb *Eco* RI insert of the genomic clone
pDC5A1 and was labeled by nick-translation.   The
migration positions of carrot 25S and 18S rRNA and
the sizes of the major hybridizing transcripts are
indicated.

Using a carrot HRGP gene as a probe, we have also
isolated and characterized a tomato genomic clone
for HRGP by screening a tomato genomic library (12).
The sequence of this clone (Tom 5) appears in Fig.2
and encodes a polypeptide with numerous Ser-(Pro)$_4$
repeats which are frequently followed by Val-His or
Val-Ala.   This tomato HRGP sequence was used to
monitor HRGP mRNA levels in unwounded and wounded
tomato stem and leaf tissue (Fig. 3A).   Unlike the
wounding response in the carrot, wounding of tomato
stems resulted in the expression of a new and larger
size HRGP transcript (1.7 kb) concomitant with the
disappearance of two HRGP mRNA species (both about

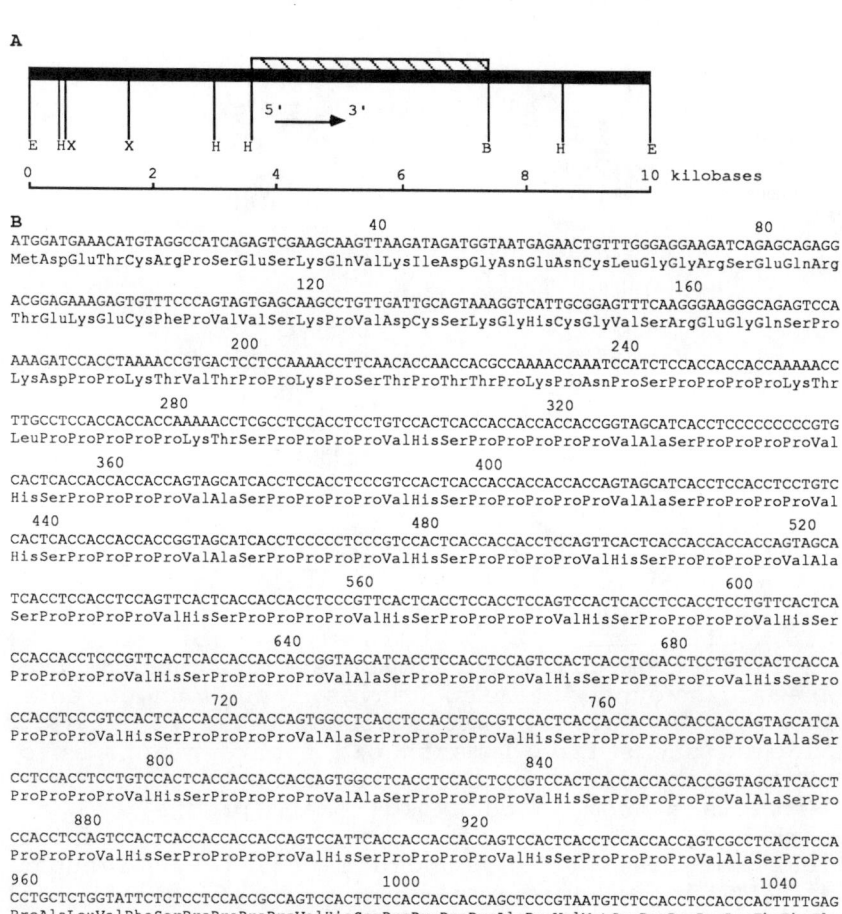

Figure 2. Restriction map (A) and DNA sequence (B) of a portion of the 10 kb *Eco* RI fragment of the tomato genomic HRGP clone Tom 5. The region of hybridization to the carrot HRGP gene sequence was determined by Southern blot analysis and is indicated by ▨▨▨▨▨. The putative direction of transcription and the region of DNA sequence shown in B are indicated by ———▶. Restriction sites for *Bam* HI (B), *Eco* RI (E), *Hind* III (H), and *Xba* I (X) are shown. Translation of this nucleotide sequence was performed using the first in-frame initiation codon 5' to the Ser-(Pro)$_4$ sequences.

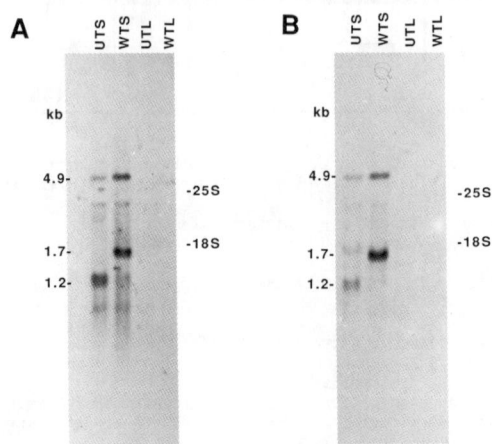

Figure 3.   RNA blot hybridization of unwounded
and wounded tomato RNA to radiolabeled tomato (A)
and carrot (B) genomic HRGP sequences.   Lanes UTS,
WTS, UTL, and WTL contain 10 μg of total RNA
isolated from unwounded tomato stems, wounded tomato
stems, unwounded tomato leaves, and wounded tomato
leaves respectively.   Stem wounding was performed by
slicing stems from 1 month old tomato plants into 3
mm sections; leaf wounding was done by punching out
leaf disks from 1 month old tomato plants with a #1
cork borer and using both the disks and the residual
leaf material.   The wounded material was incubated
at 25°C in a humid environment for 24 hours.   The
tomato (the 10 kb *Eco* RI fragment of Tom 5) and the
carrot (the 4 kb *Eco* RI insert of pDC5A1) HRGP
probes were labeled by nick translation.   The
migration positions of tomato 25S and 18S rRNA and
the sizes of the major hybridizing transcripts are
indicated.

1.2 kb) which were detected in the unwounded stems.
The accumulation of a particularly large size
transcript (4.9 kb), presumably analogous to that
seen in the carrot RNA blots, was also observed.
Leaf tissue had low levels of HRGP mRNA, even after
wounding; these data are consistent with the
abundance of wall-bound hydroxyproline in stem but

not leaf tissue (unpublished results). These hybridizations were also performed using the carrot genomic clone as a probe; they demonstrated a pattern identical to that of the tomato genomic clone (Fig. 3B).

In an effort to elucidate the nature of the wound-regulated changes observed in the tomato stems, we have constructed, screened, isolated, and partially characterized a number of cDNA clones from unwounded and wounded tomato stem poly A+ RNA. One of these cDNAs (pTom 17-1) isolated from the wounded cDNA library with the tomato and carrot HRGP gene clones, is approximately 450 base pairs in size and encodes Ser-(Pro)$_4$-Ser-Pro-Ser-(Pro)$_4$-(Tyr)$_3$-Lys hexadecapeptide repeat units (Fig. 4). These repeat units, or more precisely their hydroxylated analogs, have been previously identified by amino acid sequence analysis of tomato cell wall HRGP glyco-peptides generated by acid/protease hydrolysis of the wall (13). When this cDNA was used as a probe in RNA blots of unwounded and wounded tomato stems (Fig. 5), a pattern distinct from that of the genomic clones appeared. Here, the major hybrid-izing transcript was the large size transcript discussed above; this transcript increased approx-imately 3-fold in response to wounding in the stem and was barely detected in leaf tissue (data not shown), again consistent with protein accumulation and tissue distribution data.

Changes in HRGP mRNA levels in response to fungal elicitor and infection

In collaboration with Chris Lamb and his group at the Salk Institute and John Bailey at the Long Ashton Research Station in the United Kingdom, we have used the cloned genomic HRGP sequences as probes to monitor changes in the level of HRGP mRNAs in elicitor-treated bean cell suspension cultures and in race:cultivar-specific interactions between bean hypocotyls and the partially biotrophic fungus, *Colletotrichum lindemuthianum*, the causal agent of anthracnose. A previous report of this work appears elsewhere (12).

```
                 40                                              80
CCCATCACCTCCACCTCCCTACTACTACAAGTCTCCTCCACCACCATCGCCATCTCCTCCACCACCATACTACTACAAATCCCCACCA
 ProSerProProProProTyrTyrTyrLysSerProProProProSerProSerProProProProTyrTyrTyrLysSerProPro

                 120                                            160
CCACCATCCCCATCTCCTCCTCCACCATACTACTACAAATCACCACCTCCACCATCGCCATCTCCTCCTCCACCATACTACTACAAG
 ProProSerProSerProProProProTyrTyrTyrLysSerProProProProSerProSerProProProProTyrTyrTyrLys

                 200                                            240
TCTCCACCACCACCAGACCCATCTCCCCCACCACCATACTACTACAAGTCTCCTCCACCACCATCACCATCACCACCTCCACCATCA
 SerProProProProAspProSerProProProProTyrTyrTyrLysSerProProProProSerProSerProProProProSer

                 280                                            320
CCATCACCTCCCCCGCCCACTTACTCTTCTCCGCCACCACCACCACCATTTTACGAAAATATTCCTCTCCCACCGGTAATCGGAGTC
 ProSerProProProProThrTyrSerSerProProProProProProPheTyrGluAsnIleProLeuProProValIleGlyVal

                 360                                            400
TCCTACGCATCTCCACCACCACCAGTCATTCCATATTACTGAACAAGGGTTATCAATTAGCCTTTGATTTTCCATTATC
 SerTyrAlaSerProProProProValIleProTyrTyr*op
```

Figure 4. Nucleotide and deduced amino acid sequence of the tomato HRGP cDNA insert of pTom 17-1.

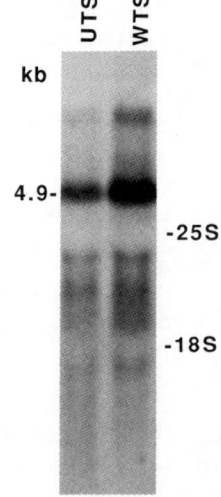

Figure 5. RNA blot hybridization of unwounded and wounded tomato stem RNA with the tomato HRGP cDNA pTom 17-1. Lanes UTS and WTS contain 10 µg of total RNA isolated from unwounded and wounded tomato stems respectively. Wounding was done as described in Figure 3. The HRGP cDNA insert of pTom 17-1 was labeled by nick translation and used as a probe. The migration positions of tomato 25S and 18S rRNA and the size of the major hybridizing transcript are indicated.

Elicitor was isolated as a high molecular weight fraction released by heat treatment of mycelial cell walls and incubated with bean cell cultures for various lengths of time ranging from 0 to 36 hours. RNA was isolated from these cultures and subjected to RNA blot hybridization analysis with the tomato HRGP gene clone (Tom 5) (Fig. 6) and the carrot gene clone (unpublished) with identical results. In particular, a marked accumulation of three HRGP transcripts of approximately 5.6, 2.7, and 1.6 kb in size was observed with increasing time of elicitor treatment. Untreated control cultures incubated over the same time period but without elicitor treatment failed to show this accumulation (data not shown). Similar experiments were carried out in soybean cell cultures treated with *Phytophthora megasperma* glucan elicitor in collaboration with Jürgen Ebel at Albert-Ludwigs-Universität in Freiburg with similar results (unpublished data).

RNA from bean hypocotyls infected with incompatible or compatible races of *Colletotrichum lindemuthianum* was also isolated and subjected to blot hybridization analysis with the tomato HRGP gene clone. This RNA was isolated at various times after fungal spore inoculation on the unwounded hypocotyl surface and from different sites on the hypocotyl as described in Fig. 7. In the incompatible interaction (host-resistant), HRGP mRNAs of 1.6 and 2.7 kb in size began to accumulate approximately 52 hours after inoculation; this lag period corresponds to the time required for spore germination, penetration of the cuticle, and initiation of the hypersensitive response (14). The accumulation kinetics of the 2.7 kb transcript is presented in Fig. 8A. An especially marked accumulation of the 2.7 kb (and 1.6 kb) transcript was observed in tissue directly underlying the site of spore inoculation (i.e. site 1); the level of hybridizable HRGP mRNA remained 10- to 20-fold higher than in equivalent uninfected control hypocotyls. Tissue distant from the site of spore inoculation (i.e. sites 2 and 3) also noticeably accumulated HRGP mRNA over control values.

In the compatible interaction (host-susceptible), the infected cells remain alive and the fungus undergoes substantial biotrophic growth. Subse-

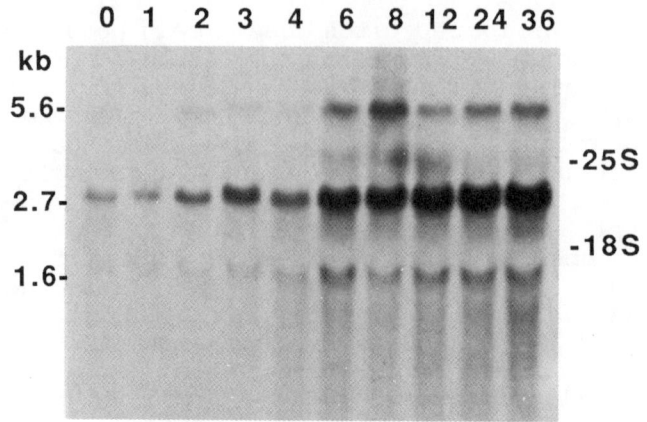

Figure 6.   RNA blot hybridization of elicitor-
treated bean cell RNA with the tomato HRGP genomic
clone (the 10 kb *Eco* RI insert of Tom 5).   Total RNA
was isolated from cells at the times (in hours)
following elicitor treatment as indicated; each lane
contains 15 µg of total cellular RNA.   The migration
positions of bean 25S and 18S rRNA and the sizes of
the major hybridizing transcripts are shown.

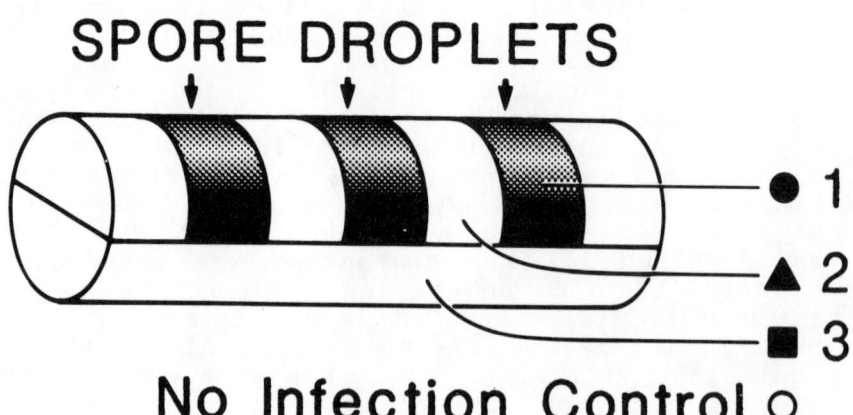

Figure 7.   Diagrammatic representation of an
infected bean hypocotyl with sites 1, 2, and 3 as
indicated.   Sites 1 and 2 were 10 mm in length.

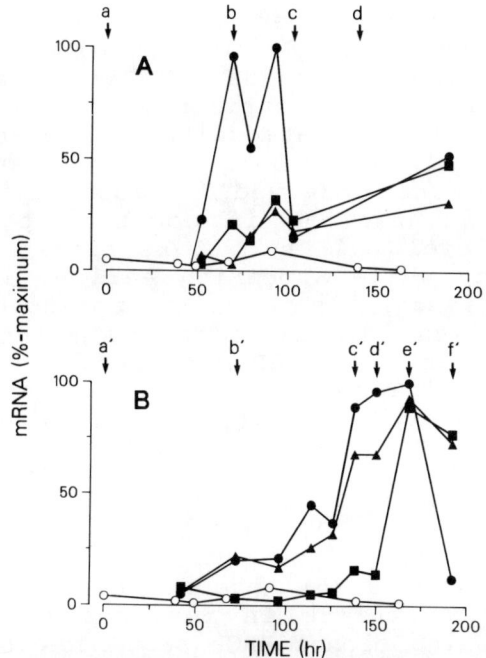

Figure 8. Kinetics of accumulation of the 2.7 kb HRGP mRNA species in bean hypocotyls in an incompatible (A) and compatible (B) interaction with two different races of *Colletotrichum lindemuthianum*. RNA was isolated from site 1 (●), site 2 (▲), site 3 (■) and equivalent uninfected hypocotyls (○). Arrows in A denote events in the expression of hypersensitive resistance at site 1: a, spore inoculation; b, onset of hypersensitive flecking in a few sites; c, hypersensitive flecking apparent at most sites; d, very dense brown flecking at all sites. No visible changes occurred in sites 2 and 3 and control hypocotyls throughout the time course. Arrows in B denote events in lesion development at site 1: a', spore inoculation; b', no visible symptoms (cf. incompatible interaction); c', onset of symptom development at a few sites; d', pale to medium brown lesions apparent at most sites; e', onset of water soaking and development of spreading lesions; f', extensive water soaking and spreading of lesions from site 1, some browning at site 2.

quently, extensive host cell death occurs and spreading anthracnose lesions develop (14). Here too, marked accumulation of HRGP mRNAs was observed but was delayed in comparison to the incompatible interaction and correlated with the onset of lesion formation(Fig. 8). Both the 1.6 and 2.7 kb transcripts were observed to accumulate in this compatible interaction in addition to weak but detectable levels of the 5.6 kb transcript. This accumulation, while appearing earlier in directly inoculated tissue, also occurred to especially high levels in tissue distant from the infection site. The maximum level of HRGP mRNA accumulation in the incompatible interaction (site 1, 93 hr.) was approximately 80% of the level attained in the compatible response (site 1, 168 hr.).

DISCUSSION

Previous studies (5) have demonstrated that cell wall HRGPs accumulate in response to mechanical wounding. Using cloned HRGP sequences as probes in RNA blot hybridizations, we have demonstrated that wounding in carrot roots (11) and tomato stems results in the increased accumulation of HRGP mRNA (Figs. 1,3,and 5). In carrot roots, this response involves accumulation of HRGP transcripts which are also present in unwounded tissues, albeit at relatively low levels. These low endogenous levels may reflect the subdued metabolic state of the carrot tissue after storage and in a relative sense enhance the degree of the wound response. In tomato stems, however, the wound response is different from that in carrot roots; a new, larger size transcript is produced in the wounded tissue concomitant with the disappearance of two smaller size transcripts present in the unwounded tissue. We have not yet determined whether this new transcript reflects accumulation from a newly transcribed gene or from differential processing of an HRGP gene(s) transcribed in the unwounded state. Both tissues investigated here contain a particularly large size HRGP transcript which accumulates during wounding and may reflect the accumulation of an unusually large and extremely tightly wall-bound HRGP as discussed below.

Chen and Varner (11) have previously reported on the characterization of the carrot genomic HRGP clone utilized in these experiments, while Showalter *et al.* (12) have previously described the partial characterization of a tomato HRGP gene clone (Tom 5). Here, additional information on the Tom 5 clone is presented and the characterization of a recently isolated tomato HRGP cDNA clone (pTom 17-1) is also described. In the case of the Tom 5 clone, its sequence encodes numerous Ser-(Pro)$_4$ repeats which are frequently followed by Val-His or Val-Ala. These dipeptide sequences have not yet been found on the protein level; however, Herrera-Estrella has recently observed such sequences in a potato tuber HRGP cDNA clone (personal communication), indicating that such sequences are transcribed. We have not yet been able to identify a signal peptide encoded by the Tom 5 sequence, and the possiblities remain that the signal peptide is still 5' of the sequenced region shown in Fig. 2 or that this gene truly lacks such a sequence. Currently, S1 nuclease mapping experiments with Tom 5 are being performed to determine whether this gene is transcribed. In contrast, the tomato cDNA sequence (pTom 17-1) encodes the hexadecapeptide sequence Ser-(Pro)$_4$-Ser-Pro-Ser-(Pro)$_4$-(Tyr)$_3$-Lys (Fig. 4), which has previously been elucidated by amino acid sequence analysis of tomato HRGP glycopeptides solubilized by acid/protease treatment of tomato cell walls (13). Smith *et al.* (15) have also isolated and sequenced three soluble tomato HRGPs; these sequences, while encoding Ser-(Hyp)$_4$ and some other repeating sequences, do not encode (Tyr)$_3$-Lys repeats. Thus, it appears that the Ser-(Pro)$_4$-Ser-Pro-Ser-(Pro)$_4$-(Tyr)$_3$-Lys repeats, or more correctly their post-translationally modified counterparts, are characteristic of a tightly wall-bound HRGP and, based on the transcript size observed in RNA blots with the cDNA probe (Fig. 5), are constituents of an exceptionally large HRGP. This interpretation would explain why this particular HRGP has not been solubilized; its participation in intermolecular cross-links, possibly by isodityrosine formation (16,17) between $(Tyr)_3$-Lys sequences, and/or its extensive interaction with other cell wall components over its long, repetitive sequence would render it insoluble in the wall.

Previous research (6-8) has also documented the accumulation of cell wall HRGPs in response to fungal infection, and in the case of cucumber plants, resistant cultivars accumulated hydroxyproline in their cell walls (indicative of HRGPs) more rapidly than susceptible cultivars. Our molecular analyses of elicitor-treated bean cell cultures and particularly of incompatible (host-resistant) and compatible (host-susceptible) interactions are consistent with these earlier findings. In these analyses, the accumulation of three different size HRGP transcripts is observed as illustrated in Fig. 6. Moreover, HRGP mRNA accumulation occurs more rapidly in the incompatible interaction than in the compatible one and is correlated with the onset of hypersensitive cell death (Fig. 8), indicating some role for these polypeptide products in disease resistance. Accumulation also occurs in tissue distant from the site of infection, indicating the presence of some intercellular elicitation signal. Such a signal may be involved with the phenomenon of induced systemic resistance and allow for the preactivation of defense mechanisms in tissue distant from the infection site. In the compatible interaction, HRGP mRNA accumulation may be the result of extensive tissue wounding as opposed to a direct host-pathogen response since accumulation here is correlated with the formation of disease lesions.

In addition to HRGP, a number of other proteins (referred to as defense response proteins) have been shown to accumulate during infection including phenylalanine ammonia-lyase (PAL) and chalcone synthase (CHS) (18). The mRNA accumulation patterns of PAL and CHS are broadly similar to those of HRGP but exhibit some differences which may reflect different regulatory circuits and induction schemes (18).

Lawton and Lamb (19) have recently demonstrated that mRNA accumulation of HRGP as well as PAL and

CHS is controlled at a transcriptional level. Specifically, they have performed nuclear run-off experiments which indicate that transcription of these genes is stimulated during wounding and infection of bean plants. The precise mechanisms of gene activation here, however, remain to be elucidated.

What is the role of HRGPs in wounding, infection, and plant disease resistance? While the answer to this question is not yet conclusive, the unique biochemical and physical properties of these molecules offer us several clues to their function(s). First, these glycoproteins are major structural components of the cell wall, which is one of the first barriers to pathogen ingress, and their accumulation during wounding and infection may serve to form a more dense and impenetrable structural barrier to potential pathogens. Additionally, Whitmore (20) has suggested that HRGPs could serve as nucleation sites for lignin deposition, again resulting in a dense structural barrier. These glycoproteins also have been shown to have the ability to agglutinate bacteria (21). Presumably this ability, which is also exhibited by polylysine, is attributable to the highly basic character of the protein backbone (HRGPs studied to date have isoelectric points in the range of 10-12 due to their high lysine content). Thus, the positively charged wall-bound HRGPs may serve to bind and immobilize pathogens in the cell wall and thereby prevent their entry into or between plant cells. In such a scenario, the HRGPs act in a manner analogous to an ion exhange resin.

It should, however, be noted that cell wall HRGPs are present in relatively high levels in unwounded and uninfected tissue and in our view were not originally or specifically designed for plant defense but rather adapted to it. We envision their primary function as a structural constituent of the cell wall and believe that it was only over the course of evolutionary time that this protein was adapted to play a secondary and somewhat general (i.e. non-specialized) role in plant defense. This view is supported to some degree by what has apparently occurred in *Chlamydomonas* in which an HRGP, different from the other cell wall HRGPs in this organism, now functions as a sexual agglutinin responsible for bringing together cells of opposite mating types (22).

The multigenic nature of HRGPs raises the question of whether one or more HRGPs are better adapted to wound healing and infection than others. While we do not yet have enough molecular or protein data to answer this question unequivocally, it is interesting to note the occurrence of HRGP polymorphism on the mRNA level in both wounding and infection. This polymorphism may indeed reflect functional differences in the resulting translation products allowing for fine tuning of the system to the particular biological stress situation. As more HRGPs and their genes are isolated and characterized, it will soon be possible to access their particular functions. In the case of disease resistance, this could be done by genetically engineering these genes so as to alter their expression either by manipulating regulatory sequences or by inactivation of transcripts with antisense RNA. In this way it should be possible to begin to dissect the contribution of HRGPs, or any plant defense response gene product, to disease resistance.

ACKNOWLEDGMENTS

We gratefully acknowledge the enormous contributions made by Susan Worst and Dat Ngo in the characterization of the tomato HRGP cDNA, by Jychian Chen in characterizing the carrot wound response, and by John Bell, Carole Cramer, John Bailey, and Chris Lamb in the fungal elicitor and infection studies. We also thank R. W. Breidenbach for his generous gift of a tomato genomic library and Cathy Hironaka and Dilip Shah for technical assistance in the construction of tomato cDNA libraries.

REFERENCES

1. Bell AA (1981). Biochemical mechanisms of disease resistance. Annu Rev Plant Physiol 32:21.

2. Sequeira, L (1983). Mechanisms of induced resistance in plants. Ann Rev Microbiol 37:51.
3. Cooper JB, Chen JA, Varner JE (1984). The glycoprotein component of plant cell walls. In Dugger WM, Bartnicki-Garcia S (eds): "Structure, Function and Biosynthesis of Plant Cell Walls," Rockville, MD: American Society of Plant Physiologists, p 75.
4. Lamport DTA, Catt JW (1981). Glycoproteins and enzymes of the cell wall. In Tanner W, Loewus FA (eds): "Encyclopedia of Plant Physiology," New York: Springer-Verlag, Vol 13B, p 133.
5. Sadava D, Chrispeels MJ (1978). Synthesis and secretion of cell wall glycoprotein in carrot root disks. In Kahl G (ed): "Biochemistry of Wounded Plant Tissues," Berlin: Walter de Gruyter & Co, p 85.
6. Esquerré-Tugayé MT, Lamport DTA (1979). Cell surfaces in plant-microorganism interactions. Plant Physiol 64:314.
7. Hammerschmidt R, Lamport DTA, Muldoon EP (1984). Cell wall hydroxyproline enhancement and lignin deposition as an early event in the resistance of cucumber *Cladosporium cucumerinum*. Physiol Plant Pathol 24:43.
8. Bolwell GP, Robbins MP, Dixon RA (1985) Metabolic changes in elicitor-treated bean cells. Eur J Biochem 148:571.
9. Esquerré-Tugayé MT, Lafitte C, Mazau D, Toppan A, Touzé A (1979). Cell surfaces in plant-microorganism interactions. Plant Physiol 64:320.
10. Chen J, Varner JE (1985). Isolation and characterization of cDNA clones for carrot extensin and a proline-rich 33-kDa protein. Proc Natl Acad Sci USA 82:4399.
11. Chen J, Varner JE (1985). An extracellular matrix protein in plants: characterization of a genomic clone for carrot extensin. EMBO J 4:2145.
12. Showalter AM, Bell JN, Cramer CL, Bailey JA, Varner JE, Lamb CJ (1985). Accumulation of hydroxyproline-rich glycoprotein mRNAs in response to fungal elicitor and infection. Proc Natl Acad Sci USA 82:6551.

13. Lamport DTA (1977). Structure, biosynthesis and significance of cell wall glycoproteins. In Loewus FA, Runeckles VC (eds): "Recent Advances in Phytochemistry," New York: Plenum Press, Vol II, p 79.
14. Bailey JA (1982). Physiological and biochemical events associated with the expression of resistance to disease. In Wood RKS (ed): "Active Defense Mechanisms in Plants," New York: Plenum Press, p 39.
15. Smith JJ, Muldoon EP, Willard JJ, Lamport DTA (1986). Tomato extensin precursors P1 and P2 are highly periodic structures. Phytochem, in press.
16. Fry SC (1982). Isodityrosine, a new cross-linking amino acid from plant cell-wall glycoprotein. Biochem J 204:449.
17. Cooper JB, Varner JE (1983). Insolubilization of hydroxyproline-rich cell wall glycoprotein in aerated carrot root slices. Biochem Biophys Res Comm 112:161.
18. Lamb CJ, Bell JN, Cramer CL, Dildine SL, Grand C, Hendrick SA, Ryder TB, Showalter AM (1985). Molecular response of plants to infection. In Augustine PC, Danforth HD, Bakst MR (eds): "Biotechnology for Solving Agricultural Problems," Totowa, NJ: Rowman & Allanheld.
19. Lawton MA, Lamb CJ (1986). Transcriptional activation of plant defense genes by fungal elicitor and infection. Manuscript submitted.
20. Whitmore FW (1978). Lignin-protein complex catalyzed by peroxidase. Plant Sci Lett 13:241.
21. Leach JE, Cantrell MA, Sequeira L (1982). A hydroxyproline-rich bacterial agglutinin from potato: its localization by immunofluorescence. Physiol Plant Pathol 21:319.
22. Goodenough UW (1985). An essay on the origins and evolution of eukaryotic sex. In Halvorson HO, Monroy A (eds): "The Origin and Evolution of Sex," New York: Alan R Liss Inc, p 123.

Molecular Strategies for Crop Protection, pages 393–400
© 1987 Alan R. Liss, Inc.

PHYSICAL AND FUNCTIONAL CHARACTERIZATION OF A GENE
FROM POTATO ENCODING PROTEINASE INHIBITOR II

Jose Sanchez-Serrano, Michael Keil
Jeff Schell and Lothar Willmitzer

Max-Planck-Institut für Züchtungsforschung,
5000 Köln 30, FRG

ABSTRACT Proteinase inhibitor II encoding cDNA's and
genomic clones have been isolated from Solanum tuberosum.
The physical structure of the genomic clone was determined by a comparison of the sequences of both the genomic as well as the cDNA revealing the presence of one
intron in the genomic clone. In the 5'-upstream and the
3'-downstream region the genomic clone displays typical
features of an eucaryotic gene. The expression of the
proteinase inhibitor II gene was found to be both under
developmental (being tuber-specific) as well as environmental (being wound-induced in leaves) control. The intact proteinase inhibitor II gene after transfer into
tobacco using Agrobacterium/Ti-plasmid derived vectors
was found to be wound-inducible in leaves of transgenic
tobacco plants.

INTRODUCTION

Protein inhibitors of proteases represent essential compounds found in a wide range of tissues of animals, plants and
microorganisms. Whereas there can be no doubt about their gross
physiological function, i.e. preventing unwanted proteolysis,
their detailed physiological function is rarely understood.
In plants proteinaceous proteinase inhibitors are most com-
monly found in seeds being especially well studied in a number of Leguminoseae and Gramineae (1). Another rich source of
proteinase inhibitors are the tubers especially of potato.
Data available in the literature indicate at least seven
different types of proteinase inhibitors as based on differences in molecular weights, N- and C-terminal amino acids or total amino acid composition (2). Concerning their physiological
function proteinase inhibitors of plants are thought to serve
a role in the defence strategy of plants against insect or

microbial attack by inhibiting the respective proteinases.
This assumption is supported by the following three observations:
- Plant proteinase inhibitors inhibit the proteolytic activity of a wide range of microorganisms and insects but only
  rarely inhibit proteinases of plant origin (1).
- Plant proteinase inhibitors preferentially accumulate in
  organs important for either vegetative or sexual propaga-
  tion. The expression of proteinase inhibitors is therefore
  in most cases under developmental control.
- The expression of at least some proteinase inhibitors can
  be induced in non-storage organs by wounding, mimicing the
  attack of a chewing insect (3).

    In addition to their important biological role, protein-
ase inhibitors therefore also represent a very interesting
system of differentially expressed genes of higher plants being both developmentally as well as environmentally controlled.
We have therefore recently started a program in order to isolate and characterize proteinase inhibitor II encoding genes
of potato with the aim to understand both their biological
function as well as the mechanism controlling the expression
of this gene family.

RESULTS

Isolation and Characterization of cDNA and Genomic Clones Encoding Proteinase Inhibitor II of Potato

    The screening of about 5000 cDNA clones prepared from
tuber A$^+$-RNA led to the isolation of three cDNA clones encoding proteinase inhibitor II. The two longer clones containing inserts of 512 and 571 nucleotides respectively were
characterized by sequencing. The results are shown in figure 1.
    cDNA 1 contains an open reading frame of 462 nucleotides
allowing for a protein of 154 amino acids. This open reading
frame is partially shared by cDNA 2 which however shows an
extension at the 3'-end all through the 3'-untranslated region
into the poly-A tail. In contrast cDNA 1 shows an extension
all into the 5'-untranslated region ending only three nucleotides in front of the transcription initiation site determined for the genomic clone (cf. below). In the overlapping
part the cDNAs differ in four positions by single nucleotide
exchanges indicating that in tubers of the potato variety
used (HH80 1201/7, a diploid line of the collection of the
Max-Planck-Institut, Köln) probably at least two genes are
expressed.

```
  1                          ACCCCAAAATTAAAAGAAAAAGAGGCAGTACTAATTAATTATCCATC    cDNA 1

 48       ATGGATGTTCACAAGGAAGTTAATTTCGTTGCTTACCTACTAATTGTTCTTGGATTATTG    cDNA 1

          MetAspValHisLysGluValAsnPheValAlaTyrLeuLeuIleValLeuGlyLeuLeu

108       GTACTTGTAAGCGCGATGGAGCATGTTGATGCGAAGGCTTGCACTTTAGAATGTGGTAAT    cDNA 1
          ***********************************************************    cDNA 2
          ValLeuValSerAlaMetGluHisValAspAlaLysAlaCysThrLeuGluCysGlyAsn

168       CTTGGGTTTGGGATATGCCCACGTTCAGAAGGAAGTCCGGAAAATCGCATATGCACCAAC    cDNA 1
          ****************************************C******************    cDNA 2
          LeuGlyPheGlyIleCysProArgSerGluGlySerProGluAsnArgIleCysThrAsn
                                                            Pro

228       TGTTGTGCAGGTTATAAAGGTTGCAATTATTATAGTGCAAATGGGGCTTTCATTTGTGAA    cDNA 1
          ***********************************************************    cDNA 2
          CysCysAlaGlyTyrLysGlyCysAsnTyrTyrSerAlaAsnGlyAlaPheIleCysGlu

288       GGACAATCTGACCCAAAAAAACCAAAAGCATGCCCCCTAAATTGCGATCCACATATTGCC    cDNA 1
          *******************************************T***************    cDNA 2
          GlyGlnSerAspProLysLysProLysAlaCysProLeuAsnCysAspProHisIleAla

348       TACTCAAAGTGTCCCCGTTCAGAAGGAAAATCGCTAATTTATCCCACCGGATGTACCACA    cDNA 1
          ***********************************************************    cDNA 2
          TyrSerLysCysProArgSerGluGlyLysSerLeuIleTyrProThrGlyCysThrThr

408       TGCTGCACAGGGTACAAGGGTTGCTACTATTTCGGTAAAAATGGCAAGTTTGTATGTGAA    cDNA 1
          ********T********************A*****************************    cDNA 2
          CysCysThrGlyTyrLysGlyCysTyrTyrPheGlyLysAsnGlyLysPheValCysGlu
                                             Ser

468       GGAGAGAGTGATGAGCCCAAGGCAAATATGTACCCTGCAATGTGA                 cDNA 1
          ********************************************CCCTAGACTTGTCCA    cDNA 2
          GlyGluSerAspGluProLysAlaAsnMetGlyProAlaMetEnd

528                                                    TCTTCTGGATTGGCCAACTTAATTAATGTATGAAATAAAAGGATGCACACATAGTGACAT    cDNA 2

588       GCTAATCACTATAATGTGGGCATCAAAGTTGTGTGTTATGTGTAATTACTAGTTATCTGA    cDNA 2

648       ATAAAAGAGAAAGAGGTCATCCATATTTCTTTTCCTAAAAAAAAAAAAAAAA          cDNA 2
                                                             701
```

FIGURE 1.   Nucleotide sequence and deduced amino acid sequence of two proteinase inhibitor II encoding cDNAs derived from potato tuber poly A$^+$ RNA. Conserved sequences between both cDNAs are indicated by asterik. The putative polyadenylation signal is underlined.

In the 3'-untranslated region of cDNA 2 the sequence AATAAA is found 38 nucleotides in front of the poly A addition site. This sequence is in complete agreement with the consensus sequence AATAAA found 25-30 nucleotides in front of the poly-A addition site in most eucaryotic genes.

The N-terminal segment (amino acids 1-30) of the protein-coding region of cDNA 1 sticks out in a hydropathy plot due to its high hydrophobicity. It displays typical features of a signal peptide containing a hydrophobic core which is flanked by more hydrophilic and charged regions (4).

The cDNA clones described above were used for screening a genomic library comprising the nuclear genome of the haploid potato line AM 80/5793 (derived from the collection of the Max-Planck-Institut) which was established in the lambda phage replacement vector EMBL 4. Four independent recombinant clones derived from two different loci were obtained by screening about 800.000 plaques. One of these clones was chosen for further characterization because under stringent conditions it showed the strongest hybridization to the cDNA clones.

The primary structure of this proteinase inhibitor II gene was determined by sequencing. The comparison of this sequence with the cDNA sequences revealed the presence of one short intron of 117 nucleotides length (cf. figure 2).

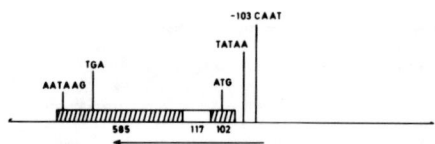

FIGURE 2.   Structure of a proteinase inhibitor II gene. Black boxes indicate the two exons, the white box indicates the intron. The direction of transcription is given by the arrow.

The genomic sequence contains an open reading frame of 462 nucleotides which therefore is in agreement with the cDNA data. The coding region as well as the 5'-untranslated leader of cDNA 1 agrees completely with the genomic clone, whereas there are seven single nucleotide exchanges as well as a deletion of three nucleotides in the genomic clone compared to cDNA 2 (5). The transcription initiation site of the genomic clone was determined by RNAse protection experiments and found

to be located at a distance of 50 nucleotides in front of the
ATG serving as the translation initiation site (5). The 5'-
upstream region of the gene contains the sequence TATAAA at
the position -26 and the sequence CAAAT at position -103.
In the 3'-region the sequence AATAAG is located 33 nucleotides
in front of the poly-A site.

Expression of Proteinase Inhibitor II Genes Is Under Both Developmental as well as Environmental Control

     The proteinase inhibitor II encoding cDNAs described
above were isolated from a collection of cDNA clones established from tuber A$^+$-RNA. Northern experiments performed with
RNA isolated from leaves, stems, roots, tubers as well as developing seeds of potato plants grown in the green-house
showed that under these conditions the expression of proteinase inhibitor II genes is under strict developmental control.
RNA hybridizing to the proteinase inhibitor II cDNA's could
only be detected in tubers of these potato plants (4,6).
     As outlined in the introduction, the expression of at
least some proteinase inhibitors has been shown to be inducible in non-storage organs of the plant by e.g. wounding (3).
Northern experiments performed using RNA from wounded and non-
wounded leaves of potato showed that the expression of proteinase inhibitor II genes can be induced in leaves of potato
plants after mechanical wounding too (4).
     The developmental control of the expression of the pro-
teinase inhibitor II can therefore be overcome by a drastic
change in the environmental conditions such as wounding. Most
interestingly however, a reciprocal situation is found in tubers of potato, where the proteinase inhibitor II is normally
expressed. Cutting potato tubers into slices and incubating
them for 16 hours at 26°C leads to a complete disappearance
of RNA homologous to the proteinase inhibitor II (4). Thus the
expression of the proteinase inhibitor II genes in potato
seems to be controlled by a rather complex interplay of both
developmental as well as environmental conditions.

Intact Proteinase Inhibitor II Gene Is Expressed in Leaves
of Transgenic Tobacco Plants after Wounding

     In order to test whether the genomic clone encoding pro-
teinase inhibitor II from potato described above contains all
signals necessary for the induction of its expression by wounding, the gene as well as 3.0 kb of its 5'-upstream and 1.45 kb
of its 3'-downstream sequences were introduced into tobacco

using Agrobacterium/Ti-plasmid mediated gene transfer. Southern analysis of the transformed and regenerated plants showed the presence of at least one copy of the intact gene, whereas no sequences homologous to the proteinase inhibitor II gene can be detected in nontransformed tobacco plants.

RNA isolated from unwounded leaves of seven independent transgenic tobacco plants did not show any detectable homology to the proteinase inhibitor II gene in a Northern type experiment, whereas an RNA of 800 nucleotides, i.e. of the expected length, clearly could be detected in two out of the seven tobacco plants after mechanical wounding of the leaves. These data therefore demonstrate that the cis-regulatory sequences necessary for wound-induction of this proteinase inhibitor II gene of potato are all contained on the genomic fragment transferred. In addition these sequences must be recognized by (probably) trans-acting factors present in a heterologous plant (tobacco) which does not contain the corresponding gene in its own genome.

DISCUSSION

The gene family encoding proteinase inhibitor II of potato represents an exciting system for studying differential gene expression in higher plants. The expression of the proteinase inhibitor II genes is under the control of both developmental as well as environmental factors. In green-house grown potato plants, RNA homologous to the proteinase inhibitor II gene is only found in tubers of the potato plant. Thus under these conditions the expression of the gene is under strict developmental control. Upon a drastic change in the environmental conditions such as mechanical wounding of leaves (mimicing pest attack), proteinase inhibitor II RNA accumulates in the leaves of the potato plant. The same change in the environment (wounding) however leads to the disappearance of the RNA in potato tubers.

As proteinase inhibitor II is encoded by a small gene family (6), its differential expression might be due to activation of different members of this gene family under the different conditions described. Another possibility would be that there are at least two different cis-acting regulatory regions in the surrounding of each gene which respond to its activation by either wounding or tuberization. The whole situation is even furthermore complicated by 'run-off'-transcription experiments performed in isolated nuclei of different organs of a non-wounded potato plant which indicate that at least the developmental-specific expression of this gene fa-

mily might be also controlled at a post-transcriptional level (6).

As indicated in the results section the N-terminal part
of the proteinase inhibitor protein displays typical features
of a signal peptide. A vacuolar localization of the corresponding proteinase inhibitor II from tomato as well as for the
carboxypeptidase inhibitor from potato tubers has been demonstrated (7,8). It therefore seems to be reasonable to speculate that the N-terminal segment of the proteinase inhibitor
II of potato might serve for co-translational uptake into the
endoplasmic reticulum from where it might be sequestered into
the vacuole.

The deduced amino acid sequence of potato proteinase inhibitor II shows 100% homology to the amino acid sequence of
a small proteinase inhibitor from potato tubers, PCI-I (9).
It has already been speculated earlier that PCI-I might be
derived from proteinase inhibitor II by a post-translational
process (10). This possibility is furthermore supported by
our failure to detect an RNA of 400-500 nucleotides length
(which would be the expected size for the 50 amino acid long
PCI-I peptide) in the RNA population of potato tubers (4).

The intact potato proteinase inhibitor II gene described
was transferred into tobacco plants using Agrobacterium mediated gene transfer techniques. In several transgenic tobacco
plants analyzed the expression of the proteinase inhibitor
II gene was found to be wound-inducible. This result indicates
that although tobacco does not contain an endogeneous proteinase inhibitor II gene the signal used for wound-induction of
the potato derived gene must be well conserved in between at
least these two very related plant species. Further experiments
using different fragments of the proteinase inhibitor II gene
will hopefully lead to the identification of the cis-regulatory sequences responsible for its environmentally induced expression.

ACKNOWLEDGEMENTS

We thank Wolfgang Schmalenbach and Eileen O'Connor San-
chez for expert technical assistance and Jutta Freitag for edi-
ting the manuscript.

REFERENCES

1.   Richardson M (1976). The proteinase inhibitors of plants
     and micro-organisms. Phytochemistry 16:159.
2.   Laskowski M, Kato I (1980). Protein inhibitors of protein-
     ases. Ann Rev Biochem 49:593.

3. Green TR, Ryan CA (1972). Wound-induced proteinase inhibi-
    tor in plant leaves: a possible defense mechanism against
    insects. Science 175:776.
4. Sanchez-Serrano J, Schmidt R, Schell J, Willmitzer L
    (1986) Nucleotide sequence of proteinase inhibitor II en-
    coding cDNA of potato (Solanum tuberosum) and its mode of
    expression. Mol Gen Genet 203:15.
5. Keil M, Sanchez-Serrano J, Schell J, Willmitzer L (1986).
    Primary structure of a proteinase inhibitor II gene from
    potato (Solanum tuberosum). Submitted.
6. Rosahl S, Eckes P, Schell J, Willmitzer L (1986) Organ-
    specific gene expression in potato: isolation and charac-
    terization of tuber-specific cDNA sequences. Mol Gen Genet
    202:368.
7. Walker-Simmons M, Ryan CA (1977) Immunological identifi-
    cation of proteinase inhibitors I and II in isolated to-
    mato leaf vacuoles. Plant Physiology 60:61.
8. Hollaender-Czytko H, Anderson JK, Ryan CA (1985). Vacuolar
    localization of wound-induced caboxypeptidase inhibitor
    in potato leaves. Plant Physiology 78:76.
9. Hass G, Hermodson M, Ryan CA, Gentry L (1982). Primary
    structures of two low molecular weight proteinase inhibi-
    tors from potatoes. Biochemistry 21:752.
10. Pearce G, Sy L, Russell C, Ryan C-A, Hass GM (1982) Iso-
    lation and characterization from potato tubers of two poly-
    peptide inhibitors of serine proteinases. Arch Biochem
    Biophys 213:456.

Molecular Strategies for Crop Protection, pages 401–413
© 1987 Alan R. Liss, Inc.

ANTIOXIDATIVE SYSTEMS: DEFENSE
AGAINST OXIDATIVE DAMAGE IN PLANTS[1]

Arno Schmidt and Karl Josef Kunert[2]

Lehrstuhl für Physiologie und Biochemie der Pflanzen,
Universität Konstanz, D-7750 Konstanz, W-Germany

ABSTRACT Different lines of our current research
activities are presented related to improve tolerance
of plants against oxidative damage by manipulation of
antioxidative protector systems. One line is the direct treatment of plants either with the antioxidant
vitamin C or with L-galactono-1,4-lactone, the biosynthesis precursor of vitamin C. Both compounds
protect against peroxidation of lipids. A second
line is the function of glutatione reductase in
plants. This enzyme prevents oxidation of the antioxidant glutathione. A third line is related to manipulation of the enzyme activity by gene amplification.
First results show that transfer of the glutathione
reductase gene results in tolerance against oxidative
stress in oxidant sensitive Escherichia coli cells.

INTRODUCTION

Natural aging and several xenobiotics, such as
certain diphenyl-ether herbicides, bipyridylium salts or
air pollutants, have been found to induce oxidation of
cell components in plant systems either directly, like
the pollutant ozone (1), or via free radical reactions.
Among radical reactions toxic to plants, peroxidation of
polyunsaturated fats has been identified (2, 3, 4, 5).

[1]Supported by Deutsche Forschungsgemeinschaft.
[2]Present address: Department of Virus Research,
John Innes Institute, Norwich NR4 7UH, England.

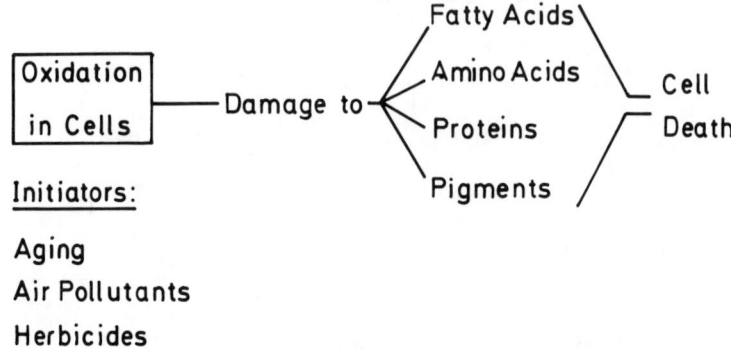

FIGURE 1.   Initiation of oxidation in cells by aging,
air pollutants, or herbicides and some of the phytotoxic
consequences observed.

If free radicals are not removed from the cell by reaction
with an antioxidative protector system, neighboring compounds, like genetic material, polyunsaturated fats, or
pigments, are rapidly damaged (6, 7, 8). Further, destruc-
tion of proteins by polymerization, chain scission of
polypeptides, and oxidation of amino acids is well-known
as a consequence of lipid peroxidation (9, 10). Figure 1
summarizes some of the phytotoxic reactions observed.
        In the subsequent chapters, a summary is given of our
current research activities related to biological protector
systems that limit the action of toxic oxidants. Finally,
we shortly present our future research strategy to improve
crop protection against oxidative damage by manipulation
of antioxidative protector systems.

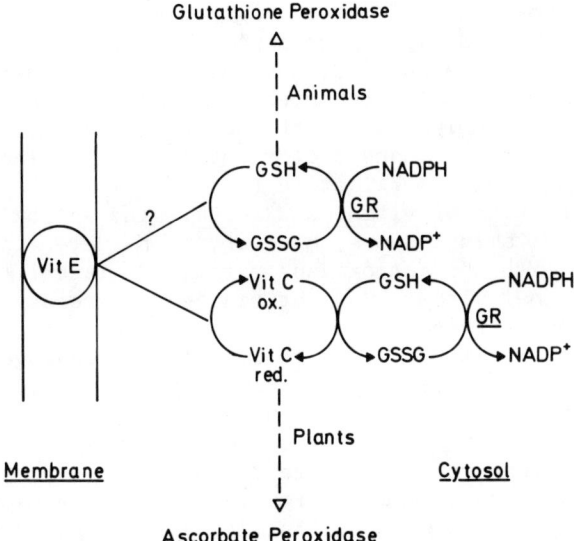

Glutathione Peroxidase

FIGURE 2.   Proposed cellular interaction of vitamin E
with vitamin C and glutathione (18, 19) to prevent peroxidation of lipids. Glutathione is reduced by the enzyme
glutathione reductase (GR). Hydroperoxides formed in cells
can be removed by glutathione peroxidase (animals) or
ascorbate peroxidase (plants).

RECENT RESULTS AND DISCUSSION

Vitamins C and E.

Protection by vitamins E and C.   Protection against
harmful oxidant reactions is provided by an array
of enzymes, such as superoxide dismutase, catalase, or
peroxidase, and a variety of small molecules, like the
vitamins E and C or glutathione (11, 12). A proposed
pathway for the antioxidant reactions partly under study
in our group is shown in Figure 2. Both vitamin C and
vitamin E are general cell constituents. In plants, a
high concentration of the lipid-soluble vitamin E has
been found in the chloroplast envelope and the osmiophilic
plastoglobuli of the plastid stroma (13). There is strong
evidence that vitamin E is the most effective radical

chain-breaking substance in biomembranes (14). Several
research groups have found strong protection against
herbicide-induced lipid peroxidation by treatment of plant
material with vitamin E (8, 15, 16). For the water-soluble
vitamin C, we recently reported a significant correlation
between the vitamin content of mustard seedlings and
herbicide-induced ethane evolution, a specific indicator
of linolenic acid peroxidation (17).

Interaction of vitamins E and C. The antioxidative
action of vitamin E seems, however, to be determined by a
synergism with the antioxidant vitamin C. Tappel (18)
suggested that vitamin C might reduce tocopheroxyl radicals formed by free-radical reactions. Therefore, one
molecule of vitamin E seems to be able to scavenge many
radicals derived from lipid peroxidation. This suggestion
has been confirmed by other research groups using in vitro
test systems, and data obtained were recently published
in an excellent review article by McCay (19).

In our group, we investigated in vivo interaction of
the vitamins E and C in plants during leaf aging (5) and
after treatment with a p-nitrodiphenyl-ether herbicide
(20). In general, the vitamin C concentration in plants
was higher than the vitamin E concentration, and vitamin C
seems to provide a reductive pool to regenerate vitamin E.
Both a substantial amount of the vitamins and a concentration ratio of vitamin C to vitamin E between 10 and 15
to 1 (wt/wt) were necessary to effectively protect plants
against diphenyl ether-induced peroxidation of lipids.
However, oxidative cell damage dramatically increased
in the presence of a much lower or a much higher concentration ratio. During leaf aging, the concentration
ratio of the vitamins declined. This decline was much
higher in leaves with a short life span than in leaves
with a long life span. Further, the decrease of the
concentration ratio was directly related to higher lipid
peroxidation.

Manipulation of the vitamin C level. Since oxidation
of cell components is toxic for plants, we expected protection against oxidative damage by increased endogenous
levels of antioxidants. By now, we are not successful to
enhance significantly the vitamin E level in plants by
either spraying vitamin E or applying different biosynthesis precursors of vitamin E.

However, experiments were recently performed in our
group in which excised cotyledons of morning glory

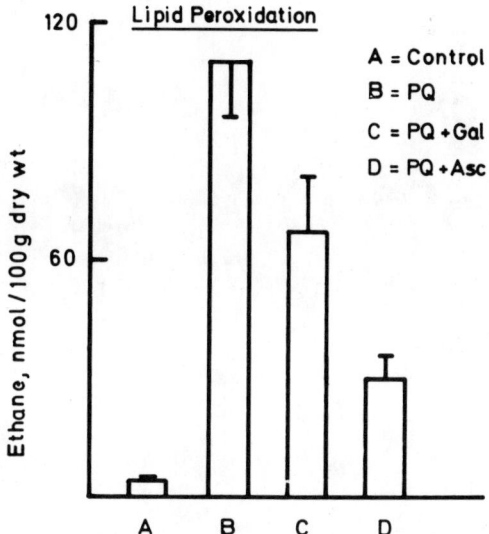

FIGURE 3. Ethane evolution, as an index of <u>in vivo</u>
lipid peroxidation, of green morning-glory cotyledons 18 h
after treatment with 0.05 mM paraquat and without the herbicide (control) in the light. To show protection against
paraquat-induced peroxidation, either 20 mM galactonolactone or 20 mM sodium ascorbate was added to the incubation medium. Before addition of the herbicide, cotyledons
were pretreated for 6 h with the protectors in the dark.

(<u>Ipomoea purpurea</u>) were treated with a relative high amount
of either sodium ascorbate or L-galactono-1,4-lactone
(21), the direct precursor of vitamin C in plants (22).
Strong oxidation was initiated by the herbicide paraquat
(1,1'-dimethyl-4,4'-bipyridinium ion). A 6-fold higher
vitamin C level was found in the cotyledons after application of ascorbate or galactonolactone that protected
against paraquat-induced oxidation of lipids (Figure 3).
In our experiments ascorbate gave better protection than
galactonolactone. Lower protection capacity of the precursor compared to ascorbate was possibly dependent on the
activity of the enzyme L-galactono-1,4-lactone oxidase.

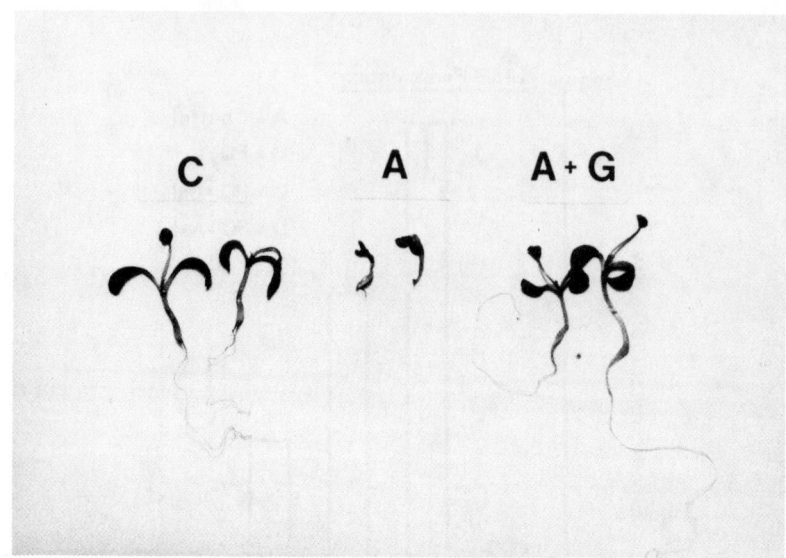

FIGURE 4. Influence of 0.5 µM acifluorfen alone (A)
and acifluorfen and 25 mM L-galactono-1,4-lactone together
(A +G) on growth of alfalfa seedlings, measured 7 d after
germination. For comparison, untreated control seedlings
(C) are shown. Cultivation was done in petri dishes on
filter paper circles.

This enzyme, which converts galactonolactone to vitamin C,
is inactivated by lipid hydroperoxides (23).
    To overcome inhibition of the galactonolactone oxidase
by lipid hydroperoxides , we used in a further study the
diphenyl-ether herbicide acifluorfen-sodium, sodium 5-[2-
chloro-4-(trifluoromethyl)phenoxy] -2-nitrobenzoic acid,
as an inducer of lipid peroxidation (15). This herbicide
is less potent than paraquat. Cultivation of alfalfa
seedlings (Medicago sativa) for 6 days in the presence of
either galactonolactone or ascorbate and acifluorfen,
growth was significantly inhibited by the herbicide
(Figure 4). Galactonolactone, however, gave strong protection against herbicidal action. There was no evidence
that ascorbate protected against growth inhibition.
Further, galactonolactone and acifluorfen-treated seedlings

had about a 20-fold higher vitamin C level than seedlings treated either with acifluorfen alone or acifluorfen and ascorbate.

Glutathione and Glutathione Reductase.

   Function and regulation of glutathione reductase. Glutathione reductase is present in cells of both plants and animals (24). In plants, we found maximal enzyme activity after germination, but enzyme activity rapidly declines with age. Glutathione reductase seems to be mainly located in the cytosol and catalyses NADPH-dependent reduction of oxidized glutathione (GSSG) to the reduced form (GSH) (24). A high GSH/GSSG ratio is necessary for several physiological functions. This includes activation and inactivation of redox-dependent enzyme systems (25), storage and transport of L-cysteine (26), detoxification of xenobiotics via conjugation (26), and regeneration of the cellular antioxidant vitamin C under oxidative conditions (27).

   Experiments were done in our group to investigate the role of glutathione and glutathione reductase in the antioxidative system of beans (Phaseolus vulgaris) (28). Glutathione synthesis, measured as total glutathione content (GSH + GSSG) of cells, and oxidation of acid-soluble SH-compounds, almost totally GSH, increased significantly after treatment of plants with the peroxidative herbicide acifluorfen. Both oxidation of GSH and enhanced glutathione synthesis indicate rapid accumulation of GSSG after herbicide treatment. When total glutathione content was at maximum, we found strong increase of glutathione reductase activity. As a consequence of higher enzyme activity, acid-soluble SH-compounds were regenerated and the total amount of glutathione declined. We propose that enhanced glutathione synthesis and activation of the enzyme glutathione reductase are possible mechanisms in biological systems to overcome oxidative stress. Our proposal is further supported by recent results of Burke et al. (29). In cotton, they observed resistance against the strong oxidative herbicide paraquat by high glutathione reductase activity induced by water deficit.

   Manipulation of glutathione reductase activity. To investigate the protective effect of glutathione reductase against oxidative damage to cells in more detail, we measured the growth of both different glutathione reductase

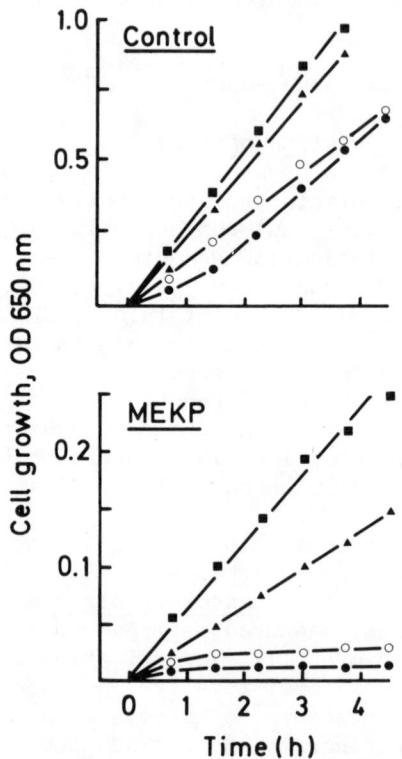

FIGURE 5. Growth of different strains of E. coli under oxidative conditions. Cells were cultivated until exponential growth (OD 650 nm = 0.1), and MEKP was added to the medium with a final concentration of 0.065%. K38 (■) and JA200/8b (▲) represent gor(+) strains; SG9 (○) and JM2267 (●) gor(-) strains.

negative (gor-) and different positive (gor+) strains of Escherichia coli in the presence of methyl ethyl ketonperoxide (MEKP), a strong inducer of lipid peroxidation (30). Growth of the gor(-) strains were strongly inhibited by MEKP (Figure 5). The gor(+) strains, however, showed much better growth under the same conditions. Transformation of the gor(-) strain SG9 with a ColE1 plasmid, carrying the

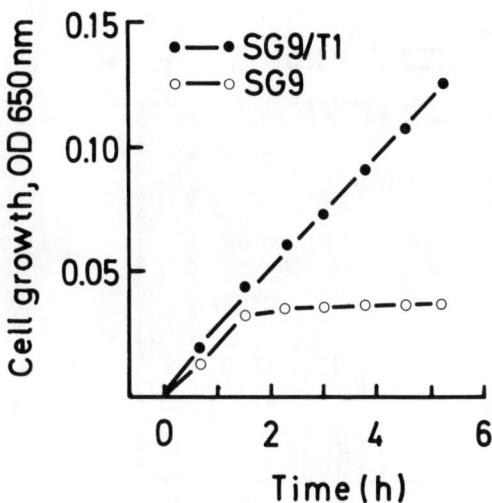

FIGURE 6. Growth of the transformed SG9/T1 gor(+)
strain compared to the parental SG9 gor(-) strain under
oxidative conditions. For experimental design see legend
of figure 5.

E. coli glutathione reductase gene, converted the gor(-)
cells from sensitivity to tolerance against MEKP (Figure
6). These results demonstrate the role of glutathione
reductase to limit the deleterious effects of oxidation
in cells.
    In a further experiment, we compared the gor(+)
strain K38, having only a single genomic gor gene, with
the transformed gor(+) strain SG9/T1 that contains multi-
copies of the plasmid carrying the gor gene (Figure 7).
The basal activity of glutathione reductase was 2-fold
higher in the strain SG9/T1 than in the strain K38. Activation of the enzyme, however, was 5-fold higher in the
strain SG9/T1 than in the strain K38 in the presence of
MEKP. From these data, we conclude that amplification of
the gor gene results in a higher activation of glutathione
reductase under oxidative stress.

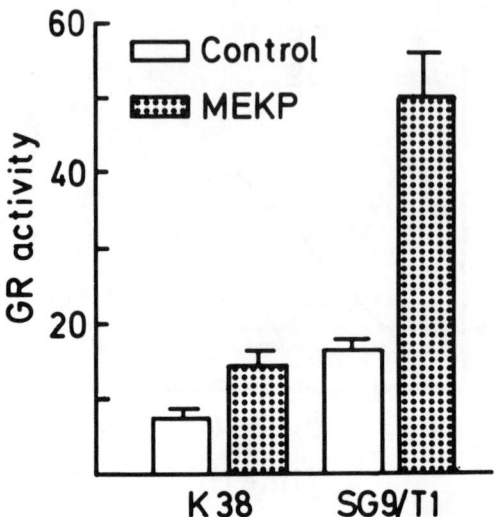

FIGURE 7. Increase of glutathione reductase activity under oxidative stress. Cells were incubated for 1h in a medium either with 0.032% MEKP or without the oxidative effector. Enzyme activity is expressed as μmol NADPH oxidized/min x g dry weight.

FUTURE RESEARCH GOALS

Increased production and/or regeneration of antioxidative protector systems are among our future research goals to protect crops against oxidative damage. Since oxidation of glutathione is among the major toxic consequences, one of our strategies to improve tolerance will be the significant increase of glutathione reductase levels in plants by gene amplification. As a consequence, we expect to maintain high GSH/GSSG ratios under oxidative stress induced by herbicides, air pollutants, and natural aging. By now, manipulation of cellular levels of nonenzymatic antioxidants, like vitamin C, via enzyme engineering seems to be less attractive for us because of the complexity of the biosynthetic pathways.

ACKNOWLEDGMENTS

Helpful discussion with Prof. P. Böger is gratefully acknowledged. We thank Prof. R. Perham (Cambridge; UK) and Prof. I. Rashed (Konstanz) for gifts of bacterial strains, and Rohm and Haas Co., Spring House, PA, for supplying us with acifluorfen-sodium.

REFERENCES

1.   Mudd JB (1982). Effects of oxidants on metabolic function. In Unsworth MH, Ormrod DP (eds): "Effects of Air Pollution in Agriculture and Horticulture," London: Butterworth Scientific, p 189.
2.   Böger P (1984). Multiple modes of action of diphenyl ethers. Z Naturforsch 39c:468.
3.   Dodge A (1982). Oxygen radicals and herbicide action. Biochem Soc Trans 10:73.
4.   Peiser GD, Yang SF (1982). Sulfite-induced lipid peroxidation in chloroplasts as determined by ethane production. Plant Physiol 70:994.
5.   Kunert KJ, Ederer M (1985). Leaf aging and lipid peroxidation: The role of the antioxidants vitamin E and C. Physiol Plant 65:85.
6 . Summerfield FW, Tappel AL (1984). Effects of dietary polyunsaturated fats and vitamin E on aging and peroxidative damage to DNA. Arch Biochem Biophys 233:408.
7.   Tappel AL (1980). Measurement of and protection from in vivo lipid peroxidation. In Pryor WA (ed): "Free Radicals in Biology, " Vol IV, New York: Academic Press, p 1.
8.   Kunert KJ, Böger P (1984). The diphenyl-ether herbicide oxyfluorfen: Action of antioxidants. J Agric Food Chem 32:725.
9.   Gardner HW (1979). Lipid hydroperoxide reactivity with proteins and amino acids: A review. J Agric Food .Chem 27:220.
10. Kunert KJ, Homrighausen C, Böhme H, Böger P (1985). Oxyfluorfen and lipid peroxidation: Protein damage as a phytotoxic consequence. Weed Sci 33:766.
11. Elstner EF (1982). Oxygen activation and oxygen toxicity. Annu Rev Plant Physiol 33:73.

12. Kunert KJ, Tappel AL (1983). The effect of vitamin C on in vivo lipid peroxidation in guinea pigs as measured by pentane and ethane production. Lipids 18:271.

13. Lichtenthaler HK, Prenzel U, Douce R, Joyard J (1981). Localization of prenylquinones in the envelope of spinach chloroplasts. Biochem Biophys Acta 641:99.

14. Burton GW, Joyce A, Ingold KU (1983). Is vitamin E the only lipid-soluble, chain-breaking antioxidant in human blood plasma and erythrocyte membranes? Arch Biochem Biophys 221:281.

15. Duke SO, Vaughn KC, Meeusen RL (1984). Mitochondrial involvment in the mode of action of acifluorfen. Pest Biochem Physiol 21:368.

16. Orr GL, Hess FD (1982). Proposed site(s) of action of new diphenyl ether herbicides. In Moreland DE, St. John JB, Hess FD (eds): "Biochemical Responses Induced by Herbicides," Washington, DC: ACS Symposium Series, No. 181, p 131.

17. Kunert KJ (1984). Herbicide-induced lipid peroxidation in higher plants: The role of vitamin C. In Bors W, Saran M, Tait D (eds): "Oxygen Radicals in Chemistry and Biology," Berlin-New York: Walter de Gruyter, p 383.

18. Tappel AL (1980). Will antioxidant nutrients slow aging process? Geriatrics 23:97.

19. McCay PB (1985). Vitamin E: Interactions with free radicals and ascorbate. Ann Rev Nutr 5:323.

20. Finckh BF, Kunert KJ (1985). Vitamins C and E: An antioxidative system against herbicide-induced lipid peroxidation in higher plants. J Agric Food Chem 33:574.

21. Kunert KJ, Müh U, Ederer M (1986). Vitamin C protects against paraquat-induced lipid peroxidation in higher plants. J Agric Food Chem submitted.

22. Baig MM, Kelly S, Loewus F (1970). L-ascorbic acid biosynthesis in higher plants from L-gulono-1,4-lactone and L-galactono-1,4-lactone. Plant Physiol 46:277.

23. Chatterjee IB, McKee RW (1965). Lipid peroxidation and biosynthesis of L-ascorbic acid in rat liver microsomes. Arch Biochem Biophys 110:254.

24. Jocelyn PC (1972). "Biochemistry of the thiol group." New York: Academic Press.

25. Ziegler DM (1985). Role of reversible oxidation-reduction of enzyme thiols-disulfides in metabolic regulation. Ann Rev Biochem 54:305.
26. Rennenberg H (1982). Glutathione metabolism and possible biological roles in higher plants. Phytochem 21:2771.
27. Foyer CH, Halliwell B (1976). The presence of glutathione and glutathione reductase in chloroplasts: A proposed role in ascorbic acid metabolism. Planta 133:21.
28. Schmidt A, Kunert KJ (1986). Glutahione reductase: A protective enzyme against oxidative stress. Arch Biochem Biophys submitted.
29. Burke JJ, Gamble PE, Hatfield JL, Quisenberry JE (1985). Plant morphological and biochemical responses to field water deficits. Plant Physiol 79:415.
30. Herschberger LA, Tappel AL (1982). Effects of vitamin E on pentane exhaled by rats treated with methyl ethyl ketone peroxide. Lipids 17:686.

Molecular Strategies for Crop Protection, pages 415–425
© 1987 Alan R. Liss, Inc.

DEVELOPING PLANT VARIETIES RESISTANT TO SULFONYLUREA HERBICIDES

R.S. Chaleff, S.A. Sebastian, G.L. Creason, B.J. Mazur, S.C. Falco, T.B. Ray, C.J. Mauvais, N.B. Yadav.

Department of Central Research and Development, Experimental Station, E.I. DuPont and Company, Wilmington, Delaware 19898

ABSTRACT Several experimental strategies are being employed to develop crop varieties resistant to sulfonylurea herbicides. Resistant tobacco mutants were first isolated by selection in cell culture. These mutants possessed an altered form of acteolactate synthase (ALS) that was less sensitive to inhibition by sulfonylurea herbicides than was the normal enzyme. ALS genes were cloned from a genomic library of a resistant tobacco mutant. Introduction of genes encoding a herbicide insensitive form of ALS offers another means by which to enhance tolerance for sulfonylurea herbicides. For species for which plant regeneration from cultured cells is difficult or has not yet been accomplished, herbicide sensitivity may be altered by mutation breeding. This last procedure has been used to generate herbicide tolerant soybean mutants.

INTRODUCTION

Herbicides have become an essential component of modern agriculture. These compounds provide an inexpensive and effective means of controlling weeds in crop cultivation. The critically important feature of herbicides that enables them to perform this function is their ability to distinguish between crop and weed species. Traditionally, this discriminatory capacity has been built into the chemical design of the herbicide. But synthesizing and testing the large numbers of compounds that must be screened to identify a useful herbicide is a process that is both expensive and time-consuming. As many as 20,000 chemicals may have to be examined to find a new selectively phytotoxic agent that merits commercial development. Such numbers provide motivation to seek other methods by which to achieve the desired specificity of herbicidal activity. One attractive alternative to chemical modification of the herbicide is genetic modification of the response of the crop. By widening the spectrum of applicability of existing compounds to include crops for which they were not originally developed, or by permitting development of new herbicides for use on several, rather than single, crops, genetic introduction of tolerance offers the potential of reducing herbicide development costs. Genetic enhancement of tolerance would also enlarge the safety margin for herbicide application and increase the flexibility of herbicide use in crop rotational systems.

The feasibility of genetically modifying the herbicide sensitivity of plants has been explored using the two DuPont herbicides, Glean® and Oust®. The active ingredients of these herbicides, chlorsulfuron (Glean®) and sulfometuron methyl (Oust®), are sulfonylurea compounds that are characterized by high herbicidal activities and low mammalian toxicities. Three different genetic strategies have been employed to modify the responses of plants to sulfonylurea herbicides: in vitro selection, transformation, and mutation breeding. No one of these approaches is universally superior to the others. Rather, each affords distinct advantages and limitations that must be considered in light of the individual features of an experimental system to determine which of these procedures is best suited to the specific application at hand.

IN VITRO SELECTION

Deliberate selection for resistant mutants is a powerful method for altering the response of a crop species to a phytotoxic agent. However, the low frequency at which such mutants are expected to appear would necessitate screening enormous numbers of plants to identify one with increased tolerance for the herbicide. It is here that the benefits of cell culture become apparent. Cell culture permits millions of cells to be grown on a chemically defined nutrient medium within the confines of a petri dish. And, as whole plants can be regenerated from cultured cells of many species, each member of this in vitro population represents a potential plant. Because herbicides that interfere with basic metabolic functions can be expected to inhibit growth of cultured cells just as they inhibit growth of the whole plant, rare resistant cells can be identified readily and unambiguously by their ability to grow on a medium supplemented with a normally toxic concentration of such a herbicide. Thus, cell culture extends to plants the experimental advantage that previously had been restricted to microbes (and that is so crucial for genetic studies) of permitting large numbers of genomes to be screened for a specific change by the direct assay of survival.

Unfortunately, plant regeneration has not yet been accomplished from cultured cells of many species. Therefore, it was because of its responsiveness to in vitro manipulation, that tobacco (Nicotiana tabacum cv. Xanthi) was used for the selection of mutants in cell culture. Mutants resistant to chlorsulfuron and sulfometuron methyl were selected by transferring callus cultures intiated from leaves of haploid plants to medium supplemented with either herbicide at a concentration of 6 nM (1). Extensive crosses with fertile, diploid plants regenerated from twelve isolates established that in all cases resistance resulted from a single dominant or semidominant nuclear mutation.

Genetic linkage studies of six mutants identified two loci, which were designated SuRA and SuRB. Although mutations at both loci confer resistance at the whole plant level, one mutation (S4) residing at the SuRB locus, was studied in greatest detail. When present in the homozygous state, the S4 mutation increased resistance to chlorsulfuron of both plants and callus cultures at least 100–fold. This level of resistance completely protected mutant plants in the field from injury by a post–emergence application of Glean® sufficient to provide excellent weed control (Fig.1).

FIGURE 1. Normal (wild type) and homozygous mutant (S4/S4) Xanthi plants and normal plants of cv. Hicks without treatment (left) and four weeks after foliar application of Glean® (formulated chlorsulfuron) at 15 gm per hectare (right).

     Passage of homozygous S4/S4 mutant callus tissue through a second cycle of selection in the presence of 600 nM sulfometuron methyl yielded an even more highly resistant cell line (2). Genetic studies with regenerated plants revealed that this enhanced level of resistance resulted from the occurrence of a second mutation (Hra), which was linked to the S4 mutation and which, therefore, resided at or near the SuRB locus. Plants homozygous for both mutations (S4 Hra/S4 Hra) were at least 1000 times more resistant to chlorsulfuron than were plants of the nonmutant parental genotype.
     Biochemical characterization of the S4 mutant led to identification of acetolactate synthase (ALS), the first enzyme specific to the isoleucine-leucine-valine biosynthetic pathway, as the site of action of chlorsulfuron and sulfometuron methyl (3). Extracts of mutant cell suspension cultures contained a form of ALS that was far less sensitive to inhibition by these two sulfonylurea herbicides than was the corresponding activity in normal cell extracts (Fig. 2). Conclusive proof that the altered ALS activity was the basis for resistance of the mutant to the two compounds and, therefore, that this enzyme must be the herbicidal site of action, was provided by the demonstration of cosegregation through genetic crosses of the herbicide insensitive ALS activity with

the resistance phenotype. Extracts of SuRA mutants also possess a resistant form of ALS activity (4). Because approximately half of the ALS activity in extracts of mutants belonging to either linkage group is resistant to chlorsulfuron, it appears that N. tabacum possesses multiple forms of ALS (isozymes), which are encoded by the SuRA and SuRB loci.

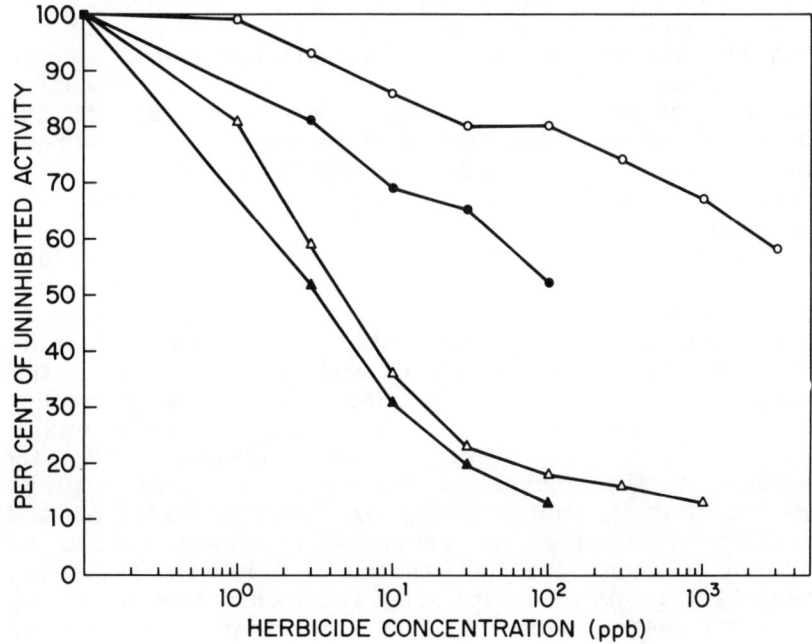

FIGURE 2. Responses to chlorsulfuron and sulfometuron methyl of ALS activities in extracts of normal and homozygous mutant (S4/S4) cell lines. Normal activity (triangles) and mutant activity (circles) were assayed in the presence of chlorsulfuron (open symbols) and sulfometuron methyl (filled symbols) (3).

GENETIC TRANSFORMATION

Although the success realized with tobacco demonstrated the power of in vitro selection as a method by which to obtain herbicide resistant varieties, this procedure would require that mutant selection be repeated de novo in each crop in which resistance is desired. Even crossing of the trait into other cultivars of the same species in which it has been isolated would require extensive and lengthy breeding programs. Herbicide resistance might be introduced more efficiently by transforming crop species with cloned genes conferring the resistance phenotype. Genetic transformation offers the additional advantage of avoiding the uncertainty of mutant selection by utilizing fully characterized mutant alleles encoding defined functions. Thus, cloning and being able to introduce that single mutant allele that confers the desired degree and specificity of response obviates the need to generate and screen large numbers of alleles repetitively in individual crops to identify a mutant form of each crop with the desired phenotype.

Because of the large size of higher plant genomes, it is not a trivial matter to clone a higher plant gene that is not either multiply represented in the genome or abundantly expressed during development. Fortunately, experimental access to the tobacco ALS gene was gained through a microbial system. Cloning of the ALS gene (ILV2) from the yeast, Saccharomyces cerevisiae, was accomplished by exploiting the increased tolerance for sulfometuron methyl that is conferred by an elevated level of ALS enzyme synthesized in response to the presence of additional copies of the structural gene (5). Herbicide resistant colonies containing increased levels of ALS activity were isolated following transformation of sensitive yeast cells with a yeast genomic library in a high copy number plasmid vector. DNA sequence analysis of the cloned yeast gene revealed an unexpectedly high degree of homology between the deduced amino acid sequences of the yeast and bacterial enzymes in three distinctly conserved domains (6). Mazur and Chui (7) reasoned that the persistence of this homology up the evolutionary ladder would permit use of the yeast gene as a heterologous hybridization probe to isolate the tobacco gene. Indeed, probing of a lambda genomic library of the S4 tobacco mutant with the cloned yeast gene produced several positive signals in a plaque filter hybridization assay. The similarity between regions of the amino acid sequence encoded by one of these cloned tobacco DNA fragments and the amino acid sequences of the conserved domains of the Escherichia

<u>coli</u> and yeast ALS genes provides strong evidence that the cloned tobacco sequence represents an ALS gene (Table 1). However, at this point it is not known which of the two <u>N</u>. <u>tabacum</u> ALS genes this cloned sequence represents: the mutant allele at one locus in the <u>S4</u> mutant that confers herbicide resistance or the normal allele of the second ALS gene. The answer, of course, will be found in the response to sulfonylurea herbicides of plants into which this putative ALS sequence has been inserted by transformation.

TABLE 1
ALS SEQUENCE HOMOLOGIES

<u>E</u>. <u>coli</u> Isozyme I	GRPGPVWI
Isozyme II	GRPGPVLV
Isozyme III	GRPGPVVV
<u>S</u>. cerevisiae	GRPGPVLV
<u>N</u>. tabacum	GRPGPILI

MUTATION BREEDING

The experimental advantages of genetic modification at the cellular level, either by mutant selection or transformation, are not to be enjoyed without certain concessions. Passage through cell culture, which is a requirement of both techniques, imposes several severe limitations on their applicability. First, <u>in</u> <u>vitro</u> selection requires expression at the cellular level of the trait to be modified (8). Fortunately, because sulfonylurea herbicides interfere with a basic biosynthetic function, both sensitive and resistant phenotypes are expressed by cultured cells. However, selection for tolerance for other herbicides that inhibit more specialized processes peculiar to the whole plant, such as photosynthesis, would not be possible with heterotrophic cultures. Second, to be useful, genetic modifications introduced in culture must be retrieved in whole plants. Yet plant regeneration has not yet been achieved from cultured cells of many important crop species. Recovery in whole plants of mutant traits introduced by <u>in</u> <u>vitro</u> selection requires that morphogenetic capacity be retained through a sufficient number of passages in culture for selection to be effective. In contrast, transformation by infection of leaf discs with <u>Agrobacterium</u> <u>tumefaciens</u>, a procedure that has been successfully applied to several dicotyledenous species(9), places a far less stringent demand on regeneration.

In addition to these experimental constraints imposed at the cellular level, introduction of resistance by genetic transformation has certain rather rigorous molecular prerequisites. Namely, that a gene conferring resistance (by encoding either an altered target site or a degradative enzyme) be identified, cloned, joined to functional regulatory sequences (if incompatibility of the native control sequences with the recipient species is anticipated), and inserted into an appropriate vector.

In cases in which the application of <u>in</u> <u>vitro</u> techniques is precluded by any of the above considerations, selection for herbicide resistance can be conducted at the whole plant level. Although somewhat less efficient than cellular manipulations, mutation breeding offers an alternative for introducing novel traits that is both effective and has the advantage of evaluating a mutant phenotype at the same level of organization and development as that at which expression of that trait is ultimately desired (8). Thus, it was for lack of an adequate cell culture system that mutation breeding was adopted as the strategy by which to increase the tolerance of soybean for sulfonylurea herbicides.

Herbicide-tolerant individuals were selected by soaking populations of mutant soybean seeds in a solution of chlorsulfuron at a concentration (0.5 ppm) sufficient to inhibit development of seedlings of the parental variety 'Williams' beyond cotyledon emergence. The mutant seed populations (M2) were produced by self fertilization of M1 plants grown from seeds treated with the chemical mutagen, ethyl methane sulfonate. Four mutants were isolated that are five to ten fold more tolerant of chlorsulfuron and DPXF6025 (the active ingredient in Classic®) than is 'Williams' (Fig. 3). In contrast to the mutations isolated in tobacco, the soybean mutations increase tolerance only modestly, are recessive, and do not affect the sensitivity of ALS to inhibition by sulfonylureas (10).

Figure 3. Normal 'Williams' (left) and mutant
1-184A (right) soybean seedlings 30 days after
planting in pots containing untreated soil (above) and
in pots in which Classic® was incorporated in the top
10 cm of soil at 400ppb (below).

FUTURE   PROSPECTS

The genetics experiments described above have added a new dimension to the design and use of chemical herbicides. First, identification of the site of action of the sulfonylurea herbicides permits the synthesis of new compounds to rely less on the laws of chance and more on the laws of physics. The traditional empirical approach to herbicide synthesis will eventually be replaced by a predictive science based on analysis of the interaction between sulfonylureas and purified ALS at the molecular level. This more direct approach to herbicide design will reduce the time and expense of herbicide development while leading to new compounds with increased activity and specificity.

The second contribution of this research has been to illustrate dramatically a concept that was perhaps accepted implicitly but never before exploited experimentally. Namely, that herbicide activity is not solely a chemical property of the herbicide, but is a property of the interaction between the herbicide and the plant. The potential utility of this realization lies in the rather obvious but important corollary that herbicide selectivity can be altered by genetic modification of the response of the plant to the chemical. Such genetic modification can be accomplished by the three procedures described in this article. But one can predict that, as techniques for introducing foreign DNA into plants are improved, genetic transformation will become the method of choice. A compelling advantage of this procedure is that it permits cloned genes to be modified in vitro before being reintroduced into the plant. Hence, in contrast to the stochastic process of in vivo mutagenesis, defined nucleotide substitutions can be made in a gene that conditions the plant's response to a herbicide and then the artificially constructed allele inserted into the plant to effect the desired resistance phenotype.

The feasibility of altering sulfonylurea sensitivity by introducing mutant ALS alleles generated in vitro was recently affirmed by the demonstration that the sensitivity of ALS to sulfonylureas can indeed be altered by replacements of individual amino acids that result from single nucleotide changes. Comparison of the DNA sequences of the normal yeast ALS gene and of a mutant derivative of this gene that encodes a herbicide insensitive form of the enzyme revealed a single nucleotide base pair difference between the two alleles. This base pair change would direct the substitution of a serine for a proline in the mutant enzyme. Similarly, DNA sequence analysis of a mutant allele of E. coli ALS isozyme II disclosed that resistance resulted from a change of the alanine at position 26 to a valine (11). It is anticipated that the characterization of additional mutant alleles will identify other amino acid subsitutions capable of conferring resistance. Eventually these studies will be coupled with X-ray diffraction analysis of the purified ALS-sulfonylurea complex to produce a complete representation of the enzyme-herbicide interaction that will allow accurate prediction of the consequences of specified amino acid substitutions to the catalytic activity and sulfonylurea sensitivity of the enzyme.

REFERENCES

1. Chaleff RS, Ray TB (1984). Herbicide-resistant mutants from tobacco cell cultures. Science223:1148.
2. Creason GL, Chaleff RS. In preparation.
3. Chaleff RS, Mauvais CJ (1984). Acetolactate synthase is the site of action of two sulfonylurea herbicides in higher plants. Science 224:1443.
4. Chaleff RS, Bascomb N. In preparation.
5. Falco SC, Dumas KS (1985). Genetic analysis of mutants of Saccharomyces cerevisiae resistant to the herbicide sulfometuron methyl. Genetics 109:21.
6. Falco SC, Dumas KS, Livak KJ (1985). Nucleotide sequence of the yeast ILV2 gene which encodes acetolactate synthase. Nucleic Acids Res. 13:4011.
7. Mazur BJ, Chui CF. In preparation.
8. Chaleff RS (1983). Isolation of agronomically useful mutants from plant cell cultures. Science 219:676.
9. Fraley RT, Rogers SG, Horsch RB (1986). Genetic transformation in higher plants. In Conger BV (ed): "CRC Critical Reviews in Plant Sciences, Vol. 4," Boca Raton:CRC Press, p 1.
10. Sebastian SA, Chaleff RS. In preparation.
11. Yadav N, McDevitt RE, Benard S, Falco SC (1986). Single amino acid substitutions in the enzyme acetolactate synthase confer resistance to the herbicide sulfometuron methyl. Proc. Natl. Acad. Sci. (USA). In press.

Index